Justification Logic

Classical logic is concerned, loosely, with the behavior of truths. Epistemic logic similarly is about the behavior of known or believed truths. Justification logic is a theory of reasoning that enables the tracking of evidence for statements and therefore provides a logical framework for the reliability of assertions. This book, the first in the area, is a systematic account of the subject, progressing from modal logic through to the establishment of an arithmetic interpretation of intuitionistic logic. The presentation is mathematically rigorous but in a style that will appeal to readers from a wide variety of areas to which the theory applies. These include mathematical logic, artificial intelligence, computer science, philosophical logic and epistemology, linguistics, and game theory.

SERGEI ARTEMOV is Distinguished Professor at the City University of New York. He is a specialist in mathematical logic, logic in computer science, control theory, epistemology, and game theory. He is credited with solving long-standing problems in constructive logic that had been left open by Gödel and Kolmogorov since the 1930s. He has pioneered studies in the logic of proofs and justifications that render a new, evidence-based theory of knowledge and belief. The most recent focus of his interests is epistemic foundations of game theory.

MELVIN FITTING is Professor Emeritus at the City University of New York. He has written or edited a dozen books and has worked in intensional logic, semantics for logic programming, theory of truth, and tableau systems for nonclassical logics. In 2012 he received the Herbrand Award from the Conference on Automated Deduction. He was on the faculty of the City University of New York from 1969 to his retirement in 2013, at Lehman College, and at the Graduate Center, where he was in the Departments of Mathematics, Computer Science, and Philosophy.

CAMBRIDGE TRACTS IN MATHEMATICS

A complete list of books in the series can be found at www.cambridge.org/mathematics.

Recent titles include the following:

181. *Totally Positive Matrices.* By A. Pinkus
182. *Nonlinear Markov Processes and Kinetic Equations.* By V. N. Kolokoltsov
183. Period Domains over Finite and *p*-adic Fields. By J.-F. Dat, S. Orlik, and M. Rapoport
184. *Algebraic Theories.* By J. Adámek, J. Rosický, and E. M. Vitale
185. *Rigidity in Higher Rank Abelian Group Actions I: Introduction and Cocycle Problem.* By A. Katok and V. Niţică
186. *Dimensions, Embeddings, and Attractors.* By J. C. Robinson
187. *Convexity: An Analytic Viewpoint.* By B. Simon
188. *Modern Approaches to the Invariant Subspace Problem.* By I. Chalendar and J. R. Partington
189. *Nonlinear Perron–Frobenius Theory.* By B. Lemmens and R. Nussbaum
190. *Jordan Structures in Geometry and Analysis.* By C.-H. Chu
191. *Malliavin Calculus for Lévy Processes and Infinite-Dimensional Brownian Motion.* By H. Osswald
192. *Normal Approximations with Malliavin Calculus.* By I. Nourdin and G. Peccati
193. *Distribution Modulo One and Diophantine Approximation.* By Y. Bugeaud
194. *Mathematics of Two-Dimensional Turbulence.* By S. Kuksin and A. Shirikyan
195. *A Universal Construction for Groups Acting Freely on Real Trees.* By I. Chiswell and T. Müller
196. *The Theory of Hardy's Z-Function.* By A. Ivić
197. *Induced Representations of Locally Compact Groups.* By E. Kaniuth and K. F. Taylor
198. *Topics in Critical Point Theory.* By K. Perera and M. Schechter
199. *Combinatorics of Minuscule Representations.* By R. M. Green
200. *Singularities of the Minimal Model Program.* By J. Kollár
201. *Coherence in Three-Dimensional Category Theory.* By N. Gurski
202. *Canonical Ramsey Theory on Polish Spaces.* By V. Kanovei, M. Sabok, and J. Zapletal
203. *A Primer on the Dirichlet Space.* By O. El-Fallah, K. Kellay, J. Mashreghi, and T. Ransford
204. *Group Cohomology and Algebraic Cycles.* By B. Totaro
205. *Ridge Functions.* By A. Pinkus
206. *Probability on Real Lie Algebras.* By U. Franz and N. Privault
207. *Auxiliary Polynomials in Number Theory.* By D. Masser
208. *Representations of Elementary Abelian p-Groups and Vector Bundles.* By D. J. Benson
209. *Non-homogeneous Random Walks.* By M. Menshikov, S. Popov, and A. Wade
210. *Fourier Integrals in Classical Analysis (Second Edition).* By C. D. Sogge
211. *Eigenvalues, Multiplicities and Graphs.* By C. R. Johnson and C. M. Saiago
212. *Applications of Diophantine Approximation to Integral Points and Transcendence.* By P. Corvaja and U. Zannier
213. *Variations on a Theme of Borel.* By S. Weinberger
214. *The Mathieu Groups.* By A. A. Ivanov
215. *Slender Modules and Rings I: Foundations.* By R. Dimitric

Justification Logic

Reasoning with Reasons

SERGEI ARTEMOV
Graduate Center, City University of New York

MELVIN FITTING
Graduate Center, City University of New York

CAMBRIDGE
UNIVERSITY PRESS

CAMBRIDGE
UNIVERSITY PRESS

University Printing House, Cambridge CB2 8BS, United Kingdom

One Liberty Plaza, 20th Floor, New York, NY 10006, USA

477 Williamstown Road, Port Melbourne, VIC 3207, Australia

314-321, 3rd Floor, Plot 3, Splendor Forum, Jasola District Centre, New Delhi - 110025, India

79 Anson Road, #06-04/06, Singapore 079906

Cambridge University Press is part of the University of Cambridge.

It furthers the University's mission by disseminating knowledge in the pursuit of education, learning and research at the highest international levels of excellence.

www.cambridge.org
Information on this title: www.cambridge.org/9781108424912
DOI: 10.1017/9781108348034

First published 2019

A catalogue record for this publication is available from the British Library

Library of Congress Cataloging in Publication data
Names: Artemov, S. N., author. | Fitting, Melvin, 1942- author.
Title: Justification logic : reasoning with reasons / Sergei Artemov (Graduate Center, City University of New York), Melvin Fitting (Graduate Center, City University of New York).
Description: Cambridge ; New York, NY : Cambridge University Press, 2019. | Series: Cambridge tracts in mathematics ; 216 | Includes bibliographical references and index.
Identifiers: LCCN 2018058431 | ISBN 9781108424912 (hardback : alk. paper)
Subjects: LCSH: Logic, Symbolic and mathematical. | Inquiry (Theory of knowledge) | Science–Theory reduction. | Reasoning.
Classification: LCC QA9 .A78 2019 | DDC 511.3–dc23
LC record available at https://lccn.loc.gov/2018058431

ISBN 978-1-108-42491-2 Hardback

To our wives, Lena and Roma.

Contents

Introduction

Why is this thus? What is the reason of this thusness?[1]

Modal operators are commonly understood to qualify the truth status of a proposition: necessary truth, proved truth, known truth, believed truth, and so on. The ubiquitous possible world semantics for it characterizes things in universal terms: $\Box X$ is true in some state if X is true in *all* accessible states, where various conditions on accessibility are used to distinguish one modal logic from another. Then $\Box(X \rightarrow Y) \rightarrow (\Box X \rightarrow \Box Y)$ is valid, no matter what conditions are imposed, by a simple and direct argument using universal quantification. Suppose both $\Box(X \rightarrow Y)$ and $\Box X$ are true at an arbitrary state. Then both X and $X \rightarrow Y$ are true at all accessible states, whatever "accessible" may mean. By the usual understanding of $\rightarrow$, Y is true at all accessible states too, and so $\Box Y$ is true at the arbitrary state we began with. Although arguments like these have a strictly formal nature and are studied as modal model theory, they also give us some insights into our informal, everyday use of modalities. Still, something is lacking.

Suppose we think of $\Box$ as epistemic, and to emphasize this we use $\mathbf{K}$ instead of $\Box$ for the time being. For some particular X, if you assert the colloquial counterpart of $\mathbf{K}X$, that is, if you say you know X, and I ask you why you know X, you would never tell me that it is because X is true in all states epistemically compatible with this one. You would, instead, give me some sort of explicit reason: "I have a mathematical proof of X," or "I read X in the encyclopedia," or "I observed that X is the case." If I asked you why $\mathbf{K}(X \rightarrow Y) \rightarrow (\mathbf{K}X \rightarrow \mathbf{K}Y)$ is valid you would probably say something like "I could use my reason for X and combine it with my reason for $X \rightarrow Y$, and infer Y." This, in effect, would be your reason for Y, given that you had reasons for X and for $X \rightarrow Y$.

[1] Charles Farrar Browne (1834–1867) was an American humorist who wrote under the pen name Artemus Ward. He was a favorite writer of Abraham Lincoln, who would read his articles to his Cabinet. This quote is from a piece called *Moses the Sassy*, Ward (1861).

Notice that this neatly avoids the *logical omniscience* problem: that we know all the consequences of what we know. It replaces logical omniscience with the more acceptable claim that there are reasons for the consequences of what we know, based on the reasons for what we know, but reasons for consequences are more complicated things. In our example, the reason for Y has some structure to it. It combines reasons for X, reasons for $X \to Y$, and inference as a kind of operation on reasons. We will see more examples of this sort; in fact, we have just seen a fundamental paradigm.

In place of a modal operator, $\Box$, justification logics have a family of *justification terms*, informally intended to represent reasons, or justifications. Instead of $\Box X$ we will see $t{:}X$, where t is a justification term and the formula is read "X is so for reason t," or more briefly, "t justifies X." At a minimum, justification terms are built up from justification variables, standing for arbitrary justifications. They are built up using a set of operations that, again at a minimum, contains a binary operation $\cdot$. For example, $x \cdot (y \cdot x)$ is a justification term, where x and y are justification variables. The informal understanding of $\cdot$ is that $t \cdot u$ justifies Y provided t justifies an implication with Y as its consequent, and u justifies the antecedent. In justification logics the counterpart of

$$\Box(X \to Y) \to (\Box X \to \Box Y)$$

is

$$t{:}(X \to Y) \to (u{:}X \to [t \cdot u]{:}Y)$$

where, as we will often do, we have added square brackets to enhance readability. Note that this exactly embodies the informal explanation we gave in the previous paragraph for the validity of $\mathbf{K}(X \to Y) \to (\mathbf{K}X \to \mathbf{K}Y)$. That is, Y has a justification built from justifications for X and for $X \to Y$ using an inference that amounts to a modus ponens application—we can think of the $\cdot$ operation as an abstract representation of this inference. Other behaviors of modal operators, $\Box X \to \Box\Box X$ for instance, will require operators in addition to $\cdot$, and appropriate postulated behavior, in order to produce justification logics that correspond to modal logics in which $\Box X \to \Box\Box X$ is valid. Examples, general methods for doing this, and what it means to "correspond" all will be discussed during the course of this book.

One more important point. Suppose X and Y are equivalent formulas, that is, we have $X \leftrightarrow Y$. Then in any normal modal logic we will also have $\Box X \leftrightarrow \Box Y$. Let us interpret the modal operator epistemically again, and write $\mathbf{K}X \leftrightarrow \mathbf{K}Y$. In fact, $\mathbf{K}X \leftrightarrow \mathbf{K}Y$, when read in the usual epistemic way, can sometimes be quite an absurd assertion. Consider some astronomically complicated tautology X of classical propositional logic. Because it is a tautology, it is equivalent to $P \vee \neg P$, which we may take for Y. Y is hardly astronomically complicated. However, because $X \leftrightarrow Y$, we will have $\mathbf{K}X \leftrightarrow \mathbf{K}Y$. Clearly, we know Y essentially by inspection and hence $\mathbf{K}Y$ holds, while $\mathbf{K}X$ on the other hand will involve an astronomical amount of work just to read it, let alone to verify it. Informally we see that, while both X and Y are tautologies, and so both are knowable in principle, any justification we might give for knowing one, *combined with quite a lot of formula manipulation*, can give us some justification for knowing the other. The two justifications may not, indeed will not, be the same. One is simple, the other very complex.

Modal logic is about *propositions*. Propositions are, in a sense, the content of formulas. Propositions are not syntactical objects. "It's good to be the king" and "Being the king is good" express the same proposition, but not in the same way. Justifications apply to *formulas*. Equivalent formulas determine the same proposition, but can be quite different as formulas. Syntax must play a fundamental role for us, and you will see that it does, even in our semantics. Consider one more very simple example. $A \rightarrow (A \wedge A)$ is an obvious tautology. We might expect $\mathbf{K}A \rightarrow \mathbf{K}(A \wedge A)$. But we should not expect $t{:}A \rightarrow t{:}(A \wedge A)$. If t does, in fact, justify A, a justification of $A \wedge A$ may involve t, but also should involve facts about the redundancy of repetition; t by itself cannot be expected to suffice.

Modal logics can express, more or less accurately, how various modal operators behave. This behavior is captured axiomatically by proofs, or semantically using possible world reasoning. These sorts of justifications for modal operator behavior are not within a modal logic, but are outside constructs. Justification logics, on the other hand, can represent the whys and wherefores of modal behavior quite directly, and from within the formal language itself. We will see that most standard modal logics have justification counterparts that can be used to give a fine-grained, internal analysis of modal behavior. Perhaps, this will help make clear why we used the quotation we did at the beginning of this Introduction.

1 What Is This Book About?

How did justification logics originate? It is an interesting story, with revealing changes of direction along the way. Going back to the days when Gödel was a young logician, there was a dream of finding a provability interpretation for intuitionistic logic. As part of his work on that project, in Gödel (1933), Gödel showed that one could abstract some of the key features of provability and make a propositional modal logic using them. Then, remarkably but naturally, one could embed propositional intuitionistic logic into the resulting system. C. I. Lewis had pioneered the modern formal study of modal logics (Lewis, 1918; Lewis and Langford, 1932), and Gödel observed that his system was equivalent to the Lewis system S4. All modern axiomatizations of modal logics follow the lines pioneered in Gödel's note, while Lewis's original formulation is rarely seen today. Gödel showed that propositional intuitionistic logic embedded into S4 using a mapping that inserted $\Box$ in front of every subformula. In effect, intuitionistic logic could be understood using classical logic plus an abstract notion of provability: a propositional formula X is an intuitionistic theorem if and only if the result of applying Gödel's mapping is a theorem of S4. (This story is somewhat simplified. There are several versions of the Gödel translation—we have used the simplest one to describe. And Gödel did not use the symbol $\Box$ but rather an operator Bew, short for beweisbar, or provability in the German language. None of this affects our main points.) Unfortunately, the story breaks off at this point because Gödel also noted that S4 does not behave like formal provability (e.g., in arithmetic), by using the methods he had pioneered in his work on incompleteness. Specifically, S4 validates $\Box X \to X$, so in particular we have $\Box\bot \to \bot$ (where $\bot$ is falsehood). This is equivalent to $\neg\Box\bot$, which is thus provable in S4. If we had an embedding of S4 into formal arithmetic under which $\Box$ corresponded to Gödel's arithmetic formula representing provability, we would be able to prove in arithmetic that falsehood was not provable. That is, we would be able to show provability of consistency, violating Gödel's second incompleteness theorem. So, work on an arithmetic semantics for propositional intuitionistic logic paused for a while.

Although it did not solve the problem of a provability semantics for intuitionistic logic, an important modal/arithmetic connection was eventually worked out. One can define a modal logic by requiring that its validities are those that correspond to arithmetic validities when reading $\Box$ as Gödel's provability formula. It was shown in Solovay (1976) that this was a modal logic already known in the literature, though as noted earlier, it is not S4. Today, the logic is called GL, standing for Gödel–Löb logic. GL is like S4 except that the T axiom $\Box X \to X$, an essential part of S4, is replaced by a modal formula abstractly representing Löb's theorem: $\Box(\Box X \to X) \to \Box X$. S4 and GL are quite different logics.

By now the project for finding an arithmetic interpretation of intuitionistic logic had reached an impasse. Intuitionistic logic embedded into S4, but S4 did not embed into formal arithmetic. GL embedded into formal arithmetic, but the Gödel translation does not embed intuitionistic logic into GL.

In his work on incompleteness for Peano arithmetic, Gödel gave a formula

$$\mathsf{Bew}(x, y)$$

that represents the relation: x is the Gödel number of a proof of a formula with Gödel number y. Then, a formal version of provability is

$$\exists x \mathsf{Bew}(x, y)$$

which expresses that there is a proof of (the formula whose Gödel number is) y. If this formula is what corresponds to $\Box$ in an embedding from a modal language to Peano arithmetic, we get the logic GL. But in a lecture in 1938 Gödel pointed out that we might work with *explicit proof representatives* instead of with provability (Gödel, 1938). That is, instead of using an embedding translating every occurrence of $\Box$ by $\exists x \mathsf{Bew}(x, y)$, we might associate with each occurrence of $\Box$ some formal term t that somehow represents a particular proof, allowing different occurrences of $\Box$ to be associated with different terms t. Then in the modal embedding, we could make the occurrence of $\Box$ associated with t correspond to $\mathsf{Bew}(\ulcorner t \urcorner, y)$, where $\ulcorner t \urcorner$ is a Gödel number for t. For each occurrence of $\Box$ we would need to find some appropriate term t, and then each occurrence of $\Box$ would be translated into arithmetic differently. The existential quantifier in $\exists x \mathsf{Bew}(x, y)$ has been replaced with a meta-existential quantifier, outside the formal language. We provide an explicit proof term, rather than just asserting that one exists. Gödel believed that this approach should lead to a provability embedding of S4 into Peano arithmetic.

Gödel's proposal was not published until 1995 when Volume 3 of his collected works appeared. By this time the idea of using a modal-like language with explicit representatives for proofs had been rediscovered independently by Sergei Artemov, see Artemov (1995, 2001). The logic that Artemov created was called LP, which stood for *logic of proofs*. It was the first example of a justification logic. What are now called justification terms were called proof terms in LP.

Crucially, Artemov showed LP filled the gap between modal S4 and Peano arithmetic. The connection with S4 is primarily embodied in a *Realization Theorem*, which has since been shown to hold for a wide range of justification logic, modal logic pairs. It will be extensively examined in this book. The connection between LP and formal arithmetic is Artemov's *Arithmetic Completeness Theorem*, which also will be examined in this book. Its range is primarily limited to the original justification logic, LP, and a few close relatives. This should not be surprising, though. Gödel's motivation for his formulation of S4 was that $\Box$ should embody properties of a formal arithmetic proof predicate. This connection with arithmetic provability is not present for almost all modal logics and is consequently also missing for corresponding justification logics, when they exist. Nonetheless, the venerable goal of finding a provability interpretation for propositional intuitionistic logic had been attained. The Gödel translation embeds propositional intuitionistic logic into the modal logic S4. The Realization Theorem establishes an embedding of S4 into the justification logic LP. And the Arithmetic Completeness Theorem shows that LP embeds into formal arithmetic.

It was recognized from the very beginning that the connection between S4 and LP could be weakened to sublogics of S4 and LP. Thus, there were justification logic counterparts for the standard modal logics, K, K4, T, and a few others. These justification logics had arithmetic connections because they were sublogics of LP. The use of *proof term* was replaced with *justification term*. Although the connection with arithmetic was weaker than it had been with LP, justification terms still had the role of supplying explicit justifications for epistemically necessary statements. One can consult Artemov (2008) and Artemov and Fitting (2012) for survey treatments, though the present book includes the material found there.

Almost all of the early work on justification logics was proof-theoretically based. Realization theorems were shown constructively, making use of a sequent calculus. The existence of an algorithm to compute what are called *realizers* is important, but this proof-theoretic approach limits the field to those logics known to have sequent calculus proof systems. For a time it was hoped that various extensions of sequent and tableau calculi would be useful and, to some extent, this has been the case. The most optimistic version of this hope was expressed in Artemov (2001) quite directly, "Gabbay's Labelled Deductive Systems, Gabbay (1994), may serve as a natural framework for LP." Unfortunately this seems to have been too optimistic. While the formats had similarities, the goals were different, and the machinery did not interact well.

A semantics for LP and its near relatives, not based on arithmetic provability, was introduced in Mkrtychev (1997) and is discussed in Chapter 3. (A constructive version of the canonical model for LP with a completeness theorem can be found already in Artemov (1995).) Mkrtychev's semantics did not use possible worlds and had a strong syntactic flavor. Possible worlds were added to the mix in Fitting (2005), producing something that potentially applied much more broadly than the earlier semantics. This is the subject of Chapter 4. Using this possible world semantics, a nonconstructive, semantic-based, proof of realization was given. It was now possible to avoid the use of a sequent calculus, though the algorithmic nature of realization was lost. More recently, a semantics with a very simple structure was created, Artemov's *basic* semantics (Artemov, 2012). It is presented in Chapter 3. Its machinery is almost minimal for the purpose. In this book, we will use possible world semantics to establish very general realization results, but basic models will often be used when we simply want to show some formula fails to be a theorem.

Though its significance was not properly realized at the time, in 2005 the subject broadened when a justification logic counterpart of S5 was introduced in Pacuit (2005) and Rubtsova (2006a, b), with a connecting realization theorem. There was no arithmetical interpretation for this justification logic. Also there is no sequent calculus for S5 of the standard kind, so the proof given for realization was nonconstructive, using a version of the semantics from Fitting (2005). The semantics needed some modification to what is called its *evidence function*, and this turned out to have a greater impact than was first realized. Eventually constructive proofs connecting S5 and its justification counterpart were found. These made use of cut-free proof systems that were not exactly standard sequent calculi. Still, the door to a larger room was beginning to open.

Out of the early studies of the logics of proofs and its variants a general logical framework for reasoning about epistemic justification at large naturally emerged, and the name, *Justification Logic*, was introduced (cf. Artemov, 2008). Justification Logic is based on justification assertions, $t{:}F$, that are read t is a *justification* for F, with a broader understanding of the word justification going beyond just mathematical proofs. The notion of justification, which has been an essential component of epistemic studies since Plato, had been conspicuously absent in the mathematical models of knowledge within the epistemic logic framework. The Justification Logic framework fills in this void.

In Fitting (2016a) the subject expanded abruptly. Using nonconstructive semantic methods it was shown that the family of modal logics having justification counterparts is infinite. The justification phenomenon is not the relatively narrow one it first seemed to be. While that work was nonconstructive, there are now cut-free proof systems of various kinds for a broader range of modal logics than was once the case, and these have been used successfully to create realization algorithms, in Kuznets and Goetschi (2012), for instance. It may be that the very general proof methodologies of Fitting (2015) and especially Negri (2005) and Negri and von Plato (2001) will extend the constructive range still further, perhaps even to the infinite family that nonconstructive methods are known to work for. This is active current work.

Work on quantified justification logics exists, but the subject is considerably behind its propositional counterpart. An important feature of justification logics is that they can, in a very precise sense, internalize their own proofs. Doing this for axioms is generally simple. Rules of inference are more of a problem. Earlier we discussed a justification formula as a simple, representative example: $t{:}(X \to Y) \to (u{:}X \to [t \cdot u]{:}Y)$. This, in effect, internalizes the axiomatic modus ponens rule. The central problem in developing quantified justification logics was how to internalize the rule of universal generalization. It turned out that the key was the clear separation between two roles played by individual variables. On the one hand, they are formal symbols, and one can simply infer $\forall x\varphi(x)$ from a proof of $\varphi(x)$. On the other hand, they can be thought of as open for substitution, that is, throughout a proof one can replace free occurrences of x with a term t to produce a new proof (subject to appropriate freeness of substitution conditions, of course). These two roles for variables are actually incompatible. It was the introduction of specific machinery to keep track of which role a variable occurrence had that made possible the internalization of proofs, and thus a quantified justification logic.

An axiomatic version of first-order LP was introduced in Artemov and Yavorskaya (Sidon) (2011) and a possible world semantics for it in Fitting (2011a, 2014b). A connection with formal arithmetic was established. There is a constructive proof of a Realization Theorem, connecting first-order LP with first-order S4. Unlike propositionally, no nonconstructive proof is currently known The possible world semantics includes the familiar monotonicity condition on world domains. It is likely that all this can be extended to a much broader range of quantified modal logics than just first-order S4, provided monotonicity is appropriate. A move to constant domain models, to quantified S5 in particular, has been made, and a semantics, but not yet a Realization Theorem, can be found in Fitting and Salvatore (2018). Much involving quantification is still uncharted territory.

This book will cover the whole range of topics just described. It will not do so in the historical order that was followed in this Introduction, but will make use of the clearer understanding that has emerged from study of the subject thus far. We will finish with the current state of affairs, standing on the edge of unknown lands. We hope to prepare some of you for the journey, should you choose to explore further on your own.

2 What Is Not in This Book?

There are several historical works and pivotal developments in justification logic that will not be covered in the book due to natural limitations, and in this section we will mention them briefly. We are confident that other books and surveys will do justice to these works in more detail.

Apart from Gödel's lecture, Gödel (1938), which remained unpublished until 1995 and thus could not influence development in this area, the first results and publications on the logic of proofs are dated 1992: a technical report, Artemov and Straßen (1992), based on work done in January of 1992 in Bern, and a conference presentation of this work at CSL'92 published in Springer Lecture Notes in Computer Science as Artemov and Straßen (1993a). In this work, the *basic logic of proofs* was presented: it had proof variables, and the format *t is a proof of F*, but without operations on proofs. However, it already had the first installment of the fixed-point arithmetical completeness construction together with an observation that, unlike provability logic, the logic of proofs cannot be limited to one standard proof predicate "from the textbook" or to any single-conclusion proof predicate.

This line was further developed in Artemov and Straßen (1993b), where the logic of single-conclusion proof predicates (without operations on proofs) was studied. This work introduced the unification axiom, which captures single-conclusioness by propositional tools. After the full-scale logic of proofs with operations had been discovered (Artemov, 1995), the logic of single-conclusion proofs with operations was axiomatized in V. Krupski (1997, 2001). A similar technique was used recently to characterize so-called sharp single-conclusion justification models in Krupski (2018).

Another research direction pursued after the papers on the basic logic of proofs was to combine provability and explicit proofs. Such a combination, with new provability principles, was given in Artemov (1994). Despite its title, this paper did not introduce what is known now as The Logic of Proofs, but rather a fusion of the provability logic GL and the basic logic of proofs without operations, but with new arithmetical principles combining proofs and provability and an arithmetical completeness theorem. After the logic of proofs paper (Artemov, 1995), the full-scale logic of provability and proofs (with operations), LPP, was axiomatized and proved arithmetically complete in Sidon (1997) and Yavorskaya (Sidon) (2001). A leaner logic combining provability and explicit proofs, GLA, was introduced and proved arithmetically complete in Nogina (2006, 2014b). Unlike LPP, the logic GLA did not use additional operations on proofs other than those inherited from LP. Later, GLA was used to find a complete classification of reflection principles in arithmetic that involve provability and explicit proofs (Nogina, 2014a).

The first publication of the full-scale logic of proofs with operations, LP, which became the first justification logic in the modern sense, was Artemov (1995). This paper contains all the results needed to complete Gödel's program of characterizing intuitionistic propositional logic IPC and its BHK semantics via proofs in classical arithmetic: internalization, the realization theorem for S4 in LP, arithmetical semantics for LP, and the arithmetical completeness theorem. It took six years for the corresponding journal paper to appear: Artemov (2001). In Goris (2008), the completeness of LP for the semantics of proofs in Peano arithmetic was extended to the semantics of proofs in Buss's bounded arithmetic S^1_2. In view of applications in epistemology, this result shows that explicit knowledge in the propositional framework can be made computationally feasible. Kuznets and Studer (2016) extend the arithmetical interpretation of LP from the original finite constant specifications to a wide class of constant specifications, including some infinite ones. In particular, this "weak" arithmetical interpretation captures the full logic of proofs LP with the total constant specification.

Decidability of LP (with the total constant specification) was also established in Mkrtychev (1997), and this opened the door to decidability and complexity studies in justification logics using model-theoretic and other means. Among the milestones are complexity estimates in Kuznets (2000), Brezhnev and Kuznets (2006), Krupski (2006a), Milnikel (2007), Buss and Kuznets (2012), and Achilleos (2014a).

The arithmetical provability semantics for the Logic of Proofs, LP, naturally generalizes to a first-order version with conventional quantifiers and to a version with quantifiers over proofs. In both cases, axiomatizability questions were answered negatively in Artemov and Yavorskaya (2001) and Yavorsky (2001). A natural and manageable first-order version of the logic of proofs, FOLP, has been studied in Artemov and Yavorskaya (Sidon) (2011), Fitting (2014a), and Fitting and Salvatore (2018) and will be covered in Chapter 10.

Originally, the logic of proofs was formulated as a Hilbert-style axiomatic system, but this has gradually broadened. Early attempts were tableau based (which could equivalently be presented using sequent calculus machinery). Tableaus generally are *analytic*, meaning that everything entering into a proof is a subformula of what is being proved. This was problematic for attempts at LP tableaus because of the presence of the $\cdot$ operation, which represented an application of modus ponens, a rule that is decidedly not analytic. Successful tableau systems, though not analytic, for LP and closely related logics can be found in Fitting (2003, 2005) and Renne (2004, 2006). The analyticity problem was overcome in Ghari (2014, 2016a). Broader proof systems have been investigated: hypersequents in Kurokawa (2009, 2012), prefixed tableaus in Kurokawa (2013), and labeled deductive systems in Ghari (2017). Indeed some of this has led to new realization results (Artemov, 1995, 2001, 2002, 2006; Artemov and Bonelli, 2007; Ghari, 2012; Kurokawa, 2012).

Finding a computational reading of justification logics has been a natural research goal. There were several attempts to use the ideas of LP for building a lambda-calculus with internalization, cf. Alt and Artemov (2001), Artemov

(2002), Artemov and Bonelli (2007), Pouliasis and Primiero (2014), and others. Corresponding combinatory logic systems with internalization were studied in Artemov (2004), Krupski (2006b), and Shamkanov (2011). These and other studies can serve as a ground for further applications in typed programming languages. A version of the logic of proofs with a built-in verification predicate was considered in Protopopescu (2016a, b).

The aforementioned intuition that justification logic naturally avoids the logical omniscience problem has been formalized and studied in Artemov and Kuznets (2006, 2009, 2014). The key idea there was to view logical omniscience as a proof complexity problem: The logical omniscience defect occurs if an epistemic system assumes knowledge of propositions, which have no feasible proofs. Through this prism, standard modal logics are logically omniscient (modulo some common complexity assumptions), and justification logics are not logically omniscient. The ability of justification logic to track proof complexity via time bounds led to another formal definition of logical omniscience in Wang (2011a) with the same conclusion: Justification logic keeps logical omniscience under control.

Shortly after the first paper on the logic of proofs, it became clear that the logical tools developed are capable of evidence tracking in a general setting and as such can be useful in epistemic logic. Perhaps, the first formal work in this direction was Artemov et al. (1999), in which modal logic S5 was equivalently modified and supplied with an LP-style explicit counterpart. Applications to epistemology have benefited greatly from Fitting semantics, which connected justification logics to mainstream epistemology via possible worlds models. In addition to applications discussed in this book, we would like to mention some other influential work. Game semantics of justification logic was studied in Renne (2008) and dynamic epistemic logic with justifications in Renne (2008) and Baltag et al. (2014). In Sedlár (2013), Fitting semantics for justification models was elaborated to a special case of the models of general awareness. Multiagent justification logic and common knowledge has been studied in Artemov (2006), Antonakos (2007), Yavorskaya (Sidon) (2008), Bucheli et al. (2010, 2011), Bucheli (2012), Antonakos (2013), and Achilleos (2014b, 2015a, b). In Dean and Kurokawa (2010), justification logic was used for the analysis of Knower and Knowability paradoxes. A fast-growing and promising area is probabilistic justification logic, cf. Milnikel (2014), Artemov (2016b), Kokkinis et al. (2016), Ghari (2016b), and Lurie (2018).

We are deeply indebted to all contributors to the exciting justification logic project, without whom there would not be this book.

Very special thanks to our devoted readers for their sharp eyes and their useful comments: Vladimir Krupski, Vincent Alexis Peluce, and Tatiana Yavorskaya (Sidon).

> I think there is no sense in forming an opinion when there is no evidence to form it on. If you build a person without any bones in him he may look fair enough to the eye, but he will be limber and cannot stand up; and I consider that *evidence* is the bones of an opinion.[2]

[2] Mark Twain (1835–1910). The quote is from his last novel, *Personal Recollections of Joan of Arc*, Twain (1896).

1
Why Justification Logic?

The formal details of justification logic will be presented starting with the next chapter, but first we give some background and motivation for why the subject was developed in the first place. We will see that it addresses, or at least partially addresses, many of the fundamental problems that have been found in epistemic logic over the years. We will also see in more detail how it relates to our understanding of intuitionistic logic. And finally, we will see how it can be used to mitigate some well-known issues that have arisen in philosophical investigations.

1.1 Epistemic Tradition

The properties of knowledge and belief have been a subject for formal logic at least since von Wright and Hintikka (Hintikka, 1962; von Wright, 1951). Knowledge and belief are both treated as modalities in a way that is now very familiar—*Epistemic Logic*. But of the celebrated three criteria for knowledge (usually attributed to Plato), *justified, true, belief*, Gettier (1963); Hendricks (2005), epistemic modal logic really works with only two of them. Possible worlds and indistinguishability model belief—one believes what is so under all circumstances thought possible. Factivity brings a trueness component into play—if something is not so in the actual world it cannot be known, only believed. But there is no representation for the justification condition. Nonetheless, the modal approach has been remarkably successful in permitting the development of rich mathematical theory and applications (Fagin et al., 1995; van Ditmarsch et al., 2007). Still, it is not the whole picture.

The modal approach to the logic of knowledge is, in a sense, built around the universal quantifier: X is known in a situation if X is true in *all* situations indistinguishable from that one. Justifications, on the other hand, bring an existential quantifier into the picture: X is known in a situation if *there exists* a justification for X in that situation. This universal/existential dichotomy is a familiar one to logicians—in formal logics there exists a proof for a formula X if and only if X is true in all models for the logic. One thinks of models as inherently nonconstructive, and proofs as constructive things. One will not go far wrong in thinking of justifications in general as much like mathematical proofs. Indeed, the first justification logic was explicitly designed to capture mathematical proofs in arithmetic, something that will be discussed later.

In justification logic, in addition to the category of formulas, there is a second category of *justifications*. Justifications are formal terms, built up from constants and variables using various operation symbols. Constants represent justifications for commonly accepted truths—axioms. Variables denote unspecified justifications. Different justification logics differ on which operations are allowed (and also in other ways too). If t is a justification term and X is a formula, $t{:}X$ is a formula, and is intended to be read

t is a justification for X.

One operation, common to all justification logics, is *application*, written like multiplication. The idea is, if s is a justification for $A \to B$ and t is a justification for A, then $[s \cdot t]$ is a justification for B.[1] That is, the validity of the following is generally assumed

$$s{:}(A \to B) \to (t{:}A \to [s \cdot t]{:}B). \tag{1.1}$$

This is the explicit version of the usual distributivity of knowledge operators, and modal operators generally, across implication

$$\mathbf{K}(A \to B) \to (\mathbf{K}A \to \mathbf{K}B). \tag{1.2}$$

How adequately does the traditional modal form (1.2) embody epistemic closure? We argue that it does so poorly! In the classical logic context, (1.2) only claims that it is impossible to have both $\mathbf{K}(A \to B)$ and $\mathbf{K}A$ true, but $\mathbf{K}B$ false. However, because (1.2), unlike (1.1), does not specify dependencies between $\mathbf{K}(A \to B)$, $\mathbf{K}A$, and $\mathbf{K}B$, the purely modal formulation leaves room for a counterexample.

The distinction between (1.1) and (1.2) can be exploited in a discussion of the paradigmatic Red Barn Example of Goldman and Kripke; here is a simplified version of the story taken from Dretske (2005).

[1] For better readability brackets will be used in terms, "[,]", and parentheses in formulas, "(,)." Both will be avoided when it is safe.

> Suppose I am driving through a neighborhood in which, unbeknownst to me, papier-mâché barns are scattered, and I see that the object in front of me is a barn. Because I have barn-before-me percepts, I believe that the object in front of me is a barn. Our intuitions suggest that I fail to know barn. But now suppose that the neighborhood has no fake red barns, and I also notice that the object in front of me is red, so I know a red barn is there. This juxtaposition, being a red barn, which I know, entails there being a barn, which I do not, "is an embarrassment."

In the first formalization of the Red Barn Example, logical derivation will be performed in a basic modal logic in which $\Box$ is interpreted as the "belief" modality. Then some of the occurrences of $\Box$ will be externally interpreted as a knowledge modality **K** according to the problem's description. Let B be the sentence "the object in front of me is a barn," and let R be the sentence "the object in front of me is red."

(1) $\Box B$, "I believe that the object in front of me is a barn." At the metalevel, by the problem description, this is not knowledge, and we cannot claim $\mathbf{K}B$.
(2) $\Box(B \wedge R)$, "I believe that the object in front of me is a red barn." At the metalevel, this is actually knowledge, e.g., $\mathbf{K}(B \wedge R)$ holds.
(3) $\Box(B \wedge R \rightarrow B)$, a knowledge assertion of a logical axiom. This is obviously knowledge, i.e., $\mathbf{K}(B \wedge R \rightarrow B)$.

Within this formalization, it appears that epistemic closure in its modal form (1.2) is violated: $\mathbf{K}(B \wedge R)$, and $\mathbf{K}(B \wedge R \rightarrow B)$ hold, whereas, by (1), we cannot claim $\mathbf{K}B$. The modal language here does not seem to help resolving this issue.

Next consider the Red Barn Example in justification logic where $t{:}F$ is interpreted as "I *believe* F for reason t." Let u be a specific individual justification for belief that B, and v for belief that $B \wedge R$. In addition, let a be a justification for the logical truth $B \wedge R \rightarrow B$. Then the list of assumptions is

(i) $u{:}B$, "u is a reason to believe that the object in front of me is a barn";
(ii) $v{:}(B \wedge R)$, "v is a reason to believe that the object in front of me is a red barn";
(iii) $a{:}(B \wedge R \rightarrow B)$.

On the metalevel, the problem description states that (ii) and (iii) are cases of knowledge, and not merely belief, whereas (i) is belief, which is not knowledge. Here is how the formal reasoning goes:

(iv) $a{:}(B \wedge R \rightarrow B) \rightarrow (v{:}(B \wedge R) \rightarrow [a{\cdot}v]{:}B)$, by principle (1.1);
(v) $v{:}(B \wedge R) \rightarrow [a{\cdot}v]{:}B$, from 3 and 4, by propositional logic;
(vi) $[a{\cdot}v]{:}B$, from 2 and 5, by propositional logic.

Notice that conclusion (vi) is $[a \cdot v]{:}B$, and not $u{:}B$; epistemic closure holds. By reasoning in justification logic it was concluded that $[a{\cdot}v]{:}B$ is a case of knowledge, i.e., "I know B for reason $a \cdot v$." The fact that $u{:}B$ is not a case of knowledge does not spoil the closure principle because the latter claims knowledge specifically for $[a{\cdot}v]{:}B$. Hence after observing a red façade, I indeed know B, but this knowledge has nothing to do with (i), which remains a case of belief rather than of knowledge. The justification logic formalization represents the situation fairly.

Tracking justifications represents the structure of the Red Barn Example in a way that is not captured by traditional epistemic modal tools. The justification logic formalization models what seems to be happening in such a case; closure of knowledge under logical entailment is maintained even though "barn" is not perceptually known.

One could devise a formalization of the Red Barn Example in a bimodal language with distinct modalities for knowledge and belief. However, it seems that such a resolution must involve reproducing justification tracking arguments in a way that obscures, rather than reveals, the truth. Such a bimodal formalization would distinguish $u{:}B$ from $[a \cdot v]{:}B$ not because they have different reasons (which reflects the true epistemic structure of the problem), but rather because the former is labeled "belief" and the latter "knowledge." But what if one needs to keep track of a larger number of different unrelated reasons? By introducing a multiplicity of distinct modalities and then imposing various assumptions governing the interrelationships between these modalities, one would essentially end up with a reformulation of the language of justification logic itself (with distinct terms replaced by distinct modalities). This suggests that there may not be a satisfactory "halfway point" between a modal language and the language of justification logic, at least inasmuch as one tries to capture the essential structure of examples involving the deductive nature of knowledge.

1.2 Mathematical Logic Tradition

According to Brouwer, truth in constructive (intuitionistic) mathematics means the existence of a proof, cf. Troelstra and van Dalen (1988). In 1931–34, Heyting and Kolmogorov gave an informal description of the intended proof-based semantics for intuitionistic logic (Kolmogoroff, 1932; Heyting, 1934), which is now referred to as the *Brouwer–Heyting–Kolmogorov* (*BHK*) *semantics*. According to the *BHK* conditions, a formula is "true" if it hasa proof. Furthermore, a proof of a compound statement is connected to proofs of its components in the following way:

- a proof of $A \wedge B$ consists of a proof of proposition A and a proof of proposition B,
- a proof of $A \vee B$ is given by presenting either a proof of A or a proof of B,
- a proof of $A \to B$ is a construction transforming proofs of A into proofs of B,
- falsehood $\bot$ is a proposition, which has no proof; $\neg A$ is shorthand for $A \to \bot$.

This provides a remarkably useful informal way of understanding what is and what is not intuitionistically acceptable. For instance, consider the classical tautology $(P \vee Q) \leftrightarrow (P \vee (Q \wedge \neg P))$, where we understand $\leftrightarrow$ as mutual implication. And we understand $\neg P$ as $P \to \bot$, so that a proof of $\neg P$ would amount to a construction converting any proof of P into a proof of $\bot$. Because $\bot$ has no proof, this amounts to a proof that P has no proof—a refutation of P.

According to *BHK* semantics the implication from right to left in $(P \vee Q) \leftrightarrow (P \vee (Q \wedge \neg P))$ should be intuitionistically valid, by the following argument. Given a proof of $P \vee (Q \wedge \neg P)$ it must be that we are given a proof of one of the disjuncts. If it is P, we have a proof of one of $P \vee Q$. If it is $Q \wedge \neg P$, we have proofs of both conjuncts, hence a proof of Q, and hence again a proof of one of $P \vee Q$. Thus we may convert a proof of $P \vee (Q \wedge \neg P)$ into a proof of $P \vee Q$.

On the other hand, $(P \vee Q) \to (P \vee (Q \wedge \neg P))$ is not intuitionistically valid according to the *BHK* ideas. Suppose we are given a proof of $P \vee Q$. If we have a proof of the disjunct P, we have a proof of $P \vee Q$. But if we have a proof of Q, there is no reason to suppose we have a refutation of P, and so we cannot conclude we have a proof of $Q \wedge \neg P$, and things stop here.

Kolmogorov explicitly suggested that the proof-like objects in his interpretation ("problem solutions") came from classical mathematics (Kolmogoroff, 1932). Indeed, from a foundational point of view this reflects Kolmogorov's and Gödel's goal to define intuitionism within classical mathematics. From this standpoint, intuitionistic mathematics is not a substitute for classical mathematics, but helps to determine what is constructive in the latter.

The fundamental value of the *BHK* semantics for the justification logic project is that informally but unambiguously *BHK* suggests treating justifications, here mathematical proofs, as objects with operations.

In Gödel (1933), Gödel took the first step toward developing a rigorous proof-based semantics for intuitionism. Gödel considered the classical modal logic **S4** to be a calculus describing properties of provability:

(1) *Axioms and rules of classical propositional logic,*
(2) $\Box(F \to G) \to (\Box F \to \Box G)$,
(3) $\Box F \to F$,
(4) $\Box F \to \Box\Box F$,
(5) *Rule of necessitation:* $\dfrac{\vdash F}{\vdash \Box F}$.

Based on Brouwer's understanding of logical truth as provability, Gödel defined a translation $tr(F)$ of the propositional formula F in the intuitionistic language into the language of classical modal logic: $tr(F)$ is obtained by prefixing every subformula of F with the provability modality $\Box$. Informally speaking, when the usual procedure of determining classical truth of a formula is applied to $tr(F)$, it will test the provability (not the truth) of each of F's subformulas, in agreement with Brouwer's ideas. From Gödel's results and the McKinsey-Tarski work on topological semantics for modal logic (McKinsey and Tarski, 1948), it follows that the translation $tr(F)$ provides a proper embedding of the Intuitionistic Propositional Calculus, IPC, into S4, i.e., an embedding of intuitionistic logic into classical logic extended by the provability operator.

$$\mathsf{IPC} \vdash F \quad \Leftrightarrow \quad \mathsf{S4} \vdash tr(F). \tag{1.3}$$

Conceptually, this defines IPC in S4.

Still, Gödel's original goal of defining intuitionistic logic in terms of classical provability was not reached because the connection of S4 to the usual mathematical notion of provability was not established. Moreover, Gödel noted that the straightforward idea of interpreting modality $\Box F$ as *F is provable in a given formal system $\mathcal{T}$* contradicted his second incompleteness theorem. Indeed, $\Box(\Box F \to F)$ can be derived in S4 by the rule of necessitation from the axiom $\Box F \to F$. On the other hand, interpreting modality $\Box$ as the predicate of formal provability in theory $\mathcal{T}$ and F as contradiction converts this formula into a false statement that the consistency of $\mathcal{T}$ is internally provable in $\mathcal{T}$.

The situation after Gödel (1933) can be described by the following figure where "$X \hookrightarrow Y$" should be read as "X is interpreted in Y":

$$\mathsf{IPC} \hookrightarrow \mathsf{S4} \hookrightarrow\ ?\ \hookrightarrow \textit{CLASSICAL PROOFS}.$$

In a public lecture in Vienna in 1938, Gödel observed that using the format of explicit proofs

$$t \textit{ is a proof of } F \tag{1.4}$$

can help in interpreting his provability calculus S4 (Gödel, 1938). Unfortunately, Gödel (1938) remained unpublished until 1995, by which time the

Gödelian logic of explicit proofs had already been rediscovered, axiomatized as the Logic of Proofs LP, and supplied with completeness theorems connecting it to both S4 and classical proofs (Artemov, 1995, 2001).

The Logic of Proofs LP became the first in the justification logic family. Proof terms in LP are nothing but *BHK* terms understood as classical proofs. With LP, propositional intuitionistic logic received the desired rigorous *BHK* semantics:

$$\mathsf{IPC} \hookrightarrow \mathsf{S4} \hookrightarrow \mathsf{LP} \hookrightarrow \textit{CLASSICAL PROOFS}\,.$$

Several well-known mathematical notions that appeared prior to justification logic have sometimes been perceived as related to the *BHK* idea: Kleene realizability (Troelstra, 1998), Curry–Howard isomorphism (Girard et al., 1989; Troelstra and Schwichtenberg, 1996), Kreisel–Goodman theory of constructions (Goodman, 1970; Kreisel, 1962, 1965), just to name a few. These interpretations have been very instrumental for understanding intuitionistic logic, though none of them qualifies as the *BHK* semantics.

Kleene realizability revealed a fundamental *computational content* of formal intuitionistic derivations; however it is still quite different from the intended *BHK* semantics. Kleene realizers are computational programs rather than proofs. The predicate "*r realizes F*" is not decidable, which leads to some serious deviations from intuitionistic logic. Kleene realizability is not adequate for the intuitionistic propositional calculus IPC. There are realizable propositional formulas not derivable in IPC (Rose, 1953).[2]

The Curry–Howard isomorphism transliterates natural derivations in IPC to typed λ-terms, thus providing a generic functional reading for logical derivations. However, the foundational value of this interpretation is limited because, as proof objects, Curry–Howard λ-terms denote nothing but derivations in IPC itself and thus yield a circular provability semantics for the latter.

An attempt to formalize the *BHK* semantics directly was made by Kreisel in his theory of constructions (Kreisel, 1962, 1965). The original variant of the theory was inconsistent; difficulties already occurred at the propositional level. In Goodman (1970) this was fixed by introducing a stratification of constructions into levels, which ruined the *BHK* character of this semantics. In particular, a proof of $A \to B$ was no longer a construction that could be applied to *any* proof of A.

[2] Kleene himself denied any connection of his realizability with the *BHK* interpretation.

1.3 Hyperintensionality

Justification logic offers a formal framework for hyperintensionality. The *hyperintensional paradox* was formulated in Cresswell (1975).

> It is well known that it seems possible to have a situation in which there are two propositions p and q which are logically equivalent and yet are such that a person may believe the one but not the other. If we regard a proposition as a set of possible worlds then two logically equivalent propositions will be identical, and so if "x believes that" is a genuine sentential functor, the situation described in the opening sentence could not arise. I call this the paradox of hyperintensional contexts. Hyperintensional contexts are simply contexts which do not respect logical equivalence.

Starting with Cresswell himself, several ways of dealing with this have been proposed. Generally, these involve adding more layers to familiar possible world approaches so that some way of distinguishing between logically equivalent sentences is available. Cresswell suggested that the syntactic form of sentences be taken into account. Justification logic, in effect, does this through its mechanism for handling justifications for sentences. Thus justification logic addresses some of the central issues of hyperintensionality but, as a bonus, we automatically have an appropriate proof theory, model theory, complexity estimates, and a broad variety of applications.

A good example of a hyperintensional context is the informal language used by mathematicians conversing with each other. Typically when a mathematician says he or she knows something, the understanding is that a proof is at hand, but this kind of knowledge is essentially hyperintensional. For instance Fermat's Last Theorem, FLT, is logically equivalent to $0 = 0$ because both are provable and hence denote the same proposition, as this is understood in modal logic. However, the context of proofs distinguishes them immediately because a proof of $0 = 0$ is not necessarily a proof of FLT, and vice versa. To formalize mathematical speech, the justification logic LP is a natural choice because $t{:}X$ was designed to have characteristics of "*t is a proof of X*."

The fact that propositions X and Y are equivalent in LP, that LP $\vdash X \leftrightarrow Y$, does not warrant the equivalence of the corresponding justification assertions, and typically $t{:}X$ and $t{:}Y$ are not equivalent, $t{:}X \not\leftrightarrow t{:}Y$. Indeed, as we will see, this is the case for every justification logic.

Going further LP, and justification logic in general, is not only sufficiently refined to distinguish justification assertions for logically equivalent sentences, but it also provides flexible machinery to connect justifications of equivalent sentences and hence to maintain constructive closure properties desirable for a logic system. For example, let X and Y be provably equivalent, i.e., there is a proof u of $X \leftrightarrow Y$, and so $u{:}(X \leftrightarrow Y)$ is provable in LP. Suppose also that v is a proof of X, and so $v{:}X$. It has already been mentioned that this does not mean v is a proof of Y—this is a hyperintensional context. However within the framework of justification logic, building on the proofs of X and of $X \leftrightarrow Y$, we can *construct* a proof term $f(u, v)$, which represents the proof of Y and so $f(u, v){:}Y$ is provable. In this respect, justification logic goes beyond Cresswell's expectations: Logically equivalent sentences display different but constructively controlled epistemic behavior.

1.4 Awareness

The logical omniscience problem is that in epistemic logics all tautologies are known and knowledge is closed under consequence, both of which are unreasonable. In Fagin and Halpern (1988) a simple mechanism for avoiding the problems was introduced. One adds to the usual Kripke model structure an *awareness* function $\mathcal{A}$ indicating for each world which formulas the agent is aware of at this world. Then a formula is taken to be known at a possible world u if (1) the formula is true at all worlds accessible from u (the Kripkean condition for knowledge) and (2) the agent is aware of the formula at u. The awareness function $\mathcal{A}$ can serve as a practical tool for blocking knowledge of an arbitrary set of formulas. However, as logical structures, awareness models exhibit abnormal behavior due to the lack of natural closure properties. For example, the agent can know $A \wedge A$ but be unaware of A and hence not know it.

Fitting models for justification logic, presented in Chapter 4, use a forcing definition reminiscent of the one from awareness models: For any given justification t, the justification assertion $t{:}F$ holds at world u iff (1) F holds at all worlds v accessible from u and (2) t is an admissible evidence for F at u, $u \in \mathcal{E}(s, F)$, read as "u is a possible world at which s is relevant evidence for F." The principal difference is that postulated operations on justifications relate to natural closure conditions on admissible evidence functions $\mathcal{E}$ in justification logic models. Indeed, this idea has been explored in Sedlár (2013), which works with the language of LP and thinks of it as a multiagent modal logic, and taking justification terms as agents (more properly, actions of agents). This shows that justification logic models absorb the usual epistemic themes of awareness, group agency, and dynamics in a natural way.

1.5 Paraconsistency

Justification logic offers a well-principled approach to paraconsistency, which looks for noncollapsing logical ways of dealing with contradictory sets of assumptions, e.g.,

$$\{A, \neg A\}.$$

The following obvious observation shows how to convert any set of assumptions

$$\Gamma = \{A_1, A_2, A_3, \ldots\}$$

into a logically consistent set of sentences while maintaining all the intrinsic structure of Γ. Informally, instead of (perhaps inconsistently) assuming that Γ holds, we assume only that each sentence A from Γ has a justification, i.e.,

$$\vec{x}{:}\Gamma = \{x_1{:}A_1, x_2{:}A_2, x_3{:}A_3, \ldots\}.$$

It is easy to see that for each Γ, the set $\vec{x}{:}\Gamma$ is consistent in what will be our basic justification logic J.

For example, for $\Gamma = \{A, \neg A\}$,

$$\vec{x}{:}\Gamma = \{x_1{:}A, x_2{:}\neg A\},$$

states that x_1 is a justification for A and x_2 is a justification for $\neg A$. Within justification logic J in which no factivity (or even consistency) of justifications is assumed, the set of assumptions $\{x_1{:}A, x_2{:}\neg A\}$ is consistent, unlike the original set of assumptions $\{A, \neg A\}$.

There is nothing paraconsistent, magical, or artificial in reasoning from $\vec{x}{:}\Gamma$ in justification logic J. In practical terms, this means we gain the ability to effectively reason about inconsistent data sets, keeping track of justifications and their dependencies, with the natural possibility to draw meaningful conclusions even when some assumed justifications from $\vec{x}{:}\Gamma$ become compromised and should be discharged.

2
The Basics of Justification Logic

In this chapter we discuss matters of syntax and axiomatics. All material is propositional, and will be so until Chapter 10. Justification logics are closely related to modal logics, so we start briefly with them in order to fix the basic notation. And just as normal modal logics all extend a single simplest example, K, all justification logics extend a single simplest example, J_0. We will begin our discussion with modal logics, then we will discuss the justification logic J_0 in detail, and finally we will extend things to the most common and best-known justification logics. A much broader family of justification logics will be discussed in Chapter 8.

2.1 Modal Logics

All propositional formulas throughout this book are built up from a countable family of propositional variables. We use P, Q, ... as propositional variables, with subscripts if necessary, and we follow the usual convention that these are all distinct. As our main propositional connective we have implication, $\rightarrow$. We have negation, $\neg$, which we will take as primitive, or defined using the propositional constant $\bot$ representing falsehood, as convenient and appropriate at the time. We also use conjunction, $\wedge$, disjunction, $\vee$, and equivalence, $\leftrightarrow$, and these too may be primitive or defined depending on circumstances. We omit outer parentheses in formulas when it will do no harm.

We usually have a single modal necessity operator. It will generally be represented by $\Box$ though in epistemic contexts it may be represented by **K**. A dual operator representing possibility, $\Diamond$, is a defined operator and actually plays little role here. There is much work on epistemic logics with multiple agents, and there is some study of justification counterparts for them. When discussing these and their connections with modal logics, we will subscript the modal operators just described.

To date, no justification logic corresponding to a nonnormal modal logic has been introduced, so only normal modal logics will appear here. A *normal* modal logic is a set of modal formulas that contains all tautologies and all formulas of the form $\Box(X \to Y) \to (\Box X \to \Box Y)$ and is closed under *uniform substitution*, modus ponens, and *necessitation* (if X is present, so is $\Box X$). The smallest normal modal logic is K; it is a subset of all normal modal logics.

The logic K has a standard axiom system. Axioms are all tautologies (or enough of them) and all formulas of the form $\Box(X \to Y) \to (\Box X \to \Box Y)$. Rules are *Modus Ponens* $X, X \to Y \Rightarrow Y$ and *Necessitation* $X \Rightarrow \Box X$.

We are not actually interested in the vast collection of normal modal logics, but only in those for which a Hilbert system exists, having an axiomatization using a finite set of axiom *schemes*. In practice, this means adding axiom schemes to the axiomatization for K just given. We assume everybody knows axiom systems like T, K4, S4, and so on. We will refer to such logics as *axiomatically formulated*. Of course semantics plays a big role in modal logics, but we postpone a discussion for the time being.

2.2 Beginning Justification Logics

Justification logics, syntactically, are like modal logics except that *justification terms* take the place of $\Box$. Justification terms are intended to represent reasons or justifications for formulas. They have structure that encodes reasoning that has gone into them. We begin our formal presentation here.

Definition 2.1 (Justification Term) The set *Tm* of justification terms is built up as follows.

(1) There is a set of *justification variables*, $x, y, \ldots, x_1, y_1, \ldots$. Every justification variable is a justification term.
(2) There is a set of *justification constants*, $a, b, \ldots, a_1, b_1, \ldots$. Every justification constant is a justification term.
(3) There are binary operation symbols, $+$ and $\cdot$. If u and v are justification terms, so are $(u + v)$ and $(u \cdot v)$.
(4) There may be additional function symbols, $f, g, \ldots, f_1, g_1, \ldots$, of various arities. Which ones are present depends on the logic in question. If f is an n-place justification function symbol of the logic, and $t_1, \ldots, t_n$ are justification terms, $f(t_1, \ldots, t_n)$ is a justification term.

Neither + nor · is assumed to be commutative or associative, and there is no distributive law. We do, however, allow ourselves the notational convenience of omitting parentheses with multiple occurrences of ·, assuming associativity to the left. Thus, for instance, $a \cdot b \cdot c \cdot d$ is short for $(((a \cdot b) \cdot c) \cdot d)$. We make the same assumption concerning +, though it actually plays a much lesser role. Also we will generally assume that · binds more strongly than +, writing $a{\cdot}b{+}c$ instead of $(a \cdot b) + c$ for instance.

Definition 2.2 (Justification Formula) The set of justification formulas, *Fm*, is built in the usual recursive way, as follows.

(1) There is a set *Var* of *propositional variables*, $P, Q, \ldots, P_1, Q_1, \ldots$ (these are also known as propositional letters). Every propositional variable is a justification formula.

(2) $\bot$ (falsehood) is a justification formula.

(3) If X and Y are justification formulas, so is $(X \to Y)$.

(4) If t is a justification term and X is a justification formula, then $t{:}X$ is a justification formula.

We will sometimes use other propositional connectives, $\wedge$, $\vee$, $\leftrightarrow$, which we can think of as defined connectives, or primitive as convenient. Outer parentheses may be omitted in formulas if no confusion will result. If justification term t has a complex structure we generally will write $[t]{:}X$, using square brackets as a visual aid. Square brackets have no theoretical significance.

In a modal formula, $\Box$ is supposed to express that something is necessary, or known, or obligatory, or some such thing, but it does not say why. A justification term encodes this missing information; it provides the *why* absent from modal formulas. This is what their structure is for. Justification variables stand for arbitrary justification terms, and substitution for them is examined beginning with Definition 2.17. Justification constants stand for reasons that are not further analyzed—typically they are reasons for axioms. Their role is discussed in more detail once *constant specifications* are introduced, in Definition 10.32. The · operation corresponds to modus ponens. If $X \to Y$ is so for reason s and X is so for reason t, then Y is so for reason $s \cdot t$. (Reasons are not unique—Y may be true for other reasons too.) The + operation is a kind of weakening. If X is so for either reason s or reason t, then $s + t$ is also a reason for X. Other operations on justification terms, if present, correspond to features peculiar to particular modal logics and will be discussed as they come up.

2.3 J_0, the Simplest Justification Logic

As we will see, there are some justification logics having a version of a necessitation rule; there are others that do not. Some justification logics are closed under substitution of formulas for propositional variables, others are not. Allowing such a range of behavior is essential to enable us to capture and study the interactions of important features of modal logics that are sometimes hidden from us. But one consequence is, there is no good justification analog of the family of normal modal logics. Still, all justification logics have a common core, which we call J_0, and it is a kind of analog of the weakest normal modal logic, K, even though there is nothing structural we can point to as determining a "normal" justification logic apart from giving an axiomatization. In this section we present J_0 axiomatically; subsequently we discuss what must be added to get the general family of justification logics.

Definition 2.3 (Justification logic J_0) The language of J_0 has no justification function symbols beyond the basic two binary ones $+$ and $\cdot$. The axiom schemes are as follows.

Classical: All tautologies (or enough of them)

Application: All formulas of the form $s{:}(X \to Y) \to (t{:}X \to [s \cdot t]{:}Y)$

Sum: All formulas of the forms $s{:}X \to [s + t]{:}X$ and $t{:}X \to [s + t]{:}X$

The only J_0 rule of inference is Modus Ponens, $X, X \to Y \Rightarrow Y$.

J_0 is a very weak justification logic. It is, for instance, incapable of proving that any formula has a justification, see Section 3.2. Reasoning in J_0 is analogous to reasoning in the modal logic K without a necessitation rule! What we can do in J_0 is derive interesting facts about justifications *provided we make explicit what other formulas we would need to have justifications for*. We give an example to illustrate this. To help bring out the points we want to make, if $(X_1 \wedge \ldots \wedge X_n) \to Y$ is provable in J_0 we may write $X_1, \ldots, X_n \vdash_{\mathsf{J}_0} Y$. Order of formulas and placement of parentheses in the conjunction of the X_i don't matter because we have classical logic to work with.

In modal K, a common first example of a theorem is $\Box(X \wedge Y) \to (\Box X \wedge \Box Y)$. Here is the closest we can come to this in J_0. Our presentation is very much abbreviated.

Example 2.4 Assume u, v, and w are justification variables.

1.	$u{:}((X \wedge Y) \rightarrow X) \rightarrow (w{:}(X \wedge Y) \rightarrow [u \cdot w]{:}X)$	Application Axiom
2.	$v{:}((X \wedge Y) \rightarrow Y) \rightarrow (w{:}(X \wedge Y) \rightarrow [v \cdot w]{:}Y)$	Application Axiom
3.	$(u{:}((X \wedge Y) \rightarrow X) \wedge v{:}((X \wedge Y) \rightarrow y)) \rightarrow$ $(w{:}(X \wedge Y) \rightarrow [u \cdot w]{:}X)$	Classical Logic on 1, 2
4.	$(u{:}((X \wedge Y) \rightarrow X) \wedge v{:}((X \wedge Y) \rightarrow y)) \rightarrow$ $(w{:}(X \wedge Y) \rightarrow [v \cdot w]{:}Y)$	Classical Logic on 1, 2
5.	$(u{:}((X \wedge Y) \rightarrow X) \wedge v{:}((X \wedge Y) \rightarrow y)) \rightarrow$ $(w{:}(X \wedge Y) \rightarrow ([u \cdot w]{:}X \wedge [v \cdot w]{:}Y))$	Classical Logic on 3, 4

So we have shown that

$$u{:}((X \wedge Y) \rightarrow X), v{:}((X \wedge Y) \rightarrow Y) \vdash_{\mathsf{J}_0} (w{:}(X \wedge Y) \rightarrow ([u \cdot w]{:}X \wedge [v \cdot w]{:}Y))$$

which we can read as an analog of $\Box(X \wedge Y) \rightarrow (\Box X \wedge \Box Y)$ as follows. In J_0, for any w there are justification terms t_1 and t_2 such that $w{:}(X \wedge Y) \rightarrow (t_1{:}X \wedge t_2{:}Y)$, provided we have justifications for the tautologies $(X \wedge Y) \rightarrow X$ and $(X \wedge Y) \rightarrow Y$.

Note that $t_1 = u \cdot w$ and $t_2 = v \cdot w$ are different terms. But, making use of the Sum Axiom scheme, these can be brought together as $t_1 + t_2 = u \cdot w + v \cdot w$. It is important to understand that justifications, when they exist, are not unique.

2.4 Justification Logics in General

The core justification logic J_0 is extended to form other justification logics using two quite different types of machinery. First, one can add new operations on justification terms, besides the basic $+$ and $\cdot$, along with axiom schemes governing their use, similar to Sum and Application. This is directly analogous to the way axiom schemes are added to K to create other modal logics. Second, one can specify which truths of logic we assume we have justifications for. This is related to the roles $u{:}((X \wedge Y) \rightarrow X)$ and $v{:}((X \wedge Y) \rightarrow Y)$ play in Example 2.4. We devote most of this section to the second kind of extension. It is, in fact, the intended role for justification constants that, up to now, have not been used for anything special.

For the time being let us assume we have a justification logic resulting from the addition of function symbols and axiom schemes to J_0. The details don't matter for now, but it should be understood that our axioms may go beyond those for J_0.

Axioms of justification logics, like axioms generally, are simply assumed and are not analyzed further. The role of justification constant symbols is to serve as reasons or justifications for axioms. If A is an axiom, we can simply announce that constant symbol c plays the role of a justification for it. It may be that some axioms are assumed to have such justifications, but not necessarily all. Suppose we look at Example 2.4 again, and suppose we have decided that $(X \wedge Y) \to X$ is an axiom for which we have a specific justification, let us say the constant symbol c plays this role. Similarly let us say the constant symbol d represents a justification for $(X \wedge Y) \to Y$. Examining the derivation given in Example 2.4, it is easy to see that if we replace the variable u throughout by c, and the variable v throughout by d we still have a derivation, but one of

$$c{:}((X \wedge Y) \to X), d{:}((X \wedge Y) \to Y) \vdash_{\mathsf{J}_0} (w{:}(X \wedge Y) \to ([c \cdot w]{:}X \wedge [d \cdot w]{:}Y)).$$

If we add $c{:}((X \wedge Y) \to X)$ and $d{:}((X \wedge Y) \to Y)$ to our axioms for J_0, we can simply prove the formula

$$(w{:}(X \wedge Y) \to ([c \cdot w]{:}X \wedge [d \cdot w]{:}Y)).$$

Roughly speaking, a constant specification tells us what axioms we have justifications for and which constants justify these axioms. As we just saw, we can use a constant specification as a source of additional axioms. But there is an important complication. If A is an axiom and constant symbol c justifies it, $c{:}A$ conceptually also acts like an axiom, and it too may have its own justification. Then a constant symbol, say d, could come in here too, as a justification for $c{:}A$, and thus we might want to assume $d{:}c{:}A$. This repeats further, of course. For many purposes exact details don't matter much, so how constants are used, and for what purposes, is turned into a kind of parameter of our logics, called a constant specification.

Definition 2.5 (Constant Specification) A *constant specification* CS for a given justification logic is a set of formulas meeting the following conditions.

(1) Members of CS are of the form $c_n{:}c_{n-1}{:}\ldots c_1{:}A$ where $n > 0$, A is an axiom of JL, and each c_i is a constant symbol.
(2) If $c_n{:}c_{n-1}{:}\ldots c_1{:}A$ is in CS where $n > 1$, then $c_{n-1}{:}\ldots c_1{:}A$ is in CS too. Thus CS contains all intermediate specifications for whatever it contains.

One reason why constant specifications are treated as parameters can be discovered through a close look at Definition 2.3. It does not really provide an axiomatization for J_0, but rather a scheme for axiomatizations. The axioms called *Classical* in that definition are not fully specified, and in common practice many classical logic axiomatizations are in use. Any set sufficient to derive all tautologies will do. Then many different axiomatizations for J_0 would meet the required conditions, and similarly for any justification logic extending J_0

as well. Because constants are supposed to be associated with axioms, a variety of constant specifications come up naturally. And because details like this often matter very little, treating constant specifications as a parameter is quite reasonable.

Definition 2.6 (Logic of Justifications with a Constant Specification) Let JL be a justification logic, resulting from the addition of function symbols to the language of J_0 and corresponding axiom schemes to those of J_0. Let CS be a constant specification for JL. Then JL(CS) is the logic JL with members of CS added as axioms (not axiom schemes), still with modus ponens as the only rule of inference.

Constant specifications allow for great flexibility. A constant specification could associate many constants with a single axiom, or none at all. Allowing for many could be of use in tracking where particular pieces of reasoning come from. Allowing none might be appropriate in dealing with axioms that have some simple form, say $X \to X$, but where the size of X is astronomical. Or again we might want to use the same constant for all instances of a particular axiom schema, or we might want to keep the instances clearly distinguishable. If details don't matter at all for some particular purpose, we might want to associate a single constant symbol with every axiom, no matter what the form. Such a constant would simply be a record that a formula is an axiom, without going into particulars.

Some conditions on constant specifications have shown themselves to be of special interest and have been given names. Here is a list of the most common. There are others.

Definition 2.7 (Constant Specification Conditions) Let CS be a constant specification for a justification logic JL. The following requirements may be placed on CS.

Empty: CS $= \emptyset$. This amounts to working with JL itself. Epistemically one can think of it as appropriate for the reasoning of a completely skeptical agent.

Finite: CS is a finite set of formulas. This is fully representative because any specific derivation in a Justification Logic will be finite and so will involve only a finite set of constants.

Schematic: If A and B are both instances of the same axiom scheme, $c{:}A \in$ CS if and only if $c{:}B \in$ CS, for every constant symbol c.

Total: For each axiom A of JL and any constants $c_1, c_2, \ldots, c_n$ we have $c_n{:}c_{n-1}{:} \ldots c_1{:}A \in$ CS.

Axiomatically Appropriate: For every axiom A and for every $n > 0$ there are constant symbols c_i so that $c_n{:}c_{n-1}{:}\ldots c_1{:}A \in \mathsf{CS}$.

The working of justification axiom systems is specified as follows.

Definition 2.8 (Consequence) Suppose JL is a justification logic, CS is a constant specification for JL, S is an arbitrary set of formulas (not schemes), and X is a single formula. By $S \vdash_{\mathsf{JL(CS)}} X$ we mean there is a finite sequence of formulas, ending with X, in which each formula is either a instance of an axiom scheme of JL, a member of CS, a member of S, or follows from earlier formulas by modus ponens.

If $\{Y_1, \ldots, Y_k\} \vdash_{\mathsf{JL(CS)}} X$ we will simplify notation and write $Y_1, \ldots, Y_k \vdash_{\mathsf{JL(CS)}} X$. If $\emptyset \vdash_{\mathsf{JL(CS)}} X$ we just write $\vdash_{\mathsf{JL(CS)}} X$, or sometimes even $\mathsf{JL(CS)} \vdash X$.

When presenting examples of axiomatic derivations using a constant specification CS, we will write $c \overset{\mathsf{CS}}{\to} X$ as a suggestive way of saying that $c{:}X \in \mathsf{CS}$, and we will say "c justifies X".

We conclude this section with some examples of theorems of justification logics. For these we work with $\mathsf{JL(CS)}$ where JL is any justification logic and CS is any constant specification for it *that is axiomatically appropriate*. We assume JL has been axiomatized taking all tautologies as axioms, though taking "enough" would give similar results once we have Theorem 2.14.

Example 2.9 $\Box P \to \Box P$ is a theorem of any normal modal logic. It has more than one proof. We could simply note that it is an instance of a tautology, $X \to X$. Or we could begin with $P \to P$, a simpler instance of this tautology, apply necessitation getting $\Box(P \to P)$, and then use the K axiom $\Box(P \to P) \to (\Box P \to \Box P)$ and modus ponens to conclude $\Box P \to \Box P$. While these are different modal derivations, the result is the same. But when we mimic the steps in $\mathsf{JL(CS)}$, they lead to different results.

Let t be an arbitrary justification term. Then $t{:}P \to t{:}P$ is a theorem of $\mathsf{JL(CS)}$ because it is an instance of a tautology. But also $P \to P$ is an instance of a tautology and so, because $\mathsf{JL(CS)}$ is assumed axiomatically appropriate, the constant specification assigns some constant to it; say $c{:}(P \to P) \in \mathsf{CS}$. Because $c{:}(P \to P) \to (t{:}P \to [c \cdot t]{:}P)$ is an axiom, $t{:}P \to [c \cdot t]{:}P$ follows by modus ponens.

In justification logic, instead of a single formula $\Box P \to \Box P$ with two proofs we have two different theorems that contain traces of their proofs. Both $t{:}P \to t{:}P$ and $t{:}P \to [c \cdot t]{:}P$ say that if there is a reason for P, then there is a reason for P, but they give us different reasons.

One of the first things everybody shows axiomatically when studying modal logic is that $\Box(P \wedge Q) \leftrightarrow (\Box P \wedge \Box Q)$ is provable in K, and thus is provable in every axiom system for a normal modal logic. But the argument from left to right is quite different from the argument from right to left. Because justification theorems contain traces of their proofs, we should not expect a single justification analog of this modal equivalence, but rather separate results for the left–right implication and for the right–left implication.

Example 2.10 Here is a justification derivation corresponding to the usual modal argument for $\Box(P \wedge Q) \to (\Box P \wedge \Box Q)$.

1.	$(P \wedge Q) \to P$	tautology
2.	$c{:}((P \wedge Q) \to P)$	cons spec
3.	$c{:}((P \wedge Q) \to P) \to (t{:}(P \wedge Q) \to [c \cdot t]{:}P)$	Application Axiom
4.	$t{:}(P \wedge Q) \to [c \cdot t]{:}P$	mod pon on 2, 3
5.	$(P \wedge Q) \to Q$	tautology
6.	$d{:}((P \wedge Q) \to Q)$	cons spec
7.	$d{:}((P \wedge Q) \to Q) \to (t{:}(P \wedge Q) \to [d \cdot t]{:}Q)$	Application Axiom
8.	$t{:}(P \wedge Q) \to [d \cdot t]{:}Q$	mod pon on 6, 7
9.	$t{:}(P \wedge Q) \to ([c \cdot t]{:}P \wedge [d \cdot t]{:}Q)$	class log on 4, 8

Then $t{:}(P \wedge Q) \to ([c \cdot t]{:}P \wedge [d \cdot t]{:}Q)$ is a theorem of JL(CS) where $c \overset{\mathsf{CS}}{\to} ((P \wedge Q) \to P)$ and $d \overset{\mathsf{CS}}{\to} ((P \wedge Q) \to Q)$.

Example 2.11 A justification counterpart of the modal theorem $(\Box P \wedge \Box Q) \to \Box(P \wedge Q)$ follows.

1.	$P \to (Q \to (P \wedge Q))$	tautology
2.	$c{:}(P \to (Q \to (P \wedge Q)))$	cons spec
3.	$c{:}(P \to (Q \to (P \wedge Q))) \to$ $(t{:}P \to [c \cdot t]{:}(Q \to (P \wedge Q)))$	Application Axiom
4.	$t{:}P \to [c \cdot t]{:}(Q \to (P \wedge Q))$	mod pon on 2, 3
5.	$[c \cdot t]{:}(Q \to (P \wedge Q)) \to$ $(u{:}Q \to [c \cdot t \cdot u]{:}(P \wedge Q))$	Application Axiom
6.	$(t{:}P \wedge u{:}Q) \to [c \cdot t \cdot u]{:}(P \wedge Q)$	class log on 4, 5

So $(t{:}P \wedge u{:}Q) \to [c \cdot t \cdot u]{:}(P \wedge Q)$ is a theorem of JL(CS) where $c \overset{\mathsf{CS}}{\to} (P \to (Q \to (P \wedge Q)))$.

Our final example illustrates the use of +, which has not come up so far. It is for handling situations where there is more than one explanation needed for something, as in a proof by cases. At first glance this seems rather minor, but + turns out to play a vital role when we come to realization results.

Example 2.12 $(\Box X \vee \Box Y) \to \Box(X \vee Y)$ is a theorem of K with an elementary proof that we omit. Let us construct a counterpart in JL(CS), still assuming that CS is axiomatically appropriate and all tautologies are axioms.

1.	$X \to (X \vee Y)$	tautology
2.	$c{:}(X \to (X \vee Y))$	cons spec
3.	$c{:}(X \to (X \vee Y)) \to$ $(t{:}X \to [c \cdot t]{:}(X \vee Y))$	Application Axiom
4.	$t{:}X \to [c \cdot t]{:}(X \vee Y)$	mod pon on 2, 3
5.	$Y \to (X \vee Y)$	tautology
6.	$d{:}(Y \to (X \vee Y))$	cons spec
7.	$d{:}(Y \to (X \vee Y)) \to (u{:}Y \to [d \cdot u]{:}(X \vee Y))$	Application Axiom
8.	$u{:}Y \to [d \cdot u]{:}(X \vee Y)$	mod pon on 6, 7
9.	$[c \cdot t]{:}(X \vee Y) \to [c \cdot t + d \cdot u]{:}(X \vee Y)$	Sum Axiom
10.	$[d \cdot u]{:}(X \vee Y) \to [c \cdot t + d \cdot u]{:}(X \vee Y)$	Sum Axiom
11.	$t{:}X \to [c \cdot t + d \cdot u]{:}(X \vee Y)$	clas log on 4, 9
12.	$u{:}Y \to [c \cdot t + d \cdot u]{:}(X \vee Y)$	clas log on 8, 10
13.	$(t{:}X \vee u{:}Y) \to [c \cdot t + d \cdot u]{:}(X \vee Y)$	class log on 11, 12

The consequents of 4 and 8 both provide reasons for $X \vee Y$, but the reasons are different. We have used + to combine them, getting a justification analog of $(\Box X \vee \Box Y) \to \Box(X \vee Y)$.

2.5 Fundamental Properties of Justification Logics

All justification logics have certain common and useful properties. Some features are identical with those of classical logic; others have twists that are special to justification logics. This section is devoted to ones we will use over and over. *Throughout this section let* JL *be a justification logic and* CS *be a constant specification for it.*

Because the only rule of inference is modus ponens the classical proof of the deduction theorem applies. We thus have $S, X \vdash_{\mathsf{JL(CS)}} Y$ if and only if $S \vdash_{\mathsf{JL(CS)}} X \to Y$. Because formal proofs are finite we have compactness that, combined with the deduction theorem, tells us: $S \vdash_{\mathsf{JL(CS)}} X$ if and only if $\vdash_{\mathsf{JL(CS)}} Y_1 \to (Y_2 \to \ldots \to (Y_n \to X)\ldots)$ for some $Y_1, Y_2, \ldots, Y_n \in S$. These are exactly like their classical counterparts. Furthermore, the following serves as a replacement for the modal Necessitation Rule.

Definition 2.13 (Internalization) JL *has the internalization property relative to constant specification* CS provided, if $\vdash_{\mathsf{JL(CS)}} X$ then for some justification term t, $\vdash_{\mathsf{JL(CS)}} t{:}X$. In addition we say that JL has the *strong* internalization property if t contains no justification variables and no justification operation or function symbols except $\cdot$. That is, t is built up from justification constants using only $\cdot$.

Theorem 2.14 *If* CS *is an axiomatically appropriate constant specification for* JL *then* JL *has the strong internalization property relative to* CS.

Proof By induction on proof length. Suppose $\vdash_{\mathsf{JL(CS)}} X$ and the result is known for formulas with shorter proofs. If X is an axiom of JL or a member of CS, there is a justification constant c such that $c{:}X$ is in CS, and so $c{:}X$ is provable. If X follows from earlier proof lines by modus ponens from $Y \rightarrow X$ and Y then, by the induction hypothesis, $\vdash_{\mathsf{JL(CS)}} s{:}(Y \rightarrow X)$ and $\vdash_{\mathsf{JL(CS)}} t{:}Y$ for some s, t containing no justification variables, and with $\cdot$ as the only function symbol. Using the J_0 Application Axiom $s{:}(Y \rightarrow X) \rightarrow (t{:}Y \rightarrow [s \cdot t]{:}X)$ and modus ponens, we get $\vdash_{\mathsf{JL(CS)}} [s \cdot t]{:}X$. □

If X is provable using an axiomatically appropriate constant specification so is $t{:}X$, and the term t constructed in the preceding proof actually internalizes the steps of the axiomatic proof of X, hence the name *internalization*. Of course different proofs of X will produce different justification terms. Here is an extremely simple example, but one that is already sufficient to illustrate this point.

Example 2.15 Assume JL is a justification logic, CS is an axiomatically appropriate constant specification for it, and all tautologies are axioms of JL.

$P \rightarrow P$ is a tautology so $c \overset{\mathsf{CS}}{\rightarrow} (P \rightarrow P)$ for some c. Then $c{:}(P \rightarrow P)$ is a theorem, and we have the justification term c internalizing a proof of $P \rightarrow P$.

Here is a more roundabout proof of $P \rightarrow P$, giving us a more complicated internalizing term. Following the method in the proof of Theorem 2.14, we construct the internalization simultaneously.

1.	$(P \rightarrow (P \rightarrow P)) \rightarrow$ $((P \rightarrow P) \rightarrow (P \rightarrow P))$	tautology	d (cons spec)
2.	$P \rightarrow (P \rightarrow P)$	tautology	e (cons spec)
3.	$(P \rightarrow P) \rightarrow (P \rightarrow P)$	mod pon on 1, 2	$d \cdot e$
4.	$P \rightarrow P$	tautology	c (cons spec)
5.	$P \rightarrow P$	mod pon on 3, 4	$d \cdot e \cdot c$

This time we get a justification term $d \cdot e \cdot c$, or more properly $(d \cdot e) \cdot c$, internalizing a proof of $P \rightarrow P$, where $c \overset{\mathsf{CS}}{\rightarrow} ((P \rightarrow (P \rightarrow P)) \rightarrow ((P \rightarrow P) \rightarrow (P \rightarrow P)))$, $e \overset{\mathsf{CS}}{\rightarrow} (P \rightarrow (P \rightarrow P))$, and $c \overset{\mathsf{CS}}{\rightarrow} (P \rightarrow P)$.

The problem of finding a "simplest" justification term is related to the problem of finding the "simplest" proof of a provable formula. It is not entirely clear what this actually means.

Corollary 2.16 (Lifting Lemma) *Suppose* JL *is a justification logic that has the internalization property relative to* CS *(in particular, if* CS *is axiomatically appropriate). If* $X_1, \ldots, X_n \vdash_{\mathsf{JL(CS)}} Y$ *then for any justification terms* $t_1, \ldots, t_n$ *there is a justification term* u *so that* $t_1{:}X_1, \ldots, t_n{:}X_n \vdash_{\mathsf{JL(CS)}} u{:}Y$.

Proof The proof is by induction on n. If $n = 0$ this is simply the definition of Internalization.

Suppose the result is known for n, and we have $X_1, \ldots, X_n, X_{n+1} \vdash_{\mathsf{JL(CS)}} Y$. We show that for any $t_1, \ldots, t_n, t_{n+1}$ there is some u so that $t_1{:}X_1, \ldots, t_n{:}X_n, t_{n+1}{:}X_{n+1} \vdash_{\mathsf{JL(CS)}} u{:}Y$.

Using the deduction theorem, $X_1, \ldots, X_n \vdash_{\mathsf{JL(CS)}} (X_{n+1} \to Y)$. By the induction hypothesis, for some v we have $t_1{:}X_1, \ldots, t_n{:}X_n \vdash_{\mathsf{JL(CS)}} v{:}(X_{n+1} \to Y)$. Now $v{:}(X_{n+1} \to Y) \to (t_{n+1}{:}X_{n+1} \to [v \cdot t_{n+1}]{:}Y)$ is an axiom hence $t_1{:}X_1, \ldots, t_n{:}X_n \vdash_{\mathsf{JL(CS)}} (t_{n+1}{:}X_{n+1} \to [v \cdot t_{n+1}]{:}Y)$. By modus ponens, $t_1{:}X_1, \ldots, t_n{:}X_n, t_{n+1}{:}X_{n+1} \vdash_{\mathsf{JL(CS)}} [v \cdot t_{n+1}]{:}Y$, so take u to be $v \cdot t_{n+1}$. □

Next we move on to the role of justification variables. We said earlier, rather informally, that variables stood for arbitrary justification terms. In order to make this somewhat more precise, we need to introduce substitution.

Definition 2.17 (Substitution) A substitution is a function σ mapping some set of justification variables to justification terms, with no variable in the domain of σ mapping to itself. We are only interested in substitutions with finite domain. If the domain of σ is $\{x_1, \ldots, x_n\}$, and each x_i maps to justification term t_i, it is standard to represent this substitution by $(x_1/t_1, \ldots, x_n/t_n)$, or sometimes as $(\vec{x}/\vec{t})$. For a justification formula X the result of applying a substitution σ is denoted $X\sigma$; likewise $t\sigma$ is the result of applying substitution σ to justification term t.

Substitutions map axioms of a justification logic into axioms (because axiomatization is by schemes), and they preserve modus ponens applications. But one must be careful because the role of constants changes with a substitution. Suppose CS is a constant specification, A is an axiom, and $c{:}A$ is added to a proof where this addition is authorized by CS. Because axiomatization is by schemes $A\sigma$ is also an axiom, but if we add $c{:}A\sigma$ to a proof this may no longer meet constant specification CS. A new constant specification, call it $(\mathsf{CS})\sigma$, can be computed from the original one: put $c{:}A\sigma \in (\mathsf{CS})\sigma$ just in case $c{:}A \in \mathsf{CS}$, for any c. If CS was axiomatically appropriate, $\mathsf{CS} \cup (\mathsf{CS})\sigma$ will also be. So, if X is provable using an axiomatically appropriate constant specification CS, the same will be true for $X\sigma$, not using the original constant specification but rather using $\mathsf{CS} \cup (\mathsf{CS})\sigma$. But this is more detail than we generally need to care about. The following suffices for much of our purposes.

Theorem 2.18 (Substitution Closure) *Suppose* JL *is a justification logic and X is provable in* JL *using some (axiomatically appropriate) constant specification. Then for any substitution σ, $X\sigma$ is also provable in* JL *using some (axiomatically appropriate) constant specification.*

We introduce some special notation that suppresses details of constant specifications when we don't need to care about these details.

Definition 2.19 Let JL be a justification logic. We write $\vdash_{\mathsf{JL}} X$ as short for: there is some axiomatically appropriate constant specification CS so that $\vdash_{\mathsf{JL(CS)}} X$.

Theorem 2.20 *Let* JL *be a justification logic.*

(1) If $\vdash_{\mathsf{JL}} X$ then $\vdash_{\mathsf{JL}} X\sigma$ for any substitution σ.
(2) If $\vdash_{\mathsf{JL}} X$ and $\vdash_{\mathsf{JL}} X \to Y$ then $\vdash_{\mathsf{JL}} Y$.

Proof Item (1) is directly from Theorem 2.18. For item (2), suppose $\vdash_{\mathsf{JL}} X$ and $\vdash_{\mathsf{JL}} X \to Y$. Then there are axiomatically appropriate constant specifications CS_1 and CS_2 so that $\vdash_{\mathsf{JL(CS_1)}} X$ and $\vdash_{\mathsf{JL(CS_2)}} X \to Y$. Now $\mathsf{CS}_1 \cup \mathsf{CS}_2$ will also be an axiomatically appropriate constant specification and $\vdash_{\mathsf{JL(CS_1 \cup CS_2)}} X$ and $\vdash_{\mathsf{JL(CS_1 \cup CS_2)}} X \to Y$, so $\vdash_{\mathsf{JL(CS_1 \cup CS_2)}} Y$ and hence $\vdash_{\mathsf{JL}} Y$. □

In fact, it is easy to check that $\vdash_{\mathsf{JL}} X$ if and only if $\vdash_{\mathsf{JL(T)}} X$, where T is the total constant specification. This gives an alternate, and easier, characterization.

2.6 The First Justification Logics

In this section and the next we present a number of specific examples of justification logics. We have tried to be systematic in naming these justification logics. Of course modal logic is not entirely consistent in this respect, and justification logic inherits some of its quirks, but we have tried to minimize anomalies.

Naming Conventions: It is common to name modal logics by stringing axiom names after K; for instance KT, K4, and so on, with K itself as the simplest case. When we have justification logic counterparts for such modal logics, we will use the same name except with a substitution of J for K; for instance JT, J4, and so on. There is a problem here because a modal logic generally has more than one justification counterpart (if it has any). We will specify which one we have in mind. Formally, JT, J4, and so on result from the addition of axiom schemes, justification function symbols, and a constant specification to

J_0. When details of a constant specification matter, we will write things like $\mathsf{JT(CS)}$, $\mathsf{J4(CS)}$, and so on, making the constant specification explicit.

We will rarely refer to J_0 again because its definition does not actually allow for a constant specification. From now on we will use J for J_0 extended with some constant specification, and we will write $\mathsf{J(CS)}$ when explicitness is called for. Note that J_0 can be thought of as $\mathsf{J}(\emptyset)$, where $\emptyset$ is the empty constant specification.

The general subject of justification logics evolved from the aforementioned Gödel–Artemov project, which embeds intuitionistic logic into the modal logic $\mathsf{S4}$, which in turn embeds into the justification logic known as LP (for *logic of proofs*). It is with LP and its standard sublogics that we are concerned in this section. These are the best-known justification logics, just as K, T (or sometimes KT), $\mathsf{S4}$ (or sometimes $\mathsf{KT4}$), and a few others are the best-known modal logics. For the time being the notion of a justification logic being a counterpart of a modal logic will be an intuitive one. A proper definition will be given in Section 7.2. With two exceptions, the justification logics examined here arise by adding additional operations to the $+$ and $\cdot$ common to all justification logics. The first exception involves *factivity*, with which we begin.

Factivity for modal logics is represented by the axiom scheme $\Box X \to X$. If we think of the necessity operator epistemically, this would be written $\mathbf{K}X \to X$. It asserts that if X is known, then X is so. The justification counterpart is the following axiom scheme.

Factivity $t{:}X \to X$

Factivity is a strong assumption: justifications cannot be wrong. Nonetheless, if the justification is a mathematical proof, factivity is something mathematicians are generally convinced of. If we think of knowledge as justified, true belief, factivity is built in. Philosophers generally understand justified, true belief to be inherent in knowledge, but not sufficient, see Gettier (1963).

The modal axiom scheme $\Box X \to X$ is called T. The weakest normal modal logic including all instances of this scheme is KT, sometimes abbreviated simply as T. We use JT for J plus Factivity and, as noted earlier, we use $\mathsf{JT(CS)}$ when a specific constant specification is needed. Note that the languages of JT and J are the same. There is one more such example, after which additional operation symbols must be brought in.

Consistency is an important special case of Factivity. Modally it can be represented in several ways. One can assume the axiom scheme $\Box X \to \Diamond X$. In any normal modal logic this turns out to coincide with assuming $\neg\Box\bot$ (where $\bot$ represents falsehood), or equivalently $\Box\bot \to \bot$, which is a very special instance of $\Box X \to X$. If one thinks of $\Box$ as representing provability, $\neg\Box\bot$ says falsehood is not provable—consistency. Suppose one thinks of $\Box$ deontically, so that $\Box X$ is read that X is obligatory, or perhaps that it is obligatory to bring about a state in which X. Then $\Box X \to \Diamond X$, or equivalently $\Box X \to \neg\Box\neg X$ says that if X is obligatory, then $\neg X$ isn't—a plausible condition on obligations. It is because of this interesting deontic reading that any of the equivalent versions is commonly called D, standing for *deontic*. Any of these has a justification counterpart. We adopt the following version.

Consistency $t{:}\bot \to \bot$

JD is J plus Consistency. Note that JT extends JD.

Positive Introspection is a common assumption about an agent's knowledge: If an agent knows something, the agent knows that it is known; an agent can introspect about the knowledge he or she possesses. In logics of knowledge it is formulated as $\mathbf{K}X \to \mathbf{KK}X$. If one understands $\Box$ as representing provability in formal arithmetic, it is possible to prove that a proof is correct: $\Box X \to \Box\Box X$. To formulate a justification logic counterpart, Artemov introduced a one-place function symbol on justification terms, denoted ! and written in prefix position. The intuitive idea is that if t is a justification of something, $!t$ is a justification that t is, indeed, such a justification. Note that the basic language of justification logics has been extended, and this must be reflected in any constant specifications being considered.

Positive Introspection $t{:}X \to\ !t{:}t{:}X$

As part of the intuition behind the program to create an arithmetic semantics for intuitionistic logic, justification terms were thought of as representing particular formal proofs. In this context justification terms were called *proof terms*, and ! was called *Proof Checker*. Later, when the idea of justification logics had broadened, ! was called *Fact Checker*. Typical everyday examples of Fact Checker applications are a referee report certifying that a proof in a paper is correct, or a computer output showing a verification that a formal proof is correct.

If Positive Introspection is among the axioms of a justification logic, constant specifications can be considerably simplified. For example, if CS is axiomatically appropriate and A is an axiom, we have been requiring that there be some constant symbol a with $a{:}A \in$ CS, and also some constant symbol b with $b{:}a{:}A \in$ CS, and also some constant symbol c with $c{:}b{:}a{:}A \in$ CS, and so on. But with Positive Introspection we really only need the first of these: there is some a with $a{:}A \in \mathsf{CS}$. We don't need to postulate the existence of b because we have that $a{:}A \rightarrow\ !a{:}a{:}A$, and so we can use $!a$ where we would have used b. Similarly we can use $!(!a)$ where we would have used c. And so on. The formulation and proof of Theorem 2.14 need some obvious modification, but it is easy. In fact, the first justification logic LP had just such a formulation. The simplification is nice, but is not a deep issue.

The addition of $\Box X \rightarrow \Box\Box X$ to K is known as K4. We use J4 for the justification counterpart, similar to what we did with JD and KD earlier.

LP is a justification logic that has been mentioned many times so far. We are finally in a position to say what it is. Axiomatically it is JT4, that is, J plus Factivity (and hence Consistency) plus Positive Introspection (and so with ! added to the language). For historical reasons instead of JT4 the name LP is commonly used standing, as we have already noted, for *logic of proofs*. It is a counterpart of the modal logic S4, sometimes known as KT4.

In Examples 2.10–2.12 we gave some instances of modal theorems and justification counterparts. Here is one more such example. This time the modal logic is S4, and the justification logic is LP(CS) where CS is an axiomatically appropriate constant specification. The example is taken from Artemov (2001), and like Example 2.12 it illustrates the uses of the + operator, as well as !.

Example 2.21 We begin by deriving $(\Box X \vee \Box Y) \rightarrow \Box(\Box X \vee \Box Y)$ in S4. Then we carry out a similar derivation in LP(CS) of a formula that, in a sense, embodies the modal argument. Here is the S4 derivation.

1.	$\Box X \rightarrow (\Box X \vee \Box Y)$, $\Box Y \rightarrow (\Box X \vee \Box Y)$	classical axioms
2.	$\Box(\Box X \rightarrow (\Box X \vee \Box Y))$, $\Box(\Box Y \rightarrow (\Box X \vee \Box Y))$	Necessitation, on 1
3.	$\Box X \rightarrow \Box\Box X$, $\Box Y \rightarrow \Box\Box Y$	axioms
4.	$\Box\Box X \rightarrow \Box(\Box X \vee \Box Y)$, $\Box\Box Y \rightarrow \Box(\Box X \vee \Box Y)$	from 2
5.	$\Box X \rightarrow \Box(\Box X \vee \Box Y)$, $\Box Y \rightarrow \Box(\Box X \vee \Box Y)$	from 3, 4
6.	$(\Box X \vee \Box Y) \rightarrow \Box(\Box X \vee \Box Y)$	from 5

And now the corresponding derivation in LP(CS). In 2 it is assumed that a and b are constants that are assigned to the axioms shown in 1 by the constant specification we are using for LP. Also x and y are justification variables.

1.	$x{:}X \to (x{:}X \lor y{:}Y),\ y{:}Y \to (x{:}X \lor y{:}Y)$	tautologies
2.	$a{:}(x{:}X \to (x{:}X \lor y{:}Y)),\ b{:}(y{:}Y \to (x{:}X \lor y{:}Y))$	cons spec
3.	$x{:}X \to !x{:}x{:}X,\ y{:}Y \to !y{:}y{:}Y$	Pos Intro Axioms
4.	$!x{:}x{:}X \to [a\cdot!x]{:}(x{:}X \lor y{:}Y),$ $!y{:}y{:}Y \to [b\cdot!y]{:}(x{:}X \lor y{:}Y)$	from 2
5.	$x{:}X \to [a\cdot!x]{:}(x{:}X \lor y{:}Y),\ y{:}Y \to [b\cdot!y]{:}(x{:}X \lor y{:}Y)$	from 3, 4
5′.	$[a\cdot!x]{:}(x{:}X \lor y{:}Y) \to [a\cdot!x + b\cdot!y]{:}(x{:}X \lor y{:}Y)$	Sum Axiom
5″.	$[b\cdot!y]{:}(x{:}X \lor y{:}Y) \to [a\cdot!x + b\cdot!y]{:}(x{:}X \lor y{:}Y)$	Sum Axiom
6.	$x{:}X \to [a\cdot!x + b\cdot!y]{:}(x{:}X \lor y{:}Y)$	from 5, 5′
6′.	$y{:}Y \to [a\cdot!x + b\cdot!y]{:}(x{:}X \lor y{:}Y)$	from 5, 5″
6″.	$(x{:}X \lor y{:}Y) \to [a\cdot!x + b\cdot!y]{:}(x{:}X \lor y{:}Y)$	from 6, 6′

Informally this says that, in LP, if we have a justification of one or X or Y, then we have a justification of that fact.

2.7 A Handful of Less Common Justification Logics

In Section 2.6 we looked at the oldest and most familiar of the justification logics. Because those early justification logics are sublogics of LP, and that is interpretable in formal arithmetic, all justification logics considered in that section are also interpretable in arithmetic. The possibility of direct arithmetic interpretability disappears as the family of justification logics grows. In this section we take a look at a broader and less familiar group of logics. For each we begin with a modal logic, then introduce a justification counterpart. Fitting Semantics for these will be discussed in Chapter 4. In Section 4.3 specific models will be presented and soundness shown. In Section 4.5 completeness will be proved. Several interesting issues are illustrated by our particular logics, and these issues will be discussed at appropriate points.

2.7.1 $\mathsf{K4}^3$ and $\mathsf{J4}^3$

Modal logic $\mathsf{K4}^3$ is one of a family, $\mathsf{K4}^n$, from Chellas (1980). $\mathsf{K4}^3$ extends K with the schema $\Box X \to \Box\Box\Box X$. (The familiar K4 is $\mathsf{K4}^2$ and is in this family.) There is more than one way of constructing a justification counterpart for $\mathsf{K4}^3$, something that is also true of K4 itself. This is a point that will be discussed later. Here we adopt the following justification version. Let ! and !! be one-place function symbols. (Please note that we are using !! as a single symbol, and not as an iteration of !.) Our justification counterpart for $\mathsf{K4}^3$ results from adding to J the following axiom schema

$$t{:}X \to\ !!t{:}!t{:}t{:}X$$

We call this justification logic $\mathsf{J4}^3$.

Every theorem of $\mathsf{K4}^3$ is a theorem of K4 because $\Box X \to \Box\Box\Box X$ is a theorem of K4 by the following simple proof.

1.	$\Box X \to \Box\Box X$	K4 axiom
2.	$\Box\Box X \to \Box\Box\Box X$	another K4 axiom instance
3.	$\Box X \to \Box\Box\Box X$	classical reasoning from 1 and 2

This argument that K4 extends $\mathsf{K4}^3$ can be adapted to J4(CS), where CS is an axiomatically complete constant specification. Here is a proof in J4(CS).

1.	$t{:}X \to\ !t{:}t{:}X$	J4 axiom
2.	$!t{:}t{:}X \to\ !(!t){:}!t{:}t{:}X$	another J4 axiom instance
3.	$t{:}X \to\ !(!t){:}!t{:}t{:}X$	classical reasoning from 1 and 2

Unlike with K4 and $\mathsf{K4}^3$, this does not say that J4 extends $\mathsf{J4}^3$. What it does say is that $\mathsf{J4}^3$ translates into, or embeds into, J4 by replacing $!!t$ by $!(!t)$.

2.7.2 S5 and JT45

S5 is perhaps the most common modal logic used in applications. One standard axiomatization is to add the scheme of *negative introspection* to S4: $\neg\Box X \to \Box\neg\Box X$. Written epistemically we have $\neg\mathbf{K}X \to \mathbf{K}\neg\mathbf{K}X$. It is the strong assumption that if you don't know something, you know you don't know it. Despite its strength, it is commonly assumed in epistemic logic, perhaps because it has the effect of simplifying the corresponding possible world semantics.

A justification counterpart for S5, now called JT45, was introduced independently in Pacuit (2005, 2006) and Rubtsova (2006b). A one-place function symbol ?, whose role is analogous to that of !, is added to the LP language, and like !, it is written in prefix position. The justification axiom scheme added to LP is the following.

$$\neg t{:}X \to\ ?t{:}\neg t{:}X$$

We will see that the semantics created for JT45 played an important role in understanding the behavior of the *evidence function*, which will be introduced in Chapter 4. We also note that while the behavior of "?" doesn't allow for a direct arithmetic interpretation, in Artemov et al. (1999) an alternative formulation of modal S5 was used to produce a different explicit counterpart, in which "?" was emulated by other means, and which does have an arithmetical provability interpretation.

2.7.3 Sahlqvist Examples

Sahlqvist formulas are important in modal logic, but we do not discuss their general significance here; an excellent treatment can be found in Blackburn et al. (2001). In example 3.5.5, section 3.6 of that book a few simple instances are discussed. We begin with the first of these examples.

Example One The formula $\Box(X \rightarrow \Diamond X)$ is a Sahlqvist scheme. An equivalent axiomatization for the modal logic it gives uses the scheme $\Box(\Box X \rightarrow X)$, and we adopt that here because of the minor role that the possibility operator plays in justification logics. We propose the following scheme for a corresponding justification logic, where f is a one-place function symbol.

$$f(t){:}(t{:}X \rightarrow X)$$

Example Two $(X \wedge \Diamond\neg X) \rightarrow \Diamond X$ is a Sahlqvist schema, with an equivalent logic axiomatization using $\Box X \rightarrow (X \vee \Box\neg X)$. We suggest the following justification axiom scheme, where g is a one-place function symbol.

$$t{:}X \rightarrow (X \vee g(t){:}\neg X)$$

2.7.4 S4.2 and JT4.2

The modal logic S4.2 is from Geach (1973). It is the paradigm example of a broader class of logics that we will examine in detail in Chapter 8. Axiomatically S4.2 extends S4 with the scheme $\Diamond\Box X \rightarrow \Box\Diamond X$. Because the possibility operator does not play much of a role in justification logics, we reformulate this scheme as $\neg\Box\neg\Box X \rightarrow \Box\neg\Box\neg X$. For a justification counterpart we build on LP, our counterpart of S4. We add two function symbols, f and g, not the same as in the Sahlqvist examples, each two-place, and adopt the following axiom scheme.

$$\neg f(t, u){:}\neg t{:}X \rightarrow g(t, u){:}\neg u{:}\neg X$$

We call this justification logic JT4.2.

Although we primarily have a formal interest here, a few motivating words about this justification scheme above might be in order. In LP, because of the Factivity axiom scheme $t{:}X \rightarrow X$, we have provability of $(t{:}X \wedge u{:}\neg X) \rightarrow \bot$, for any t and u, and thus provability of $\neg t{:}X \vee \neg u{:}\neg X$. In any context one of the disjuncts must hold. The axiom scheme we adopted is equivalent to $f(t,u){:}\neg t{:}X \vee g(t,u){:}\neg u{:}\neg X$. Very informally, this says that in any context we have means for computing a justification for the disjunct that holds. It is a strong assumption, but not implausible under some circumstances. We do not pursue the point further here.

2.7.5 KX4 and JX4

KX4 is a weakening of S4 in which factivity, $\Box X \rightarrow X$, is replaced with the axiom scheme $\Box\Box X \rightarrow \Box X$. This scheme has sometimes been called C4 but more recently X has come into use, and we adopt this name. The logic KX4 is axiomatized using modus ponens and the following axiom schemes.

Classical All (or enough) tautologies
K $\Box(X \rightarrow Y) \rightarrow (\Box X \rightarrow \Box Y)$
X $\Box\Box X \rightarrow \Box X$
4 $\Box X \rightarrow \Box\Box X$

For a justification counterpart we start with LP again. We drop the Factivity axiom scheme, $t{:}X \rightarrow X$. To replace it we introduce a binary function symbol c, which we write in infix position, and we add the following scheme.

$$s{:}t{:}X \rightarrow [s \mathsf{c} t]{:}X$$

We call the resulting justification logic JX4. This logic has interesting features, which are explored in Fitting (2017).

3
The Ontology of Justifications

So far we have seen justification logics axiomatically presented—it was with axiomatic LP that the subject began. But proof machinery is not enough—one also needs semantics. Technically, without a semantics it is difficult to show something is not provable. But more deeply, a semantics gives us some precise idea of what our proofs are about—what our ontological assumptions are. Over the years, apart from the original provability semantics for the logic of proofs, LP, and related systems, three semantics have been introduced for justification logics. All three are very general and are discussed in this chapter and the next.

Historically, the first nonprovability semantics for justification logic (then the logic of proofs LP) is from Mkrtychev (1997). It is called *Mkrtychev semantics* here. This was followed by *Fitting semantics*, which combined Mkrtychev's machinery with that of Kripke's possible world semantics (Fitting, 2003, 2005). Chapter 4 is devoted to it. The third semantics is due to Artemov, and is called *modular semantics* (which includes the so-called *basic semantics*), see Artemov (2012), Kuznets and Studer (2012), and Artemov (2018). In many ways it is the simplest and most basic of the three, though it was the last to be created. Mkrtychev and modular semantics are covered in this chapter.

3.1 Generic Logical Semantics of Justifications

What kind of logical objects are justifications? When asked in a mathematical context, "What is a predicate?," we have a ready answer: a subset of a Cartesian product of the domain set. We know a reasonable mathematical answer (within Kolmogorov's model) to the question, "What is probability?": a function from a σ-algebra of events to $[0, 1]$, with some natural conditions imposed. Within an exact mathematical theory, there should be a similar kind of answer to the question, "What is a justification?" In addition to its conceptual value, clarity on this issue should also lead to cleaner mathematical models.

We consider this question in its full generality, which, surprisingly, yields a clean and meaningful answer. As in Chapter 2 we assume the language of justification logic consists of two disjoint sets of syntactic objects:

(1) a set of justification terms *Tm*. Naturally, the set of terms of any Justification Logic considered in this book is covered by this (Definition 2.1).
(2) a set of formulas *Fm*, built inductively from propositional variables, *Var*, using Boolean connectives and the justification formula formation rule: if F is a formula and t a justification term ($t \in Tm$), then $t{:}F$ is again a formula (Definition 2.2).

In classical logic an assignment of truth values to the propositional variables, *Var*, completely determines the truth values for all propositional formulas, using the familiar Boolean conditions. For a justification language more is needed because the Boolean conditions don't supply truth values for formulas of the form $t{:}F$. To handle this, a semantic meaning must be supplied for justification terms, members of *Tm*. The following places minimal requirements on such a semantics, giving us the simplest ontology for justification logic. The meaning assigned to formulas is a classical truth value, and we retain classical logic behavior for propositional connectives. We write 0 for *false* and 1 for *true*, so that our space of truth values is $\{0, 1\}$. The key item is the meaning of justification terms, and this will be a *set of formulas*. Each justification term is simply interpreted as *the set of formulas for which it is a justification*. There are loose similarities between this and neighborhood models for modal logics, in which one simply announces arbitrarily how necessitated formulas are to behave.

Definition 3.1 (Basic Model) A *basic model*, typically called $*$, consists of an interpretation of the members of *Var*, propositional variables, and an interpretation of the members of *Tm*, meeting conditions given later. Overloading notation, we use $*$ for a basic model, and also for the interpretations of *Var* and of *Tm* in that model.

The interpretation of a propositional in a basic model is a truth value. That is,

$$*{:}Var \mapsto \{0, 1\}.$$

The interpretation of each justification term is a *set of formulas*. That is,

$$*{:}Tm \mapsto 2^{Fm}.$$

For a justification term t, we write the simpler expression t^* for $*(t)$, the interpretation of term t in basic model $*$.

Just as a truth value assignment to propositional variables extends to a valuation on all classical formulas, a basic model and its mappings do the same for justification formulas.

Definition 3.2 (Evaluation in a Basic Model) Let $*$ be a basic model. It is easy to show that $*$ determines a unique mapping from all formulas to truth values, $*{:}Fm \mapsto \{0, 1\}$, meeting the following conditions. (As we did with terms, we write the simpler X^* instead of $*(X)$.)

(1) $P^* = *(P)$ for $P \in Var$.
(2) $\bot^* = 0$.
(3) $(X \to Y)^* = 1$ if and only if $X^* = 0$ or $Y^* = 1$ (and similar for the other classical connectives).
(4) $(t{:}X)^* = 1$ if and only if $X \in t^*$.

If $*$ is a basic model, we will often use the suggestive notation $\models_* X$ as another way of writing $X^* = 1$, and $\not\models_* X$ for $X^* = 0$.

Note that while propositions are interpreted semantically, as truth values, justifications are interpreted syntactically, as sets of formulas. This is a principal *hyperintensional* feature: a basic model may treat distinct formulas F and G as equal, i.e., $F^* = G^*$, but still be able to distinguish justification assertions $t{:}F$ and $t{:}G$, e.g., when $F \in t^*$, but $G \notin t^*$, yielding $\models_* t{:}F$ but $\not\models_* t{:}G$.

So far, a basic model is nothing but a classical propositional model in which justification assertions $t{:}F$ are treated as independent propositional atoms. This analogy will be abandoned when we turn to models for specific justification logics in which the various $t{:}F$ are no longer independent and have to obey some constraints called *closure conditions*. In the meantime, this minimal construction already admits a generic justification semantics.

Definition 3.3 Let $S \subseteq Fm$ and $X \in Fm$. We write $S \vdash_{\mathsf{CL}} X$ if formula X is derivable from the set S of formulas in classical logic, CL, using (say) a standard axiomatization of classical logic, treating justification assertions $t{:}F$ as propositional atoms, and with modus ponens as the only rule of inference. We say that S is *classically consistent*, if $S \nvdash_{\mathsf{CL}} \bot$.

Obviously if X is a classical tautology and $*$ is a basic model, $\models_* X$, or briefly, all classical tautologies are true in all basic models.

Theorem 3.4 (Basic Model Existence) *Any classically consistent set of formulas S has a basic model, that is, there is a basic model in which its members are true.*

Proof Assume S is classically consistent. Let Γ be any set that extends S and is maximally classically consistent. Define a (canonical) basic model $*$ by stipulating

(1) $P^* = 1$ if and only if $P \in \Gamma$ for propositional atoms P;
(2) $t^* = \{F \mid t{:}F \in \Gamma\}$ for justification terms t.

Now we need The Truth Lemma: *for each formula X we have $\models_* X$ iff $X \in \Gamma$.* This is proved by induction on the complexity of X. The base case where X is atomic follows by the definition of $*$. The Boolean cases are standard and are omitted. If X is $t{:}Y$ for some t and Y, by the definitions,

$$\models_* t{:}Y \quad \text{iff} \quad Y \in t^* \quad \text{iff} \quad t{:}Y \in \Gamma.$$

Because $S \subseteq \Gamma$, we have $\models_* S$, that is, all members of S evaluate to 1, so $*$ defines a basic model of S. □

Definition 3.5 (Set of Basic Models) For $S \subseteq Fm$, by $BM(S)$ we mean the class of all basic models of S.

Using Theorem 3.4, S is classically consistent if and only if $BM(S) \neq \emptyset$.

Theorem 3.6 (Generic Completeness) *Each set of formulas S is sound and complete with respect to its class of basic models $BM(S)$. In other words, $S \vdash_{\mathsf{CL}} F$ if and only if F is true in each basic model of S.*

Proof The theorem vacuously holds for an inconsistent S. Now assume that S is consistent.

Soundness. Let $S \vdash_{\mathsf{CL}} F$ and $*$ be a basic model of S. By induction on derivation length we show that any formula derivable from S is true in $*$, and hence $\models_* F$. All formulas of S are true. The only rule of inference in CL, modus ponens, is truth preserving.

Completeness. Suppose $S \nvdash_{\mathsf{CL}} F$. Then $S \cup \{\neg F\}$ is classically consistent. Otherwise, $S \cup \{\neg F\} \vdash_{\mathsf{CL}} \bot$, and, by the Deduction Theorem for classical logic, $S \vdash_{\mathsf{CL}} \neg\neg F$, which is impossible because $S \nvdash_{\mathsf{CL}} F$. Using Theorem 3.4, let $*$ be a basic model of $S \cup \{\neg F\}$. Obviously, this is a basic model of S. Furthermore, because $\models_* \neg F$, then $\not\models_* F$. □

Example 3.7 We work with classical logic CL over Fm, as in Definition 3.3, taking $S = \emptyset$.

(1) For every justification term t,

$$\nvdash_{\mathsf{CL}} t{:}F.$$

For a countermodel put $t^* = \emptyset$ for each term $t \in Tm$, and on each propositional variable P, P^* is arbitrary. In this basic model, all $t{:}F$ are false, and so $\not\vdash_{\mathsf{CL}} t{:}F$ by Theorem 3.6.

(2) For any propositional variable P, and term t,

$$\not\vdash_{\mathsf{CL}} t{:}P \to P.$$

Put $t^* = Fm$ for each term t, and $P^* = 0$, with other propositional variable assignments being arbitrary. In this model, all justification assertions are true, but $t{:}P \to P$ is false.

(3) For any propositional variable P, and term t,

$$\not\vdash_{\mathsf{CL}} P \to t{:}P.$$

Put $t^* = \emptyset$ for each term t, and $P^* = 1$. In this model, all justification assertions are false as well as $P \to t{:}P$.

(4) Take $x^* = \{P\}$ for justification variable x, and $t^* = Fm$ for any other t. Propositional variable assignments are arbitrary. In this model, $x{:}P$ holds, but $x{:}(P \wedge P)$ does not. By soundness,

$$\not\vdash_{\mathsf{CL}} x{:}P \to x{:}(P \wedge P)$$

which demonstrates hyperintensionality of our classical logic base because P and $P \wedge P$ are provably equivalent there, but not $x{:}P$ and $x{:}(P \wedge P)$.

Discussion

(1) Informally, a basic model of a set of formulas S is nothing but a maximal consistent extension $\widetilde{S}$ of S over the classical logic base. What makes $\widetilde{S}$ a model rather than a syntactic logical structure is the set-theoretical reading of justification assertions $t{:}F$ as $F \in t^*$. In the justification logic context, this replaces logical identities with set-theoretic closure conditions. Instead of reasoning about logical connectives and axioms, we begin by reasoning about nonlogical objects. Specifically, we begin with sets (of formulas) and operations on these sets, with their own set-theoretical intuitions. This proves to be useful.

(2) A special case that falls under the scope of Theorems 3.4 and 3.6 is the conventional multimodal language. Each modality $\Box_i$ there can be viewed as a justification with reading $\Box_i F$ as $i{:}F$.

In classical logic CL without specific modal axioms, basic models for a modal language are the same as for any justification language with the matching set of justifications. However, adding modal and justification axioms lead to different classes of basic models.

As an example, consider the family of basic models for the multimodal version of K, $BM(\mathsf{K}_m)$, where these models are in a language with modalities $\Box_1, \Box_2, \ldots$ and are required to satisfy the following conditions:

- deductive closure for each i:

$$\models_* \Box_i(F \to G) \textit{ and } \models_* \Box_i F \quad \Rightarrow \ \models_* \Box_i F;$$

- "common knowledge of axioms of K":

$$\models_* \Box_{i_1}\Box_{i_2}\ldots\Box_{i_n}A$$

  for each string $i_1, i_2, \ldots, i_n$ and axiom A.

This is a non-Kripkean semantics for modal logic K, with easy soundness and completeness theorems. It is unclear to what extent this semantics would be useful for modal logic. However, such basic models prove to be a potent tool in justification logics because the latter enjoy evidence tracking capabilities that provide more control over a model.

3.2 Models for J_0 and J

So far basic models may seem to have only a tenuous connection with justification logics. Theorem 3.6 gives completeness with respect to classical logic, for instance. But the machinery we have provided is, in fact, sufficiently general. Recall Definition 2.6—to build a justification logic axiomatically, we start with J_0, in which only two axiom schemes are added to classical logic. Then other justification logics can be constructed from J_0 with the addition of further axiom schemes. Finally a constant specification, CS, is added whose members act like specific axioms. In this way we get a justification logic with a constant specification, JL(CS). But instead of thinking of JL(CS) as built on J_0 we can think of it as built on classical logic itself, by the addition of the J_0 axiom schemes, the axiom schemes special to J, and members of CS taken as axioms. Then $\vdash_{\mathsf{JL(CS)}} X$, that is, provability in a justification logic (Definition 2.8), can be thought of as derivability in classical logic, $S \vdash_{\mathsf{CL}} X$, where S is the entire set of added axiom schemes and axioms. This way of viewing things, at least potentially, turns any justification logic into something a basic model might be useful for. In this section we start the process of showing just how useful the result can be.

Recall Definition 2.3, of justification logic J_0, in which the set Tm of terms is built up from variables and constants constructed using the operations application, $\cdot$, and sum, $+$. The set Fm of formulas is defined accordingly. The axioms of J_0 are those of classical logic CL enhanced by two principles

$$s{:}(F \to G) \to (t{:}F \to [s \cdot t]{:}G) \quad \text{and} \quad s{:}F \vee t{:}F \to [s + t]{:}F.$$

Then, basic J_0-models are the basic models in which both of these principles hold. Importantly, we will soon see that a more structural characterization is possible.

Definition 3.8 For sets of formulas X and Y, we define

$$X \rhd Y = \{G \mid F \to G \in X \text{ and } F \in Y \text{ for some } F\}.$$

Informally, $X \rhd Y$ is the result of applying modus ponens once to all members of X and of Y (in a given order).

Theorem 3.9 *$BM(\mathsf{J}_0)$ is the class of basic models in which the following closure conditions are met*

$$s^* \rhd t^* \subseteq (s{\cdot}t)^* \quad \textit{and} \quad s^* \cup t^* \subseteq (s + t)^*.$$

Proof Let us first assume the closure conditions and check the validity of the application and sum axioms. Indeed, $\models_* s{:}(F \to G)$ and $\models_* t{:}F$ yield $(F \to G) \in s^*$ and $F \in t^*$ respectively. By the closure conditions, $G \in [s \cdot t]^*$, i.e., $\models_* [s \cdot t]{:}G$. Likewise, $\models_* s{:}F$ or $\models_* t{:}F$ yield $F \in s^*$ or $F \in t^*$. Then by the closure conditions, $F \in [s + t]^*$, i.e., $\models_* [s + t]{:}F$.

Next we assume the application and sum axioms hold in basic model $*$ and we derive the closure conditions. Let $(F \to G) \in s^*$ and $F \in t^*$. By definition, $\models_* s{:}(F \to G)$ and $\models_* t{:}F$. By the application axiom, $\models_* [s \cdot t]{:}G$, hence $G \in [s \cdot t]^*$. Similarly let $F \in s^*$ or $F \in t^*$, i.e., $\models_* s{:}F$ or $\models_* t{:}F$. By the sum axiom, $\models_* [s + t]{:}F$, hence $F \in [s + t]^*$. □

Definition 3.10 Let CS be a constant specification. *A basic* CS*-model* is a basic model in which all formulas from CS hold.

Recall from Section 2.6 that we use $\mathsf{J(CS)}$ when we are dealing with J_0 extended with constant specification CS.

Corollary 3.11 *The basic models for* $\mathsf{J(CS)}$*, that is* $BM(\mathsf{J(CS)})$*, are the basic* CS*-models for* J_0*.* $\mathsf{J(CS)}$ *is sound and complete with respect to* $BM(\mathsf{J(CS)})$*.*

Note that in the closure conditions of Theorem 3.9, one cannot replace either of inclusions $\subseteq$ by equalities $=$ without violating the completeness Theorem 3.6. A verification for each of the two cases follows.

Example 3.12 (Application closure) We designate a justification constant to play a special role. Suggestively we use 0 as this constant. Consider the justification logic

$$\mathcal{L} = \mathsf{J}_0 \cup \{\neg 0{:}F \mid F \in Fm\}.$$

Informally, the justification term 0 is interpreted as the empty evaluation in any basic model, that is, $0^* = \emptyset$. We claim that formula $F = \neg[0 \cdot 0]{:}P$ is not derivable in $\mathcal{L}$, but is true in any basic model of $\mathcal{L}$ that meets the closure condition $s^* \triangleright t^* = (s{\cdot}t)^*$.

To show that $\mathcal{L} \nvdash_{\mathsf{CL}} F$, it suffices to find a basic model for $\mathcal{L}$ in which F is false. Consider a basic model $\bullet$ such that $0^\bullet = \emptyset$ and $t^\bullet = Fm$ for any other justification term t. Obviously, the closure conditions from Theorem 3.9 together with $0^\bullet = \emptyset$ are met. Therefore, $\bullet$ is a basic model of $\mathcal{L}$. It is immediate that F is false in $\bullet$, because $[0 \cdot 0]^\bullet = Fm$. On the other hand, for any basic model $*$ of $\mathcal{L}$ with the closure condition $0^* \triangleright 0^* = (0{\cdot}0)^*$ we have $(0{\cdot}0)^* = \emptyset$ because $0^* = \emptyset$ and $\emptyset \triangleright \emptyset = \emptyset$.

Example 3.13 (Sum closure) We have that $[x + y]{:}P \to (x{:}P \vee y{:}P)$ holds in any basic model with $s^* \cup t^* = (s + t)^*$ but it is not derivable in any J(CS). The former is obvious. To show the latter, let x and y be justification variables and set $x^* = y^* = \emptyset$ and $t^* = Fm$ for any other t. The closure conditions, as well as validity of any constant specification CS are straightforward. In this model, $[x + y]{:}P$ is true, but both $x{:}P$ and $y{:}P$ are false. Therefore,

$$\nvdash_{\mathsf{J(CS)}} [x + y]{:}P \to (x{:}P \vee y{:}P).$$

Examples 3.7. all hold for J_0 instead of CL because all the interpretations $*$ enjoy the J closure conditions from Theorem 3.9. In particular, in 1, $t^* = \emptyset$ for all t and the closure conditions hold vacuously.

3.3 Basic Models for Positive and Negative Introspection

Basic models handle justification logics whose postulates are conditions on justification operations, such as J, J4, J45, etc. For such models justification term evaluation conditions can be expressed as closure conditions, and then any propositional atom valuation and any justification term valuation can be extended inductively to all formulas. In this section we sketch basic models for J4 and J45.

Recall, J4 is J with an additional "proof checking" axiom $t{:}F \to\, !t{:}t{:}F$ in the extended language of J with a new unary operation ! on justifications. We begin with the empty constant specification, and then full generality is immediate. We remind the reader of Definitions 3.5 and 3.10.

Theorem 3.14 *$BM(\mathsf{J4}(\emptyset))$ is the class of basic models for J_0 with the following closure condition for !*

$$F \in t^* \quad \Rightarrow \quad t{:}F \in (!t)^*. \tag{3.1}$$

Proof If this !-closure condition holds in basic model $*$, then $\models_* t{:}F \to !t{:}t{:}F$. Indeed, if $\models_* t{:}F$, then $F \in t^*$. By the !-closure, $t{:}F \in (!t)^*$ hence $\models_* !t{:}t{:}F$. Conversely, suppose $\models_* t{:}F \to !t{:}t{:}F$. If $F \in t^*$, then $\models_* t{:}F$, hence $\models_* !t{:}t{:}F$ and $t{:}F \in (!t)^*$. □

Corollary 3.15 *Let CS be an arbitrary constant specification. Basic models for $\mathsf{J4(CS)}$, that is $BM(\mathsf{J4(CS)})$, are the basic CS-models for $\mathsf{J4}$. $\mathsf{J4(CS)}$ is sound and complete with respect to $BM(\mathsf{J4(CS)})$.*

The justification logic J45 is J4 with an additional "negative proof checking" axiom

$$\neg t{:}F \to ?t{:}\neg t{:}F$$

in the extension of the language of J4 with a new unary operation ? on justifications.

Theorem 3.16 *$BM(\mathsf{J45}(\emptyset))$ is the class of basic models for $\mathsf{J4}(\emptyset)$ with the ?-condition*

$$F \notin t^* \quad \Rightarrow \quad \neg t{:}F \in (?t)^*.$$

Proof If the ?-condition holds in basic model $*$, then $\models_* \neg t{:}F \to ?t{:}t{:}F$. Indeed, if $\models_* \neg t{:}F$, then $\not\models_* t{:}F$ and $F \notin t^*$. By the ?-condition, $\neg t{:}F \in (?t)^*$ hence $\models_* ?t{:}\neg t{:}F$. Conversely, suppose $\models_* \neg t{:}F \to ?t{:}\neg t{:}F$. If $F \notin t^*$, then $\models_* \neg t{:}F$, hence $\models_* ?t{:}\neg t{:}F$ and $\neg t{:}F \in (?t)^*$. □

Corollary 3.17 *Let CS be an arbitrary constant specification. Basic models for $\mathsf{J45(CS)}$, $BM(\mathsf{J45(CS)})$, are the basic CS-models for $\mathsf{J45}$. $\mathsf{J45(CS)}$ is sound and complete with respect to $BM(\mathsf{J45(CS)})$.*

3.4 Adding Factivity: Mkrtychev Models

Things change with the factivity axiom.

$$t{:}F \to F$$

Atomic and justification evaluations are no longer sufficient to define a model.

Consider the simplest justification logic with factivity,

$$\mathsf{JT} = \mathsf{J} + (t{:}F \to F).$$

Theorem 3.18 *$BM(\mathsf{JT}(\emptyset))$ is the class of basic* J*-models in which the following factivity condition holds:*

$$\models_* t{:}F \quad \Rightarrow \quad \models_* F.$$

Proof Axiom $t{:}F \to F$ is true in each basic J-model with the factivity condition. Conversely, if $t{:}F \to F$ is true in a basic model, then $\models_* t{:}F$ yields $\models_* F$. □

Corollary 3.19 *Basic models for* $\mathsf{JT}(\mathsf{CS})$*, that is, members of the family* $BM(\mathsf{JT}(\mathsf{CS}))$*, are the basic* CS*-models for* $\mathsf{JT}(\emptyset)$*.* $\mathsf{JT}(\mathsf{CS})$ *is sound and complete with respect to* $BM(\mathsf{JT}(\mathsf{CS}))$.

So, to be a JT-model evaluations should satisfy factivity, and this adds a potentially infinite number of things to check in order to determine whether $t{:}F$ indeed yields F in order to certify that a given evaluation is indeed a model. A modification of the truth condition for justification assertions $t{:}F$, first suggested by Mkrtychev in Mkrtychev (1997), eliminates this problem.

Definition 3.20 An *Mkrtychev model* is an evaluation $*$ satisfying the closure conditions for basic models but with a different truth assignment for justification assertions

$$\models_* t{:}F \quad \text{iff} \quad F \in t^* \textit{ and } \models_* F.$$

The advantage of Mkrtychev models is that they allow the extension of any atomic and justification assignment to all formulas in an inductive way: the truth value of $t{:}F$ is assigned after F gets its truth value.

Theorem 3.21 *Each basic* JT*-model is an Mkrtychev model. For every Mkrtychev model $*$ there is a basic* JT*-model $\bullet$ such that for each formula F,*

$$\models_* F \textit{ if and only if } \models_\bullet F.$$

Proof For any JT-model $*$, $\models_* t{:}F$ yields $\models_* F$ hence $\models_* t{:}F$ is equivalent to $F \in t^*$ *and* $\models_* F$, i.e., $*$ is an Mkrtychev model.

Consider an Mkrtychev JT-model $*$. Define an evaluation $\bullet$ on propositional variables as $P^\bullet = P^*$ and on justification assertions as

$$F \in t^\bullet \quad \textit{iff} \quad F \in t^* \textit{ and } F^* = 1$$

which means

$$t{:}F \textit{ is true in } \bullet \quad \textit{iff} \quad t{:}F \textit{ is true in } *.$$

We check the J closure conditions.

Application: Suppose $(F \to G) \in s^\bullet$ and $F \in t^\bullet$. Then $(F \to G) \in s^*$,

$(F \to G)^* = 1$, $F \in t^*$, and $F^* = 1$. By the classical truth tables and application closure for $*$, $G \in [s \cdot t]^*$ and $G^* = 1$, hence $G \in [s \cdot t]^\bullet$.

Sum: Suppose $F \in s^\bullet$ or $F \in t^\bullet$. Then ($F \in s^*$ and $F^* = 1$) or ($F \in t^*$ and $F^* = 1$), and hence $F \in s^* \cup t^*$ and $F^* = 1$. By the sum closure of $*$, $F \in [s+t]^*$ and $F^* = 1$, hence $F \in [s+t]^\bullet$.

Next we establish that $\models_\bullet F$ iff $\models_* F$ by induction on the complexity of F. For atomic formulas, $P^\bullet = P^*$, which yields the claim. The Boolean cases are straightforward. For $F = t{:}X$,

$$\models_\bullet t{:}X \;\; \textit{iff} \;\; X \in t^\bullet \;\; \textit{iff} \;\; X \in t^* \text{ and } X^* = 1 \;\; \textit{iff} \;\; \models_* t{:}X.$$

Finally, we check the factivity of $\bullet$. By definitions, $t{:}X$ is true in $\bullet$ yields $X^* = 1$, hence, by the equivalence of $*$ and $\bullet$, $X^\bullet = 1$, i.e., X holds in $\bullet$. □

Mkrtychev models offer a practical semantic tool for justification logic with factivity. Let us consider an example comparing basic and Mkrtychev models.

Example 3.22 Take as a base logic a version of JT with an axiomatically appropriate constant specification CS over a language with propositional variables P, Q, and R and two justification variables x and y. We assume that P is false, Q and R are true, and x and y are justifications for P and Q respectively, and for nothing else. Proposition R though true is left without justification. This is a reasonable epistemic scenario with factive and nonfactive justifications, known and unknown truths, etc.

The natural Mkrtychev model $*$ for this story has

(1) $P^* = 0, \quad Q^* = R^* = 1$;

(2) $x^* = \{P\}, \;\; y^* = \{Q\}, \;\; 0^* = \emptyset$, all other justification constants are evaluated according to CS.

In addition, we assume the "minimal" closure conditions:

$$(s{\cdot}t)^* = s^* \triangleright t^* \;\text{ and }\; (s+t)^* = s^* \cup t^*.$$

The Mkrtychev model is now completely defined.

To define an equivalent basic model $\bullet$ we have to specify $\bullet$ on all justification terms. This requires an additional induction on formulas with the justification assertion steps corresponding to Definition 3.20. However, this essentially is a redundant technical step that does not enhance our understanding of the model $\bullet$ beyond what we know from the Mkrtychev model $*$. In this case, as well as in many other cases, the natural format of specifying a model is Mkrtychev's.

3.5 Basic and Mkrtychev Models for the Logic of Proofs LP

The Logic of Proofs LP is historically the first justification logic. It occupies a prominent position in this area due to its arithmetical semantics and connections to the Brouwer–Heyting–Kolmogorov semantics for intuitionistic logic. This warrants a special section devoted to basic and Mkrtychev semantics for LP.

As we have seen in Section 2.6, LP is J4 with factivity $t{:}F \to F$. The following is yet another Corollary of Theorem 3.14. showing that basic models for LP are J4-models with the factivity condition.

Corollary 3.23 *The basic models for* LP(CS), *members of* BM(LP(CS)), *are the basic* CS-*models for* J4 *with factivity.* LP(CS) *is sound and complete with respect to* BM(LP(CS)).

Mkrtychev models for LP are those for JT that are closed under the proof checking condition (3.1). An analogue of Theorem 3.21 holds for LP as well.

Theorem 3.24 *For every Mkrtychev* LP-*model* $*$ *there is a basic* LP-*model* $\bullet$ *such that for each formula* F,

$$F \text{ is true in } \bullet \quad \textit{iff} \quad F \text{ is true in } *.$$

Proof The proof is the same as the proof of Theorem 3.21, but in addition we have to verify the closure of $\bullet$ with respect to the proof checking condition (3.1). Suppose $F \in t^\bullet$. By our definitions, this yields $F \in t^*$ and $F^* = 1$ hence $t{:}F$ is true in $*$, $(t{:}F)^* = 1$. Furthermore, by !-closure of $*$, $t{:}F \in (!t)^*$ hence $(t{:}F)^\bullet = 1$. □

3.6 The Inevitability of Possible Worlds: Modular Models

Basic and Mkrtychev models were originally confined to a single possible world, which is sufficient to provide a countermodel for any nontheorem of a given justification logic (cf. the completeness Theorem 3.6). But this does not suffice for capturing the intrinsic epistemic structure of situations that naturally include *possible worlds*. We look at some examples involving possible worlds. These examples can, for the time being, be understood informally.

Example 3.25 Suppose a proposition P holds at Γ, but is not known to an agent due to lack of evidence. In addition, the agent deems possible a world Δ at which P is true and supported by a convincing justification, which we designate by x. A natural epistemic model could be $\mathcal{M}_1$, shown in Figure 3.1, with two possibly reflexive worlds Γ and Δ such that Δ is accessible from Γ, with P holding at Γ and Δ. Thus we do not have $t{:}P$ holding at Γ for any t, but we do have $x{:}P$ being true at Δ.

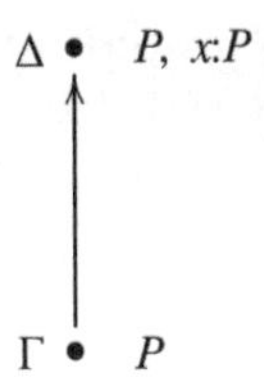

Figure 3.1 Model $\mathcal{M}_1$

Formally, $P^*_\Gamma = P^*_\Delta = 1$, $t^*_\Gamma = \emptyset$ for all t, $x^*_\Delta = \{P\}$. Other details are of no importance, but let us assume the underlying justification logic is JT. In $\mathcal{M}_1$, P holds at both Γ and Δ, and according to Kripke–Hintikka epistemology, P is *believed* in $\mathcal{M}_1$. So, at Γ, proposition P is *believed but not justified*. However, at Γ proposition P is believed without evidence, which exists only at Δ.

The worlds Γ and Δ are essentially basic models, which we can identify with maximal consistent sets $\Gamma_{\max}$ and $\Delta_{\max}$ of formulas true at the corresponding worlds, and these are classically maximal consistent sets. However, $\Gamma_{\max}$ and $\Delta_{\max}$ are not compatible, e.g., $\neg x{:}P \in \Gamma_{\max}$ but $x{:}P \in \Delta_{\max}$; there is no single basic model that represents the whole of $\mathcal{M}_1$.

So, introducing possible worlds into the semantic consideration of justification logics is inevitable. Possible world frames with accessibility relations directly relate justification logics to modal logics. Note, however, that an accessibility relation between possible worlds makes sense also outside a modal logic context. In model $\mathcal{M}_1$, there are no modalities present, but nevertheless, it makes perfect sense to model epistemic possibilities there.

How should we connect justification logic with mainstream epistemic logic, which relies heavily on possible worlds models? The standard semantics of

F is believed at world u

is

F holds at all worlds considered possible at u.

How can justifications be made to fit into this picture? The idea is to formally associate a basic model with each possible world of a frame, so that the interpretation of justification terms can vary from world to world.

Definition 3.26 A *modular model* is a structure

$$\mathcal{M} = \langle \mathcal{G}, \mathcal{R}, * \rangle$$

such that $\langle \mathcal{G}, \mathcal{R} \rangle$ is a Kripke frame ($\mathcal{G}$ is a nonempty set of possible worlds, $\mathcal{R}$ is a binary *accessibility relation* on $\mathcal{G}$), and $*$ is a mapping that assigns a basic model to each member of $\mathcal{G}$.

Let $\mathcal{M} = \langle \mathcal{G}, \mathcal{R}, * \rangle$ be a modular model, $\Gamma \in \mathcal{G}$, and $X \in Fm$ be a justification formula. We write

$$\mathcal{M}, \Gamma \Vdash X \text{ if and only if } \models_{*(\Gamma)} X,$$

that is, if X is true in the modular model associated with world Γ.

This definition of modular models combines basic models with possible worlds frames without any constraints. Basic models at worlds may have different closure conditions, or may have none. Justifications are not yet linked to the epistemic modalities, which are implicitly present in modular models via accessibility relations.

A modular model $\mathcal{M} = \langle \mathcal{G}, \mathcal{R}, * \rangle$ is also a Kripke model (with justifications as an additional feature). Therefore, it supports semantics for both justification and modalities. Even if there is no modality $\Box$ in the language, we can emulate

$$\Box F \textit{ holds at } \Gamma$$

by

$$\mathcal{M}, \Delta \Vdash F \text{ for every } \Delta \in \mathcal{G} \text{ with } \Gamma \mathcal{R} \Delta.$$

Definition 3.27 A modular model $\mathcal{M} = \langle \mathcal{G}, \mathcal{R}, * \rangle$ is a J-model if basic models $*(\Gamma)$ for every $\Gamma \in \mathcal{G}$ satisfy the J_0-closure conditions from Theorem 3.9.

It immediately follows from Corollary 3.11 that J is sound and complete with respect to modular J-models.

The notion of a modular model naturally extends to other justification logics, e.g., J4, J5, JT, LP (cf. Artemov, 2012; Kuznets and Goetschi, 2012).

Example 3.28 Figure 3.2 shows an example, of a JT-model and its epistemic reading.

In $\mathcal{M}_2$, x is a justification of P at Γ, but the agent does not believe P at Γ because there is a possible world Δ at which P does not hold. So, at Γ, P is *justified, but not believed.*

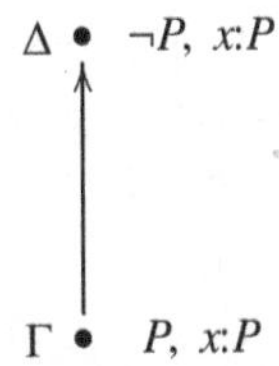

Figure 3.2 Model $\mathcal{M}_2$

3.7 Connecting Justifications, Belief, and Knowledge

We begin with a few items of notation. Suppose $\mathcal{M} = \langle \mathcal{G}, \mathcal{R}, * \rangle$ is a modular model, Definition 3.26. Then for each $\Gamma \in \mathcal{G}$, $*(\Gamma)$ is a basic model. We will informally say something is the case at possible world $\Gamma \in \mathcal{G}$ if it is so in the basic model $*(\Gamma)$. Then for a justification term t, t^Γ is the set of formulas we think of t as justifying at Γ, and for a formula X, X^Γ is the truth value of X at Γ. Recall that if $X^\Gamma = 1$ we may also write $\mathcal{M}, \Gamma \Vdash X$.

We now introduce a new piece of notation, $\square^\Gamma$, for the set of formulas true at all possible worlds accessible from Γ. That is,

$$\square^\Gamma = \{F \mid \mathcal{M}, \Delta \Vdash F \textit{ for all } \Delta \in \mathcal{G} \textit{ such that } \Gamma \mathcal{R} \Delta\}.$$

In Kripke–Hintikka epistemology, $\square^\Gamma$ is the *set of formulas believed (or known) at* Γ. Conceptually, for a given world Γ, t^Γ reflects believing with justification t, whereas $\square^\Gamma$ represents believing without providing a specific reason.

The following is a fundamental connection between justification terms and the accessibility relation in a modular model. It is not part of the general definition—we specify explicitly when we have a modular model meeting this condition.

Definition 3.29 (Justification Yields Belief) We say that *justification yields belief, JYB,* in a modular model $\mathcal{M} = \langle \mathcal{G}, \mathcal{R}, * \rangle$, if

$$t^\Gamma \subseteq \square^\Gamma$$

for each justification term t and each $\Gamma \in \mathcal{G}$. In other words, if t is a justification for F at Γ, then F is believed at Γ.

Example 3.30 Model $\mathcal{M}_1$ from Example 3.25 enjoys *JYB*, whereas $\mathcal{M}_2$ from Example 3.28 does not.

We have soundness of modular models with *JYB* for the corresponding justification logics J, J4, J45, JT, LP, etc. This is because each of the worlds is a basic model. Completeness immediately follows from the corresponding completeness theorems for basic models because each basic model may be regarded as a modular model $\langle \mathcal{G} = \langle \mathcal{W}, \mathcal{R}, * \rangle$ with a singleton $\mathcal{G}$ and empty $\mathcal{R}$.

3.8 History and Commentary

The historical order in which the variety of semantic models for justification logics were created was quite different from the order they have been presented in this chapter.

The whole justification logic project has been semantically motivated from the beginning and grew from an attempt to find the right logical format for and the basic laws of reasoning about mathematical proofs. The immediate goal of this formalization was to define Brouwer–Heyting–Kolmogorov proof objects via classical proofs. The first arithmetical semantics for Justification Logic appeared in Artemov (1995) (cf. also Artemov, 2001) as the semantics of classical proofs, in Peano Arithmetic, for the Logic of Proofs LP.

In addition, Artemov (1995) introduced a constructive version of the canonical model for LP, which was, in the current terminology, a special case of basic models and established completeness of LP. In a more general setting, the abstract logic semantics for LP was developed in Mkrtychev (1997) as basic and Mkrtychev models (the current terminology). Mkrtychev models for J, JT, and J4 were introduced in Kuznets (2000).

Fitting models for LP were introduced and developed in Fitting (2005). Fitting models for a variety of justification logics that were sublogics of LP, such as J, J4, JD, and JT were studied in Artemov (2008). Fitting models for JT45 were developed in Pacuit (2006) and Rubtsova (2006b). Extensions of Fitting models for combinations of modal logics with justifications were introduced in Artemov (2006) and Artemov and Nogina (2005). More recently, Fitting models for an infinite family of justification logics were created in Fitting (2016a). This family will be discussed in Chapter 8.

Modular models for J were introduced in Artemov (2012) and for a number of other justification logics in Kuznets and Goetschi (2012). Basic models in their current general form are due to Artemov (2018).

Conceptually, basic models and modular models, supported by soundness and completeness theorems, provide an answer to the ontological question *What kind of logical object is a justification?* However, these models might be difficult to work with, and more convenient constructions, such as Mkrtychev and Fitting models, are widely used in this area. Technically, basic models and Mkrtychev models may be regarded as special cases of Fitting models. On the other hand, Fitting models can be identified with modular models having the JYB property. This provides a natural hierarchy of the classes of models:

basic and Mkrtychev models $\subset$ *Fitting models* $\subset$ *modular models.*

Even the smallest class, basic models, is already sufficient for mathematical completeness results for justification logics, and interesting consequences can be shown, as in Example 3.7. So, the main idea of going higher to Fitting and modular models is not a pursuit of completeness but rather a desire to offer natural models for a variety of epistemic situations involving evidence, belief, and knowledge. There are also important technical advantages. Mkrtychev models have played an important role in proving results about the complexity of justification logics, and Fitting models provide a broad and uniform, though nonconstructive, treatment of realization. This will be discussed in Chapters 6 and 7.

Basic models reduce all the justification information to the syntactic evaluation of justifications, i.e., sets of formulas t^*. Conceptually, this approach reflects only one reason for *not knowing* F: there is no sufficient justification for F available. In contrast, Fitting models take into account two reasons for *not knowing* F: the *Kripkean reason*, F fails in some possible world, and the *awareness reason*, no justification for F is available. Modular models view evidence and belief as independent concepts that can help to analyze a broader class of epistemic situations.

4
Fitting Models

From the very beginning, justification logics had important connections with modal logics. Indeed, this was the main reason for studying them. For instance, as we have noted several times, the first justification logic LP connects with the modal logic S4, and this plays an important role in giving an arithmetic interpretation for intuitionistic logic. We will examine the details of this LP/S4 connection in Chapter 9. Both basic models and Mkrtychev models lack specific machinery to help bring the justification/modal connection out. Fitting semantics, introduced in Fitting (2003, 2005), is a possible world semantics. Because of this, justification/modal connections are consequently much easier to grasp. Modular models extend this use of possible world machinery to a still more general setting because *JYB*, justification yields belief, is not required though it is essentially built into Fitting models. As we will see in Chapter 7, Fitting models turned out to be the appropriate tools for establishing deep general results connecting modal and justification logics, but we can't go into details at this point.

In this chapter we first present modal semantics. This is familiar and standard, and we have been making some use of it all along. To do this we must first establish notation and set up basic machinery. Then we introduce Fitting semantics and prove general completeness results.

4.1 Modal Possible World Semantics

A *frame* is a directed graph, $\mathcal{F} = \langle \mathcal{G}, \mathcal{R} \rangle$, with $\mathcal{G}$ being the nodes, or possible worlds, or states and $\mathcal{R}$ being the directed edges, or accessibility relation. We use the modal language specified in Section 2.1. A frame becomes a modal model by specifying which atomic formulas are true at which possible worlds—there is a mapping $\mathcal{V}$ from propositional letters to subsets of

$\mathcal{G}$. We take $\mathcal{V}(\bot) = \emptyset$ by convention. A modal model is thus a structure $\mathcal{M} = \langle \mathcal{G}, \mathcal{R}, \mathcal{V} \rangle$, where $\langle \mathcal{G}, \mathcal{R} \rangle$ is a frame. We say $\mathcal{F} = \langle \mathcal{G}, \mathcal{R} \rangle$ is the *underlying frame* of the model $\mathcal{M} = \langle \mathcal{G}, \mathcal{R}, \mathcal{V} \rangle$, or that model $\mathcal{M}$ is *based on* the frame $\mathcal{F}$.

If $\mathcal{M} = \langle \mathcal{G}, \mathcal{R}, \mathcal{V} \rangle$, we write $\mathcal{M}, \Gamma \Vdash X$ to indicate that modal formula X is true at possible world $\Gamma \in \mathcal{G}$. Because there is a direct connection between $\Vdash$ and $\mathcal{V}$ it is sometimes convenient to refer to a model as $\langle \mathcal{G}, \mathcal{R}, \Vdash \rangle$ as well as $\langle \mathcal{G}, \mathcal{R}, \mathcal{V} \rangle$, We list the familiar truth conditions.

(1) if A is atomic, $\mathcal{M}, \Gamma \Vdash A$ if and only if $\Gamma \in \mathcal{V}(A)$
(2) $\mathcal{M}, \Gamma \Vdash X \to Y$ if and only if $\mathcal{M}, \Gamma \nVdash X$ or $\mathcal{M}, \Gamma \Vdash Y$
(3) $\mathcal{M}, \Gamma \Vdash \Box X$ if and only if $\mathcal{M}, \Delta \Vdash X$ for every $\Delta \in \mathcal{G}$ with $\Gamma \mathcal{R} \Delta$

Note that because $\mathcal{V}(\bot) = \emptyset$, $\mathcal{M}, \Gamma \nVdash \bot$ for every possible world and every model. Other connectives can be taken to be defined, or primitive, and similarly for the dual modal operator. Either way we have the following.

(4) $\mathcal{M}, \Gamma \Vdash X \wedge Y$ if and only if $\mathcal{M}, \Gamma \Vdash X$ and $\mathcal{M}, \Gamma \Vdash Y$
(5) $\mathcal{M}, \Gamma \Vdash X \vee Y$ if and only if $\mathcal{M}, \Gamma \Vdash X$ or $\mathcal{M}, \Gamma \Vdash Y$
(6) $\mathcal{M}, \Gamma \Vdash \neg X$ if and only if $\mathcal{M}, \Gamma \nVdash X$
(7) $\mathcal{M}, \Gamma \Vdash \Diamond X$ if and only if $\mathcal{M}, \Delta \Vdash X$ for some $\Delta \in \mathcal{G}$ with $\Gamma \mathcal{R} \Delta$

A formula X is *valid* in a modal model $\mathcal{M} = \langle \mathcal{G}, \mathcal{R}, \mathcal{V} \rangle$ if $\mathcal{M}, \Gamma \Vdash X$ for every $\Gamma \in \mathcal{G}$. A formula is *valid in a frame* if it is valid in every model based on that frame. And a formula is valid in a class of frames if it is valid in every frame in the class.

If $\mathcal{R}$ is a reflexive relation on $\mathcal{G}$ we simply say the frame $\langle \mathcal{G}, \mathcal{R} \rangle$ is reflexive, and similarly for other conditions on accessibility. Various natural classes of frames famously correspond to well-known modal logics. X is valid in the class of all frames just in case X is provable in axiomatic K. Similarly the class of reflexive frames corresponds to axiomatic T, the class of transitive frames corresponds to $\mathsf{K4}$, the class of reflexive and transitive frames corresponds to $\mathsf{S4}$, and so on.

4.2 Fitting Models

Now assume the language is not modal, but involves justification terms as in Section 2.2. Fitting models have a special piece of machinery, syntactic in nature and tracing back to Mkrtychev (1997): an *evidence function*. This function, $\mathcal{E}$, maps justification terms and formulas to sets of possible worlds. Think of $\Gamma \in \mathcal{E}(t, X)$ as saying that at possible world Γ, t is relevant evidence for the truth of X. Informally, relevant evidence need not be conclusive. For example evidence that might be admitted in court, whose truth can then be discussed during the court proceedings, is relevant evidence. As another example, consider the findings of an expert who is knowledgeable and whose opinion must be taken into account, but who is fallible. The informal idea Fitting models embody is that to say $t{:}X$ is so is to say that X is a necessary truth and t is relevant evidence for it.

Definition 4.1 (Fitting Model) A Fitting model for a justification logic is a structure $\mathcal{M} = \langle \mathcal{G}, \mathcal{R}, \mathcal{V}, \mathcal{E} \rangle$, where $\langle \mathcal{G}, \mathcal{R}, \mathcal{V} \rangle$ is as in Kripke models, Section 4.1, and $\mathcal{E}$ is an evidence function. Conditions for truth functional connectives are the same as for Kripke models—Boolean at each possible world. The possibility operator condition, (7), is dropped. The necessity condition, (3), is replaced with the following.

(3′) $\mathcal{M}, \Gamma \Vdash t{:}X$ if and only if

Modal Condition: $\mathcal{M}, \Delta \Vdash X$ for every $\Delta \in \mathcal{G}$ with $\Gamma \mathcal{R} \Delta$
Evidence Condition: $\Gamma \in \mathcal{E}(t, X)$

With modal models various *frame conditions* are imposed: transitivity, symmetry, convergence, etc. This is done with Fitting models in exactly the same way. But in addition one also puts conditions on Fitting model evidence functions. Particular justification logics can have operations on justification terms that are not shared by other logics, but we have assumed $+$ and $\cdot$ are always present axiomatically, and correspondingly we always require the following semantically:

Definition 4.2 (Minimum Evidence Conditions) In all Fitting models the evidence function must meet the following conditions.

$\cdot$ **Condition:** $\mathcal{E}(s, X \to Y) \cap \mathcal{E}(t, X) \subseteq \mathcal{E}(s \cdot t, Y)$
$+$ **Condition:** $\mathcal{E}(s, X) \cup \mathcal{E}(t, X) \subseteq \mathcal{E}(s + t, X)$

The $\cdot$ condition can be read informally as follows. Any world at which s counts as relevant evidence for an implication and t counts as relevant evidence for the antecedent is a world where $s \cdot t$ is relevant evidence for the consequent. Likewise the $+$ condition says that $s + t$ is relevant evidence for something provided one of s or t is. These general conditions are quite reasonable, but as logics get more and more esoteric, informal readings of evidence conditions are harder to come by. Ultimately they are simply mathematical requirements, of course.

Constant specifications (Definitions 10.32, 2.6, and 2.6) will be required to hold universally, whenever imposed by the constant specification.

Definition 4.3 (Constant Specification Condition) We say a Fitting model $\mathcal{M} = \langle \mathcal{G}, \mathcal{R}, \mathcal{V}, \mathcal{E} \rangle$ *meets constant specification* CS provided that $\mathcal{E}(c, X) = \mathcal{G}$ for each $c{:}X \in \mathsf{CS}$.

As in the modal case, a justification formula X is valid in a Fitting model $\mathcal{M} = \langle \mathcal{G}, \mathcal{R}, \mathcal{V}, \mathcal{E} \rangle$ provided that for each $\Gamma \in \mathcal{G}$ it holds that $\mathcal{M}, \Gamma \Vdash X$. Likewise a formula is valid in a class of Fitting models if it is valid in every member of the class. If JL is a justification logic and CS is a constant specification for it, we say a Fitting model $\mathcal{M}$ is a model for JL(CS) if all axioms of JL are valid in $\mathcal{M}$ and $\mathcal{M}$ meets CS. An induction on axiomatic proof length gives an easy verification that if $\mathcal{M}$ is a model for JL(CS) then all theorems of JL(CS) are valid in $\mathcal{M}$.

Fitting models historically preceded modular models and have a structure similar to modular models, with similar closure conditions, but they differ in how they evaluate justification assertions. In Fitting models a justification assertion $t{:}F$ is true at possible world Γ iff $F \in t^{*(\Gamma)} \cap \Box^{*(\Gamma)}$. In fact, each modular model $\mathcal{M} = \langle \mathcal{G}, \mathcal{R}, * \rangle$ with *JYB* is a Fitting model: In modular models with *JYB* the truth condition $F \in \Box^{*(\Gamma)}$ becomes redundant. Furthermore, each Fitting model encodes a modular model with *JYB* over the same frame and with the same truth evaluation of formulas at each node (Artemov, 2012; Kuznets and Goetschi, 2012).

Two special conditions are often imposed on Fitting models: being *fully explanatory*, originating in Fitting (2003, 2005) and having a *strong evidence function*, originating in Rubtsova (2006a).

Definition 4.4 (Fully Explanatory) A Fitting model $\mathcal{M} = \langle \mathcal{G}, \mathcal{R}, \mathcal{V}, \mathcal{E} \rangle$ is *fully explanatory* provided that, whenever $\mathcal{M}, \Delta \Vdash X$ for every $\Delta \in \mathcal{G}$ with $\Gamma \mathcal{R} \Delta$, there is some justification term t so that $\mathcal{M}, \Gamma \Vdash t{:}X$.

Somewhat informally, if we had a modal $\Box$ symbol in the justification language then fully explanatory would say that if $\mathcal{M}, \Gamma \Vdash \Box X$ then $\mathcal{M}, \Gamma \Vdash t{:}X$ for some justification term t. In words, anything that is necessary at a world has a reason at that world.

Definition 4.5 (Strong Evidence Function) Definition 4.1 for Fitting Models imposes two conditions: a Modal condition and an Evidence condition. A *strong evidence function* is one for which the Evidence condition implies the Modal condition. That is, if $\Gamma \in \mathcal{E}(t, X)$ for some justification term t then $\mathcal{M}, \Delta \Vdash X$ for every $\Delta \in \mathcal{G}$ with $\Gamma \mathcal{R} \Delta$.

If we have a Strong Evidence function, condition (3′) from Definition 4.1 simplifies: $\mathcal{M}, \Gamma \Vdash t{:}X$ if and only if $\Gamma \in \mathcal{E}(t, X)$. Speaking informally, strong evidence amounts to assuming that we are not dealing with mere *relevant* evidence, but with *conclusive* evidence.

Fully explanatory and strong evidence have quite different natures, and some general remarks are appropriate.

Being fully explanatory is attractive conceptually. Informally it says that necessary truths always have reasons. Still, despite its attractiveness no formal consequences of interest have yet been found. It is not a requirement on Fitting models and does not hold universally. Our completeness proofs use a justification logic analog of the familiar modal canonical model construction. These canonical justification models will always be fully explanatory provided we assume we have an axiomatically appropriate constant specification. In such cases we generally have soundness with respect to a set of Fitting models whether or not they are fully explanatory, but if X is not provable it will be falsifiable in a Fitting model that does have this feature.

Strong Evidence functions are quite another thing, however. Canonical justification models always have Strong Evidence functions, though for logics such as LP, Strong Evidence is not part of the characterization of LP model. But there are many examples of justification logics for which the natural class of Fitting models for them requires a Strong Evidence condition in order to establish soundness. A justification counterpart of S5 is such a case (Rubtsova, 2006a). This is the first such example known, but there are now many others for which strong evidence is needed to establish soundness. It can be quite difficult to actually exhibit particular models with strong evidence functions, though if such a condition is not needed, exhibiting a model is much less of a problem.

4.3 Soundness Examples

In Chapter 2 we looked at a handful of justification logics: LP and various sublogics in Section 2.6, and other less common justification logics in Section 2.7. We now examine Fitting models for these justification logics, and verify soundness.

4.3.1 J(CS)

Assume the logic J(CS) is built on J_0 from Section 2.3, with constant specification CS added. No special assumptions are placed on CS. Consider a Fitting model $\mathcal{M} = \langle \mathcal{G}, \mathcal{R}, \mathcal{V}, \mathcal{E} \rangle$ where $\langle \mathcal{G}, \mathcal{R} \rangle$ is any Kripke frame, $\mathcal{E}$ is any evidence function, and $\mathcal{V}$ is any mapping from propositional letters to subsets of $\mathcal{G}$. We assume $\mathcal{M}$ meets constant specification CS, that is, if $t{:}X \in \mathsf{CS}$ then $\mathcal{E}(t, X) = \mathcal{G}$.

We first verify soundness: if $\vdash_{\mathsf{J(CS)}} X$ then X is valid in $\mathcal{M}$, that is, $\mathcal{M}, \Gamma \Vdash X$ for every $\Gamma \in \mathcal{G}$. This is done by induction on the length of an axiomatic proof of X. It is trivial to check that applications of modus ponens preserve validity. Likewise validity of all tautologies is immediate. All members of the constant specification CS are valid because $\mathcal{M}$ meets the specification. We concentrate on the axioms for the basic justification operations Application $\cdot$ and Sum $+$.

Consider the Application axiom scheme $s{:}(X \to Y) \to (t{:}X \to [s \cdot t]{:}Y)$. Let Γ be an arbitrary member of $\mathcal{G}$ and suppose $\mathcal{M}, \Gamma \Vdash s{:}(X \to Y)$ and $\mathcal{M}, \Gamma \Vdash t{:}X$. We must show that $\mathcal{M}, \Gamma \Vdash [s \cdot t]{:}Y$. By the Modal Condition from Definition 4.1, for any $\Delta \in \mathcal{G}$ with $\Gamma \mathcal{R} \Delta$ we have $\mathcal{M}, \Delta \Vdash X$ and $\mathcal{M}, \Delta \Vdash X \to Y$. Then $\mathcal{M}, \Delta \Vdash Y$ for every accessible Δ. By the Evidence Condition, $\Gamma \in \mathcal{E}(s, X \to Y)$ and $\Gamma \in \mathcal{E}(t, X)$ and then the $\cdot$ Condition from Definition 4.2 gives us that $\Gamma \in \mathcal{E}(s \cdot t, Y)$. We now have both conditions required by Definition 4.1, and so $\mathcal{M}, \Gamma \Vdash [s \cdot t]{:}Y$. Validity of the Sum axiom is similarly established; we omit details.

We now have that all theorems of J(CS) are valid in any Fitting model meeting constant specification CS. This means we can show unprovability of a formula by constructing a suitable Fitting countermodel. The formula $x{:}P \to x{:}(P \wedge Q)$ is not provable (P and Q are atomic and x is a justification variable). We construct two different countermodels for it. These are illustrative of some of the ideas behind the semantics. A formula $t{:}X$ might fail at a possible world either because X is not believed there (it is false at some accessible world), or because t is not an appropriate reason for X. The two models illustrate these alternatives. To keep things uncluttered, we assume CS is the empty constant specification so we can ignore the role of justification constants in what follows.

The first model $\mathcal{M}_1$ has a single state, Γ, accessible to itself, with an evidence function such that $\mathcal{E}(t, Z) = \{\Gamma\}$ for every justification term t and every formula Z. In this model every t serves as *universal* evidence. Use a valuation such that $\mathcal{V}(P) = \{\Gamma\}$ and $\mathcal{V}(Q) = \emptyset$. Then we have $\mathcal{M}_1, \Gamma \Vdash x{:}P$ but $\mathcal{M}_1, \Gamma \nVdash x{:}(P \wedge Q)$ because, even though x serves as universal evidence, $P \wedge Q$ is not believed at Γ in the Hintikka/Kripke sense because Q is not true.

The second model $\mathcal{M}_2$ again has a single state Γ accessible to itself. This time use the valuation $\mathcal{V}$ that assigns $\{\Gamma\}$ to every propositional letter. It follows that both P and $P \wedge Q$ are true at Γ. The evidence function is somewhat more complicated this time. For every justification variable v set $\mathcal{E}(v, Z) = \{\Gamma\}$ if and only if Z is a propositional letter. For every justification term t that is not a justification variable, set $\mathcal{E}(t, Z) = \{\Gamma\}$ for every formula Z. This meets the Minimum Evidence Conditions from Definition 4.2 because both $\mathcal{E}(s \cdot t, Z)$ and $\mathcal{E}(s + t, Z)$ will evaluate to the whole space of possible worlds for every Z. Also $\mathcal{M}_2, \Gamma \Vdash x{:}P$ because P is true at every accessible world (namely Γ itself) and $\Gamma \in \mathcal{E}(x, P)$. But $\mathcal{M}_2, \Gamma \nVdash x{:}(P \wedge Q)$ because x is a propositional variable but $P \wedge Q$ is not a propositional letter, so $\Gamma \notin \mathcal{E}(x, P \wedge Q)$.

In Hintikka–Kripke models belief and knowledge are essentially semantic notions, but the justification logic treatment of evidence has a partly syntactic nature. For example, the model $\mathcal{M}_2$ above also invalidates $x{:}P \rightarrow x{:}(P \wedge P)$. At first glance this is surprising because in any standard logic of knowledge or belief $\Box P \rightarrow \Box(P \wedge P)$ is valid. But, just because x serves as evidence for P, it need not follow that it also serves as evidence for $P \wedge P$. The formulas are syntactically different, and effort is needed to recognize that the later formula is a redundant version of the former. To take this to an extreme, consider the formula $x{:}P \rightarrow x{:}(P \wedge P \wedge \ldots \wedge P)$, where the consequent has as many conjuncts as there are elementary particles in the universe! In brief, Hintikka–Kripke style knowledge is knowledge of propositions, but justification terms justify sentences.

4.3.2 LP

The first justification logic was LP, the *Logic of Proofs*. It is presented axiomatically in Section 2.6. To get a Fitting semantics for it we build on that just introduced for J. For the following, $\mathcal{M} = \langle \mathcal{G}, \mathcal{R}, \mathcal{V}, \mathcal{E} \rangle$ is a Fitting model for J to which we add extra conditions on accessibility and evidence.

Factivity is captured axiomatically with the scheme $t{:}X \rightarrow X$. Modally, $\Box X \rightarrow X$ corresponds to the semantic condition of reflexivity on Kripke models, and the same is the case for justification logics. It is easy to check that the Factivity scheme is valid in all Fitting models in which accessibility, $\mathcal{R}$, is a reflexive relation. We use JT when referring to this logic, semantically or axiomatically (a completeness theorem will be shown soon, so there is no ambiguity between axiomatics and semantics here).

Positive Introspection modally is given by the axiom scheme $\Box X \rightarrow \Box\Box X$, which corresponds to a transitivity requirement on accessibility. The modal logic thus characterized axiomatically or semantically is called K4. For a justification counterpart a unary operation symbol, !, is added to the language and one adopts the axiom scheme $t{:}X \rightarrow\ !t{:}t{:}X$. For Fitting models transitivity is needed but is not enough. The justification analog of K4 is called J4, and semantically one requires that accessibility, $\mathcal{R}$, be transitive on $\mathcal{G}$, as in the modal case, and in addition we want the following.

Monotonicity Condition: If $\Gamma\mathcal{R}\Delta$ and $\Gamma \in \mathcal{E}(t, X)$ then $\Delta \in \mathcal{E}(t, X)$.
! Condition: $\mathcal{E}(t, X) \subseteq \mathcal{E}(!t, t{:}X)$.

These conditions together entail the validity of $t{:}X \rightarrow\ !t{:}t{:}X$, by the following argument. Assume $\Gamma \in \mathcal{G}$ and $\mathcal{M}, \Gamma \Vdash t{:}X$. Breaking this down according to Definition 4.1, we have that $\mathcal{M}, \Delta \Vdash X$ for every $\Delta \in \mathcal{G}$ such that $\Gamma\mathcal{R}\Delta$ (Modal Condition) and that $\Gamma \in \mathcal{E}(t, X)$ (Evidence Condition). We will show that $\mathcal{M}, \Gamma \Vdash\ !t{:}t{:}X$. This means we must show the Modal Condition and the Evidence Condition for $!t{:}t{:}X$ at Γ.

First we establish the Modal Condition for $!t{:}t{:}X$ at Γ. Let $\Delta \in \mathcal{G}$ be such that $\Gamma\mathcal{R}\Delta$; we show that $\mathcal{M}, \Delta \Vdash t{:}X$. This means we must establish the Modal Condition and the Evidence Condition for $t{:}X$ at Δ. Beginning with the Modal Condition, let $\Omega \in \mathcal{G}$ such that $\Delta\mathcal{R}\Omega$. Because $\mathcal{R}$ is transitive, $\Gamma\mathcal{R}\Omega$, and because $\mathcal{M}, \Gamma \Vdash t{:}X$, then $\mathcal{M}, \Omega \Vdash X$. Thus every state accessible from Δ is one at which X is true, and this is the Modal Condition we need for $t{:}X$ at Δ. Next the Evidence Condition. We have that $\Gamma \in \mathcal{E}(t, X)$ so by the Monotonicity Condition, $\Delta \in \mathcal{E}(t, X)$, the Evidence Condition we need. We now have shown that $\mathcal{M}, \Delta \Vdash t{:}X$.

Next we establish the Evidence Condition for $!t{:}t{:}X$ at Γ, and this is quite direct. Because $\Gamma \in \mathcal{E}(t, X)$, by the ! Condition $\Gamma \in \mathcal{E}(!t, t{:}X)$.

We have enough to conclude $\mathcal{M}, \Gamma \Vdash\ !t{:}t{:}X$.

LP is the combination of JT and J4, and so is sometimes known as JT4. The name LP is retained for historical reasons. Semantically one assumes reflexivity and transitivity of accessibility, the Monotonicity Condition, and the ! Condition.

4.3.3 K4^3 and J4^3

Modal K4^3 and justification J4^3 were introduced axiomatically in Section 2.7.1. Following Chellas (1980), in a modal model with possible worlds Γ and Δ and an accessibility relation $\mathcal{R}$, let $\Gamma\mathcal{R}^3\Delta$ mean that there are possible worlds Δ_1 and Δ_2 so that $\Gamma\mathcal{R}\Delta_1\mathcal{R}\Delta_2\mathcal{R}\Delta$. A modal model for K4^3 must meet the condition: whenever $\Gamma\mathcal{R}^3\Delta$ then $\Gamma\mathcal{R}\Delta$, a generalization of transitivity.

Fitting models for the justification logic J4^3 are models $\mathcal{M} = \langle \mathcal{G}, \mathcal{R}, \mathcal{V}, \mathcal{E} \rangle$ where the underlying frame meets the condition for K4^3 above, and in which the evidence function meets the following conditions, generalizing from ! to !!.

! and !! Conditions:

$$\Gamma \in \mathcal{E}(t, X) \quad \Longrightarrow \quad \begin{cases} \Gamma \in \mathcal{E}(!!t, !t{:}t{:}X) & (a) \\ \Gamma \mathcal{R} \Delta \Longrightarrow \Delta \in \mathcal{E}(!t, t{:}X) & (b) \\ \Gamma \mathcal{R}^2 \Delta \Longrightarrow \Delta \in \mathcal{E}(t, X) & (c) \end{cases}$$

Condition (a) generalizes the ! Condition from J4 while (b) and (c) are Monotonicity Condition generalizations.

We verify that $t{:}X \rightarrow !!t{:}\ !t{:}t{:}X$ is valid in all J4^3 Fitting models. Let $\mathcal{M} = \langle \mathcal{G}, \mathcal{R}, \mathcal{V}, \mathcal{E} \rangle$ be such a model, $\Gamma \in \mathcal{G}$, and suppose $\mathcal{M}, \Gamma \Vdash t{:}X$ but $\mathcal{M}, \Gamma \not\Vdash !!t{:}\ !t{:}t{:}X$; we derive a contradiction

Because $\mathcal{M}, \Gamma \Vdash t{:}X$, we must have $\Gamma \in \mathcal{E}(t, X)$. Because we have this and (a), but $\mathcal{M}, \Gamma \not\Vdash !!t{:}\ !t{:}t{:}X$, there must be some $\Delta_1 \in \mathcal{G}$ with $\Gamma \mathcal{R} \Delta_1$, and $\mathcal{M}, \Delta_1 \not\Vdash !t{:}t{:}X$.

Because $\mathcal{M}, \Gamma \Vdash t{:}X$ and $\Gamma \mathcal{R} \Delta_1$, by (b) we must have $\Delta_1 \in \mathcal{E}(!t, t{:}X)$. By this, and $\mathcal{M}, \Delta_1 \not\Vdash !t{:}t{:}X$, there must be some $\Delta_2 \in \mathcal{G}$ with $\Delta_1 \mathcal{R} \Delta_2$, and $\mathcal{M}, \Delta_2 \not\Vdash t{:}X$.

Again, because $\mathcal{M}, \Gamma \Vdash t{:}X$ and $\Gamma \mathcal{R}^2 \Delta_2$, by (c) we must have $\Delta_2 \in \mathcal{E}(t, X)$. Then since $\mathcal{M}, \Delta_2 \not\Vdash t{:}X$, there must be some $\Delta \in \mathcal{G}$ with $\Delta_2 \mathcal{R} \Delta$ and $\mathcal{M}, \Delta \not\Vdash X$.

Now by assumption, $\mathcal{M}, \Gamma \Vdash t{:}X$. Also $\Gamma \mathcal{R}^3 \Delta$ so by the frame condition generalizing transitivity, $\Gamma \mathcal{R} \Delta$. It follows that $\mathcal{M}, \Delta \Vdash X$, and we have a contradiction.

4.3.4 S5 and JT45

We have seen semantics for justification logics that are sublogics of LP, corresponding to sublogics of the modal logic S4. Historically, the first justification example that went beyond LP was JT45, corresponding to modal S5. It was discussed axiomatically in Section 2.7.2 and involves a negative introspection operator, "?". There are three semantic conditions needed for this operator. First, the accessibility relation $\mathcal{R}$ should be symmetric, similar to S5 Kripke modal models. Here are the other two conditions, which have no modal analog.

Strong Evidence: As given by Definition 4.5.
? Condition: If $\Gamma \notin \mathcal{E}(t, X)$ then $\Gamma \in \mathcal{E}(?t, \neg t{:}X)$

If these conditions are added to those for LP we have a semantics for the justification logic JT45. (If the conditions are added directly to J, the corresponding logic is called J5.) Notice that the ? Condition is stated with a negative antecedent. Understanding how to handle this was a major breakthrough. Our formulation is from Rubtsova (2006a, b), with a similar approach given in Pacuit (2005, 2006). The most significant new thing is that a Strong Evidence function is needed. In order to see how this is used, we show the validity of $\neg t{:}X \to ?t{:}\neg t{:}X$ in J5 Fitting models (and hence also in JT45 models). The argument is actually quite simple.

Suppose $\mathcal{M} = \langle \mathcal{G}, \mathcal{R}, \mathcal{V}, \mathcal{E} \rangle$ is a Fitting model for J5, and it holds that both $\Gamma \in \mathcal{G}$ and $\mathcal{M}, \Gamma \Vdash \neg t{:}X$. Because $\mathcal{M}, \Gamma \not\Vdash t{:}X$ and a Strong Evidence condition is imposed, we must have that $\Gamma \notin \mathcal{E}(t, X)$ so, by the ? Condition, $\Gamma \in \mathcal{E}(?t, \neg t{:}X)$. Then by Strong Evidence again, $\mathcal{M}, \Gamma \Vdash ?t{:}\neg t{:}X$.

4.3.5 Sahlqvist Examples

We give semantic counterparts to the axiomatic presentation in Section 2.7.3.

Example One The Sahlqvist scheme $\Box(X \to \Diamond X)$, or $\Box(\Box X \to X)$ in our version, corresponds to the modal frame condition $\Gamma\mathcal{R}\Delta \implies \Delta\mathcal{R}\Delta$. For the corresponding justification logic we adopted the scheme $f(t){:}(t{:}X \to X)$, where f is a one-place function symbol. A Fitting semantics for this logic begins with the J conditions for $\mathcal{M} = \langle \mathcal{G}, \mathcal{R}, \mathcal{V}, \mathcal{E} \rangle$ and adds the requirements that the frame meet the modal condition just given, and the evidence condition be such that $\mathcal{E}(f(t), t{:}X \to X) = \mathcal{G}$.

To show soundness we must show that for every model $\mathcal{M} = \langle \mathcal{G}, \mathcal{R}, \mathcal{V}, \mathcal{E} \rangle$ meeting the preceding conditions, and for every $\Gamma \in \mathcal{G}$, it holds that $\mathcal{M}, \Gamma \Vdash f(t){:}(t{:}X \to X)$. We do this by contradiction. Suppose we had $\mathcal{M}, \Gamma \not\Vdash f(t){:}(t{:}X \to X)$. It is clearly not the Evidence Condition that is violated, so there must be some $\Delta \in \mathcal{G}$ with $\Gamma\mathcal{R}\Delta$ and $\mathcal{M}, \Delta \not\Vdash t{:}X \to X$. But because $\Gamma\mathcal{R}\Delta$, the frame condition tells us that $\Delta\mathcal{R}\Delta$, so if $\mathcal{M}, \Delta \Vdash t{:}X$ then $\mathcal{M}, \Delta \Vdash X$, and we have a contradiction.

We grant that the evidence condition seems a bit like cheating. But we will see that the Fitting semantics for this logic is a useful tool for establishing a Realization Theorem. Sometimes the obvious is what works best.

Example Two In Blackburn et al. (2001) it is shown that the Sahlqvist scheme $(X \wedge \Diamond\neg X) \to \Diamond X$ corresponds to the modal frame condition $(\Gamma\mathcal{R}\Delta \wedge \Gamma \neq \Delta) \implies \Gamma\mathcal{R}\Gamma$. Although it is not noted in Blackburn et al. (2001), this reduces to the simpler frame condition $\Gamma\mathcal{R}\Delta \implies \Gamma\mathcal{R}\Gamma$, and it is this condition that we will use. Recall, in Section 2.7.3 we revised the modal scheme to avoid $\Diamond$, obtaining $\Box X \to (X \vee \Box\neg X)$, and we proposed as a justification counterpart the axiom scheme $t{:}X \to (X \vee g(t){:}\neg X)$, where g is a one-place function symbol. For a Fitting semantics we use J models $\mathcal{M} = \langle \mathcal{G}, \mathcal{R}, \mathcal{V}, \mathcal{E} \rangle$ where the frame meets the modal condition just given, and the evidence condition has the property: $\Gamma \in \mathcal{E}(t, X) \Rightarrow \Gamma \Vdash X$ or $\Gamma \in \mathcal{E}(g(t), \neg X)$. Read informally it says that if

Γ is a possible world where t is relevant evidence for X, but X is not the case, then there is also relevant evidence for $\neg X$, namely $g(t)$.

To show soundness, let $\mathcal{M} = \langle \mathcal{G}, \mathcal{R}, \mathcal{V}, \mathcal{E} \rangle$ be a Fitting model for this justification logic, and $\Gamma \in \mathcal{G}$; we show the formula $t{:}X \to (X \vee g(t){:}\neg X)$ holds at Γ. Suppose we have $\mathcal{M}, \Gamma \Vdash t{:}X$, but $\mathcal{M}, \Gamma \not\Vdash X$; we show we must have $\mathcal{M}, \Gamma \Vdash g(t){:}\neg X$. Because $\mathcal{M}, \Gamma \Vdash t{:}X$ we have $\Gamma \in \mathcal{E}(t, X)$, and because $\mathcal{M}, \Gamma \not\Vdash X$, by the Evidence Condition we must have $\Gamma \in \mathcal{E}(g(t), \neg X)$. Then if we did not have $\mathcal{M}, \Gamma \Vdash g(t){:}\neg X$ we would have some Δ with $\Gamma \mathcal{R} \Delta$ and $\mathcal{M}, \Delta \not\Vdash \neg X$. Because $\Gamma \mathcal{R} \Delta$, by the frame condition $\Gamma \mathcal{R} \Gamma$, and since $\mathcal{M}, \Gamma \Vdash t{:}X$ we would have $\mathcal{M}, \Gamma \Vdash X$. This is a contradiction. Thus we must have $\mathcal{M}, \Gamma \Vdash g(t){:}\neg X$.

4.3.6 S4.2 and JT4.2

Axiomatic systems for S4.2 and JT4.2 were discussed in Section 2.7.4. We begin the discussion here with the modal logic S4.2, which extends S4 with the Geach axiom scheme $\Diamond\Box X \to \Box\Diamond X$. This logic is complete with respect to the family of models based on frames that are reflexive, transitive, and *convergent*, meaning that whenever $\Gamma_1 \mathcal{R} \Gamma_2$ and $\Gamma_1 \mathcal{R} \Gamma_3$, there is some Γ_4 such that $\Gamma_2 \mathcal{R} \Gamma_4$ and $\Gamma_3 \mathcal{R} \Gamma_4$.

Showing soundness for S4.2 begins by showing soundness for the S4 part. That is, we want validity of $\Box X \to \Box\Box X$ because we have transitive models, and of $\Box X \to X$ because we have reflexive models. This is standard and is omitted here. Then one shows validity of the Geach scheme in convergent models. We look at this part in some detail. Suppose we have a modal model $\mathcal{M} = \langle \mathcal{G}, \mathcal{R}, \mathcal{V} \rangle$ whose frame is convergent. Suppose that $\Gamma_1 \in \mathcal{G}$ and $\mathcal{M}, \Gamma_1 \Vdash \Diamond\Box X$ but $\mathcal{M}, \Gamma_1 \not\Vdash \Box\Diamond X$; we derive a contradiction. By the first item, for some $\Gamma_2 \in \mathcal{G}$ with $\Gamma_1 \mathcal{R} \Gamma_2$, $\mathcal{M}, \Gamma_2 \Vdash \Box X$, and by the second item, for some $\Gamma_3 \in \mathcal{G}$ with $\Gamma_1 \mathcal{R} \Gamma_3$, $\mathcal{M}, \Gamma_3 \not\Vdash \Diamond X$. Because the frame is convergent, there is some Γ_4 with $\Gamma_2 \mathcal{R} \Gamma_4$ and $\Gamma_3 \mathcal{R} \Gamma_4$. But then we would have both $\mathcal{M}, \Gamma_4 \Vdash X$ and $\mathcal{M}, \Gamma_4 \not\Vdash X$, our contradiction.

Axiomatically the justification logic JT4.2 builds on LP. In Section 2.7.4 two binary function symbols, f and g were added to LP, and we adopted the following axiom scheme.

$$\neg f(t,u){:}\neg t{:}X \to g(t,u){:}\neg u{:}\neg X$$

Semantically, $\mathcal{M} = \langle \mathcal{G}, \mathcal{R}, \mathcal{V}, \mathcal{E} \rangle$ is a J4.2 Fitting model if its frame is reflexive, transitive, and convergent, as with S4.2, we have the LP requirements, the Monotonicity Condition, and the ! Condition from Section 4.3.2, $\mathcal{E}$ is a *Strong Evidence* function, Definition 4.5, and we have the following.

Geach Condition: $\mathcal{E}(f(t,u), \neg t{:}X) \cup \mathcal{E}(g(t,u), \neg u{:}\neg X) = \mathcal{G}$

We verify soundness for $\neg f(t,u){:}\neg t{:}X \rightarrow g(t,u){:}\neg u{:}\neg X$. Suppose the formula fails at Γ in model $\mathcal{M} = \langle \mathcal{G}, \mathcal{R}, \mathcal{V}, \mathcal{E} \rangle$ meeting the conditions given above; we derive a contradiction. Because it fails at Γ then $\mathcal{M}, \Gamma \not\Vdash f(t,u){:}\neg t{:}X$ and $\mathcal{M}, \Gamma \not\Vdash g(t,u){:}\neg u{:}\neg X$. Because $\mathcal{E}$ is a strong evidence function, it must be that $\Gamma \notin \mathcal{E}(f(t,u), \neg t{:}X)$ and $\Gamma \notin \mathcal{E}(g(t,u), \neg u{:}\neg X)$. But this contradicts the Geach Condition. Note that this argument is extremely trivial. The assumption of a strong evidence function is a strong assumption indeed.

This material will be revisited in a much broader setting, in Section 8.2. A separate chapter is required because of both length and significance.

4.3.7 KX4 and JX4

Recall from Section 2.7.5 that KX4, axiomatically, is modal K4 together with the scheme $\Box\Box X \rightarrow \Box X$. Semantically this corresponds to a *density* condition: if $\Gamma\mathcal{R}\Delta$ then for some Ω, $\Gamma\mathcal{R}\Omega$ and $\Omega\mathcal{R}\Delta$. Verification of soundness is simple. Suppose $\mathcal{M} = \langle \mathcal{G}, \mathcal{R}, \mathcal{V} \rangle$ is a Kripke model for which the denseness condition holds. Suppose $\Gamma \in \mathcal{G}$ and $\mathcal{M}, \Gamma \Vdash \Box\Box X$; we show $\mathcal{M}, \Gamma \Vdash \Box X$. Let $\Delta \in \mathcal{G}$ be an arbitrary state with $\Gamma\mathcal{R}\Delta$. By density there is some Ω so that $\Gamma\mathcal{R}\Omega$ and $\Omega\mathcal{R}\Delta$. Because $\mathcal{M}, \Gamma \Vdash \Box\Box X$ then $\mathcal{M}, \Omega \Vdash \Box X$, and so $\mathcal{M}, \Delta \Vdash X$. Because Δ was arbitrary, $\mathcal{M}, \Gamma \Vdash \Box X$.

The justification counterpart JX4 of KX4 adds a binary operator c to the machinery of J4, from Section 4.3.2, along with the axiom scheme $s{:}t{:}X \rightarrow [s \mathsf{c}\, t]{:}X$. Semantically we add the same denseness condition we did earlier for the modal case. We also add the following Evidence Function requirement.

c **Condition** $\mathcal{E}(s, t{:}X) \subseteq \mathcal{E}(s \mathsf{c}\, t, X)$

Soundness now follows quite easily. The part involving ! carries over from J4 and is omitted. Suppose we have a JX4 model $\mathcal{M} = \langle \mathcal{G}, \mathcal{R}, \mathcal{V}, \mathcal{E} \rangle$ as just defined, and suppose $\mathcal{M}, \Gamma \Vdash s{:}t{:}X$ for some $\Gamma \in \mathcal{G}$. We show $\mathcal{M}, \Gamma \Vdash [s \mathsf{c}\, t]{:}X$. First, making use of density exactly as we did earlier for KX4, we can show that $\mathcal{M}, \Delta \Vdash X$ for every $\Delta \in \mathcal{G}$ with $\Gamma\mathcal{R}\Delta$. Next, we have $\Gamma \in \mathcal{E}(s, t{:}X)$, so by the c Condition $\Gamma \in \mathcal{E}(s \mathsf{c}\, t, X)$. This is enough to ensure that $\mathcal{M}, \Gamma \Vdash [s \mathsf{c}\, t]{:}X$.

4.3.8 A Remark about Strong Evidence Functions

Imposing a strong evidence assumption can sometimes make other requirements on a model redundant. For instance, if we restrict the semantics for $\mathsf{J4}^3$ from Section 4.3.3 to models with strong evidence functions, two of the three conditions on evidence functions are unnecessary. Assume, for this discussion, that $\Gamma \in \mathcal{E}(t, X)$. We retain consequence (a) that $\Gamma \in \mathcal{E}(!!t, !t{:}t{:}X)$, but we can drop (b) and (c) by the following reasoning.

Suppose that $\Gamma \in \mathcal{E}(t, X)$ and also $\Gamma\mathcal{R}\Delta$. We have $\Gamma \in \mathcal{E}(!!t, !t{:}t{:}X)$ by condition (a). Because $\mathcal{E}$ is a strong evidence function, it follows that $\mathcal{M}, \Gamma \Vdash !!t{:}!t{:}t{:}X$. Then because $\Gamma\mathcal{R}\Delta$, we must have that $\mathcal{M}, \Delta \Vdash !t{:}t{:}X$, and hence $\Delta \in \mathcal{E}(!t, t{:}X)$. Redundancy of condition (c) has a similar argument.

For J4.2 in Section 4.3.6, a similar argument shows that the Monotonicity Condition is unnecessary if we have a Strong Evidence function. Suppose $\Gamma\mathcal{R}\Delta$ and $\Gamma \in \mathcal{E}(t, X)$. By the ! Condition, $\Gamma \in \mathcal{E}(!t, t{:}X)$. Assuming we have Strong Evidence, $\mathcal{M}, \Gamma \Vdash !t{:}t{:}X$. Because $\Gamma\mathcal{R}\Delta$, $\mathcal{M}, \Delta \Vdash t{:}X$, and hence $\Delta \in \mathcal{E}(t, X)$.

Strong Evidence is, in fact, quite strong.

4.4 Canonical Models and Completeness

Canonical models for modal logics are constructed with possible worlds that are maximally consistent sets of formulas and with accessibility defined in a standard way. Logics for which canonical models suffice to establish axiomatic completeness are called canonical logics. Most common modal logics are canonical, though there are important exceptions. All the modal logics we have considered here are canonical. The methodology developed for canonical modal logics carries over to justification logics and plays a fundamental role. For reference, we begin with a review of canonical machinery in the modal setting before moving to justification logics.

4.4.1 Canonical Modal Logics

Let KL be a normal modal logic, which we assume is axiomatized by adding a set of axioms to K. Call a set S of modal formulas KL-*consistent* provided that for no $X_1, \ldots, X_n \in S$ is $(X_1 \wedge \ldots \wedge X_n) \rightarrow \bot$ provable in KL. Using Lindenbaum's construction, every KL-consistent set extends to a maximally KL-consistent set, one that is consistent but no proper extension is. Familiar arguments tell us that if S is maximally consistent then, for each formula X, exactly one of X or $\neg X$ will be in S. Likewise $X \rightarrow Y \in S$ exactly when $X \notin S$ or $Y \in S$, and similarly for the other propositional connectives.

Definition 4.6 (Sharp Operation, Modal Version) For a set S of modal formulas, let $S^{\sharp} = \{X \mid \Box X \in S\}$.

Let $\mathcal{G}$ be the set of all maximally KL-consistent sets. For $\Gamma, \Delta \in \mathcal{G}$, let $\Gamma\mathcal{R}\Delta$

provided $\Gamma^\sharp \subseteq \Delta$. This gives us a *frame*, $\langle \mathcal{G}, \mathcal{R} \rangle$. Finally for each propositional letter A, set $\Gamma \in \mathcal{V}(A)$ just if $A \in \Gamma$ (so then A is true at $\Gamma \in \mathcal{G}$ if and only if $A \in \Gamma$) . This completely specifies a Kripke model $\mathcal{M} = \langle \mathcal{G}, \mathcal{R}, \mathcal{V} \rangle$, called the *canonical* model for KL.

The key fact about modal canonical models is the following. It is very well-known, but is included for reference.

Theorem 4.7 (Modal Truth Lemma) *Let $\mathcal{M} = \langle \mathcal{G}, \mathcal{R}, \mathcal{V} \rangle$ be the canonical model for modal logic* KL*. For any $\Gamma \in \mathcal{G}$ and for any formula X,*

$$X \in \Gamma \Longleftrightarrow \mathcal{M}, \Gamma \Vdash X.$$

Proof The argument is by induction on the degree of X. If X is a propositional letter, the equivalence holds by definition of $\mathcal{V}$ in the canonical model. If X is $\bot$ both sides of the equivalence are false, the left because Γ is consistent, the right because $\mathcal{V}(\bot) = \emptyset$ in any Kripke model.

Assume the result is known for formulas simpler than $X \to Y$. Then we have the following.

$$\begin{aligned} X \to Y \in \Gamma &\Leftrightarrow X \notin \Gamma \text{ or } Y \in \Gamma \text{ (maximal consistency)} \\ &\Leftrightarrow \mathcal{M}, \Gamma \nVdash X \text{ or } \mathcal{M}, \Gamma \Vdash Y \text{ (induction hypothesis)} \\ &\Leftrightarrow \mathcal{M}, \Gamma \Vdash X \to Y \text{ (definition of } \Vdash) \end{aligned}$$

The necessitation case must be done in two parts, one of which will be seen to hold essentially by definition. Assume the result is known for formulas simpler than $\Box X$.

Suppose $\Box X \in \Gamma$. Let Δ be an arbitrary member of $\mathcal{G}$ with $\Gamma \mathcal{R} \Delta$. By definition of $\mathcal{R}$, it holds that $\Gamma^\sharp \subseteq \Delta$. So, because $X \in \Gamma^\sharp$, we know that $X \in \Delta$ and by the induction hypothesis, $\mathcal{M}, \Delta \Vdash X$. Because Δ was arbitrary, $\mathcal{M}, \Gamma \Vdash \Box X$.

For the converse direction, suppose $\Box X \notin \Gamma$. We must show $\mathcal{M}, \Gamma \nVdash \Box X$. The argument proceeds in two stages.

Using our assumption, $\neg\Box X \in \Gamma$ by maximal consistency. We first show that $\Gamma^\sharp \cup \{\neg X\}$ is consistent. This is a proof by contradiction. If $\Gamma^\sharp \cup \{\neg X\}$ is not consistent, there is some finite set $\{G_1, \ldots, G_n\} \subseteq \Gamma^\sharp$ such that $(G_1 \wedge \ldots \wedge G_n \wedge \neg X) \to \bot$ is a KL theorem. Then so is $(G_1 \wedge \ldots \wedge G_n) \to X$. By standard normal modal logic manipulation it follows that $(\Box G_1 \wedge \ldots \wedge \Box G_n) \to \Box X$ is also a theorem. Because each $G_i \in \Gamma^\sharp$, $\Box G_i \in \Gamma$, so by maximal consistency, $\Box X \in \Gamma$, but it does not.

We have shown that $\Gamma^\sharp \cup \{\neg X\}$ is consistent. By a Lindenbaum construction, it extends to a maximally consistent set, call it Δ. Then $\Delta \in \mathcal{G}$ and because $\Gamma^\sharp \subseteq \Delta$ we have $\Gamma \mathcal{R} \Delta$. Because $\neg X \in \Delta$, $X \notin \Delta$ so by the induction hypothesis, $\mathcal{M}, \Delta \nVdash X$. It follows that $\mathcal{M}, \Gamma \nVdash \Box X$. □

A canonical model is a universal countermodel. For, suppose X is not provable in KL. Then $\{\neg X\}$ is KL-consistent and so can be extended to a maximally consistent set Γ. Then $\Gamma \in \mathcal{G}$ and, by the Truth Lemma, $\mathcal{M}, \Gamma \not\Vdash X$.

What has been shown is that if X is not provable in KL, then X is falsifiable in the canonical model. Now we have a very important point. It can happen that the canonical model is not actually a model that meets the intended semantic conditions for its logic. For instance, this happens with Gödel-Löb logic, GL. This logic is characterized independently by two different sets of frames; for one of them frames must be finite. Because any canonical model is infinite, it is not built on such a frame. The other kind of frame is a bit more complicated to describe, but the canonical model is not built on one of these either. Completeness can be proved for GL, but a direct canonical model argument won't work.

An axiomatic modal logic KL is *canonical* if its canonical model meets the semantic conditions adopted for KL. This is stated loosely because it requires an independent determination of what we want to count as a model for KL. The statement can be cleaned up, but we don't need the resulting complications here. We will simply assume the modal logics we discuss come with intended semantics. Then GL is not canonical, but most common modal logics are. Indeed, all the modal logics considered in Sections 2.7 and 4.3 are canonical. For instance, KT is axiomatized by adding the scheme $\Box X \rightarrow X$ to K, and its intended semantics requires that models have accessibility relations that are reflexive. This is canonical because, if Γ is a possible world in the canonical KT model, $\Gamma^\sharp \subseteq \Gamma$ by the following argument. Suppose $X \in \Gamma^\sharp$. Then $\Box X \in \Gamma$. Because Γ is maximally consistent and we have $\Box X \rightarrow X$, then $X \in \Gamma$.

The key fact to take away is that canonical modal logics have axiomatic completeness proofs that are uniform across a broad range of modal logics.

4.4.2 Canonical Justification Models

Canonical models for justification logics were introduced in Fitting (2005). Until recently all known justification logics were counterparts of canonical modal logics, but recently Shamkanov (2016) has shown a realization theorem for the Gödel–Löb modal logic, and this is not canonical. Nonetheless, canonical justification models supply the main semantic machinery behind almost all realization proofs, as we will see in Chapters 7 and 8.

Definition 4.8 (Consistency) Let JL(CS) be some axiomatically formulated justification logic, with constant specification CS. We say a set S of formulas is JL(CS)-*inconsistent* if $S \vdash_{\mathsf{JL(CS)}} \bot$, and S is JL(CS)-*consistent* if it is not JL(CS)-inconsistent.

We need an analog of the modal sharp operation, Definition 4.6, which was used in the previous section.

Definition 4.9 (Sharp Operation, Justification Version) For a set S of formulas of a justification logic, $S^\sharp = \{X \mid t{:}X \in S \text{ for some justification term } t\}$.

The following definition is directly analogous to that for modal canonical models—only part (4) is new.

Definition 4.10 (Canonical Justification Model) The *canonical Fitting model* $\mathcal{M}_{\mathsf{JL(CS)}} = \langle \mathcal{G}, \mathcal{R}, \mathcal{V}, \mathcal{E} \rangle$ for justification logic JL(CS), is defined as follows.

(1) $\mathcal{G}$ is the set of all maximally JL(CS)-consistent sets of formulas.
(2) For $\Gamma, \Delta \in \mathcal{G}$, $\Gamma \mathcal{R} \Delta$ if $\Gamma^\sharp \subseteq \Delta$.
(3) For atomic A, $\Gamma \in \mathcal{V}(A)$ if $A \in \Gamma$.
(4) $\Gamma \in \mathcal{E}(t, X)$ if $t{:}X \in \Gamma$.

We need to verify that the canonical model meets the Minimum Evidence Conditions from Definition 4.2 and the Constant Specification Condition from Definition 4.3, but this is simple. The $\cdot$ condition on evidence functions says $\mathcal{E}(s, X \to Y) \cap \mathcal{E}(t, X) \subseteq \mathcal{E}(s \cdot t, Y)$. Well, suppose $\Gamma \in \mathcal{E}(s, X \to Y)$ and $\Gamma \in \mathcal{E}(t, X)$. By definition of $\mathcal{E}$ in the canonical model, $s{:}(X \to Y) \in \Gamma$ and $t{:}X \in \Gamma$. Because JL(CS) axiomatically extends J_0, $s{:}(X \to Y) \to (t{:}X \to [s \cdot t]{:}Y)$ is an axiom. Since Γ is maximally JL(CS) consistent, it follows that $[s \cdot t]{:}Y \in \Gamma$, and hence $\Gamma \in \mathcal{E}(s \cdot t, Y)$. The + condition is $\mathcal{E}(s, X) \cup \mathcal{E}(t, X) \subseteq \mathcal{E}(s + t, X)$, and is treated the same way using J_0 axioms $s{:}X \to [s + t]{:}X$ and $t{:}X \to [s + t]{:}X$. The Constant Specification Condition is quite immediate.

As with modal canonical models, the key item to show is a truth lemma, analogous to Theorem 4.7. Curiously, for justification logics the proof is simpler than in the modal case.

Theorem 4.11 (Justification Truth Lemma) *Let* $\mathcal{M}_{\mathsf{JL(CS)}} = \langle \mathcal{G}, \mathcal{R}, \mathcal{V}, \mathcal{E} \rangle$ *be the canonical justification model for justification logic* JL(CS). *For any* $\Gamma \in \mathcal{G}$ *and any formula* X,

$$X \in \Gamma \iff \mathcal{M}_{\mathsf{JL(CS)}}, \Gamma \Vdash X$$

Proof The proof is by induction on the degree of X. The atomic case is by definition. Propositional connective cases are the same as in the modal proof for Theorem 4.7, making use of maximal consistency of Γ. This leaves the justification case. Assume as induction hypothesis that the result holds for for the formula X.

Suppose $t{:}X \in \Gamma$. By definition of $\mathcal{E}$ we have $\Gamma \in \mathcal{E}(t, X)$. Let Δ be an arbitrary member of $\mathcal{G}$ with $\Gamma\mathcal{R}\Delta$. Then $\Gamma^\sharp \subseteq \Delta$, so $X \in \Delta$, and by the induction hypothesis $\mathcal{M}_{\mathsf{JL(CS)}}, \Delta \Vdash X$. Then $\mathcal{M}_{\mathsf{JL(CS)}}, \Gamma \Vdash t{:}X$.

Now suppose $t{:}X \notin \Gamma$. Then $\Gamma \notin \mathcal{E}(t, X)$, so $\mathcal{M}_{\mathsf{JL(CS)}}, \Gamma \nVdash t{:}X$. □

As usual, the canonical justification model $\mathcal{M}_{\mathsf{JL(CS)}}$ is a universal counter model for $\mathsf{JL(CS)}$, and by the familiar argument. Then the completeness question reduces to the following—a direct analog of what happens in modal logic. We may have some class $\mathcal{F}_{\mathsf{JL(CS)}}$ of Fitting models, and we want to prove completeness of justification logic $\mathsf{JL(CS)}$ relative to this class. If it turns out that the canonical model $\mathcal{M}_{\mathsf{JL(CS)}}$ is in $\mathcal{F}_{\mathsf{JL(CS)}}$, we have succeeded. Typically this comes down to showing the frame and the evidence function have appropriate properties. We will look at examples in Section 4.5, but first we have a bit more to extract from the canonical model construction.

4.4.3 Strong Evidence and Fully Explanatory

In Definition 4.4 *Fully Explanatory* Fitting models were characterized, and in Definition 4.5 *Strong Evidence Functions* as well. As we are about to see, the second of these comes for free, while the first requires a familiar argument already seen in the proof of Theorem 4.7. We begin with Strong Evidence. Recall that an evidence function in a Fitting model is *strong* if the Evidence condition for it implies the Modal condition.

Theorem 4.12 (Strong Evidence) *Let* $\mathsf{JL(CS)}$ *be an axiomatically presented justification logic, and let* $\mathcal{M}_{\mathsf{JL(CS)}} = \langle \mathcal{G}, \mathcal{R}, \mathcal{V}, \mathcal{E} \rangle$ *be the canonical model for it. The evidence function* $\mathcal{E}$ *is a strong evidence function.*

Proof Suppose $\Gamma \in \mathcal{E}(t, X)$. Then, by definition of $\mathcal{E}$ in canonical models, $t{:}X \in \Gamma$ so by Theorem 10.41, $\mathcal{M}_{\mathsf{JL(CS)}}, \Gamma \Vdash t{:}X$. But then $\mathcal{M}_{\mathsf{JL(CS)}}, \Delta \Vdash X$ for every $\Delta \in \mathcal{G}$ with $\Gamma\mathcal{R}\Delta$. □

We move on to Fully Explanatory. Speaking informally, suppose we say a formula is *knowable* at a possible world of a Fitting model if it is true at all accessible possible worlds. Then Fully Explanatory says: Any formula that is knowable at a possible world has a justification at that possible world.

Theorem 4.13 (Fully Explanatory) *Again let* $\mathsf{JL(CS)}$ *be an axiomatically presented justification logic, and let* $\mathcal{M}_{\mathsf{JL(CS)}} = \langle \mathcal{G}, \mathcal{R}, \mathcal{V}, \mathcal{E} \rangle$ *be the canonical model for it. Assume* JL *has the internalization property relative to constant specification* CS *(by Theorem 2.14, this is the case if* CS *is axiomatically appropriate). Then the canonical model is fully explanatory. That is, if*

$\mathcal{M}_{\mathsf{JL(CS)}}, \Delta \Vdash X$ for every $\Delta \in \mathcal{G}$ with $\Gamma\mathcal{R}\Delta$, then for some justification term t, $\mathcal{M}_{\mathsf{JL(CS)}}, \Gamma \Vdash t{:}X$

Proof We show $\mathcal{M}_{\mathsf{JL(CS)}}$ is Fully Explanatory by working with the definition in contrapositive form. Let $\Gamma \in \mathcal{G}$, and suppose X is a formula such that that $\mathcal{M}_{\mathsf{JL(CS)}}, \Gamma \not\Vdash t{:}X$ for every justification term t. We show that for some $\Delta \in \mathcal{G}$ with $\Gamma\mathcal{R}\Delta$, $\mathcal{M}_{\mathsf{JL(CS)}}, \Delta \not\Vdash X$.

Using Theorem 10.41, the Truth Lemma, $t{:}X \notin \Gamma$ for every justification term t. We begin by showing $\Gamma^\sharp \cup \{\neg X\}$ is consistent in JL(CS).

Well suppose otherwise, $\Gamma^\sharp \cup \{\neg X\}$ is inconsistent. We derive a contradiction. Assuming the inconsistency, there are $Y_1, \ldots, Y_n \in \Gamma^\sharp$ so that $Y_1, \ldots, Y_n \vdash_{\mathsf{JL(CS)}} X$. For each i, because $Y_i \in \Gamma^\sharp$ then $t_i{:}Y_i \in \Gamma$ for some t_i. Then by the Lifting Lemma, Theorem 2.16, there is some justification term u so that $t_1{:}Y_1, \ldots, t_n{:}Y_n \vdash_{\mathsf{JL(CS)}} u{:}X$. Because $t_i{:}Y_i \in \Gamma$ and Γ is maximally consistent, $u{:}X \in \Gamma$, and so by the Truth Lemma again, $\mathcal{M}_{\mathsf{JL(CS)}}, \Gamma \Vdash u{:}X$ contradicting our initial assumption that $\mathcal{M}_{\mathsf{JL(CS)}}, \Gamma \not\Vdash t{:}X$ for every justification term t.

We have shown that $\Gamma^\sharp \cup \{\neg X\}$ is consistent. Extend this set to a maximally consistent set Δ. Then $\Delta \in \mathcal{G}$ and $\Gamma\mathcal{R}\Delta$. Because $\neg X \in \Delta$, then clearly $\mathcal{M}_{\mathsf{JL(CS)}}, \Delta \not\Vdash X$. □

4.5 Completeness Examples

We resume our examination of the justification logic examples that we first looked at axiomatically in Section 2.7, then again in Section 4.3 where semantics were given and soundness was established. Now we show completeness with respect to appropriate Fitting model classes. For each example we begin with references to the earlier discussions of the logic involved.

4.5.1 LP and Sublogics

(Continuing 2.6 and 4.3.2.)

We concentrate on completeness for LP because showing it for J, JD, JT, J4, and JD4 can be done by omitting parts of the argument. Given the work in Section 4.4.2, all that needs to be shown for LP is that the canonical model $\mathcal{M}_{\mathsf{LP(CS)}} = \langle \mathcal{G}, \mathcal{R}, \mathcal{V}, \mathcal{E} \rangle$ for axiomatic LP with constant specification CS is an LP model as defined in Section 4.3.2.

We need that $\mathcal{R}$ be transitive; $\Gamma\mathcal{R}\Delta$ and $\Delta\mathcal{R}\Omega$ implies $\Gamma\mathcal{R}\Omega$. Using the accessibility definition in canonical models, we assume that $\Gamma^\sharp \subseteq \Delta$ and $\Delta^\sharp \subseteq \Omega$, and we show $\Gamma^\sharp \subseteq \Omega$. Suppose $X \in \Gamma^\sharp$. Then $t{:}X \in \Gamma$ for some t. But $t{:}X \to\, !t{:}t{:}X$ is an axiom of LP and Γ is maximally consistent, so $!t{:}t{:}X \in \Gamma$. Then $t{:}X$ is in $\Gamma^\#$ and hence in Δ, so X is in $\Delta^\#$ and hence in Ω.

We need that $\mathcal{R}$ be reflexive, that is, for $\Gamma \in \mathcal{G}$, it holds that $\Gamma\mathcal{R}\Gamma$. This is an easy consequence of the fact that we have $t{:}X \to X$ as an axiom scheme, and we omit the details.

Moving to the evidence function, we need two conditions in addition to those required for J_0. The first is a monotonicity condition on accessibility: if $\Gamma\mathcal{R}\Delta$ and $\Gamma \in \mathcal{E}(t, X)$ then $\Delta \in \mathcal{E}(t, X)$. Using the definition of canonical model this becomes: $\Gamma^\# \subseteq \Delta$ and $t{:}X \in \Gamma$ implies $t{:}X \in \Delta$. This is immediate, using the LP axiom $t{:}X \to !t{:}t{:}X$

The second evidence condition we need is: if $\Gamma \in \mathcal{E}(t, X)$ then $\Gamma \in \mathcal{E}(!t, t{:}X)$. The canonical model definition turns this into: $t{:}X \in \Gamma$ implies $!t{:}t{:}X \in \Gamma$, and the axiom $t{:}X \to !t{:}t{:}X$ gives us this as well.

4.5.2 J4^3

(Continuing 2.7.1 and 4.3.3.)

Let $\mathcal{M}_{\mathsf{J4^3(CS)}} = \langle \mathcal{G}, \mathcal{R}, \mathcal{V}, \mathcal{E}\rangle$ be the canonical J4^3 model, using CS as constant specification. We saw in Section 4.3.8 that when dealing with a strong evidence function, as we have in $\mathcal{M}_{\mathsf{J4^3(CS)}}$, only one of the three evidence function conditions for J4^3 needs to be established. We must show $\Gamma \in \mathcal{E}(t, X)$ implies $\Gamma \in \mathcal{E}(!!t, !t{:}t{:}X)$. For canonical models this is equivalent to $t{:}X \in \Gamma$ implies $!!t{:}\ !t{:}t{:}X \in \Gamma$. But $t{:}X \to !!t{:}\ !t{:}t{:}X$ is an axiom of J4^3, and we have maximal consistency of Γ.

We must also show that in $\mathcal{M}_{\mathsf{J4^3(CS)}}$, if $\Gamma\mathcal{R}^3\Delta$ then $\Gamma\mathcal{R}\Delta$, and this can be done by an extension of the proof that the canonical model for LP is transitive, given earlier. We recall the details, for convenience. Making use of the accessibility definition in canonical models, we assume that $\Gamma^\# \subseteq \Delta_1$, $\Delta_1^\# \subseteq \Delta_2$, and $\Delta_2^\# \subseteq \Omega$, and we show $\Gamma^\# \subseteq \Omega$. Suppose $X \in \Gamma^\#$. Then $t{:}X \in \Gamma$ for some t. Because $t{:}X \to !!t{:}!t{:}t{:}X$ is an axiom of LP and Γ is maximally consistent, then $!!t{:}!t{:}t{:}X \in \Gamma$. But then $!t{:}t{:}X \in \Gamma^\# \subseteq \Delta_1$, so $t{:}X \in \Delta_1^\# \subseteq \Delta_2$, and so $X \in \Delta_2^\# \subseteq \Omega$.

4.5.3 JT45

(Continuing 2.7.2 and 4.3.4.)

Assume $\mathcal{M}_{\mathsf{JT45(CS)}} = \langle \mathcal{G}, \mathcal{R}, \mathcal{V}, \mathcal{E}\rangle$ is the canonical model for JT45 with a constant specification of CS. Because this logic extends LP, we have the items shown in Section 4.5.1. The additional items we need are that we have a symmetric accessibility relation, the Strong Evidence condition, and the ? condition: if $\Gamma \notin \mathcal{E}(t, X)$ then $\Gamma \in \mathcal{E}(?t, \neg t{:}X)$.

Strong Evidence is from Theorem 4.12. For the ? condition on evidence, it must be shown that if $t{:}X \notin \Gamma$ then $?t{:}\neg t{:}X \in \Gamma$. Suppose $t{:}X \notin \Gamma$. Because Γ is maximally consistent, $\neg t{:}X \in \Gamma$. Then the JT45 axiom $\neg t{:}X \to ?t{:}\neg t{:}X$, and maximal consistency of Γ gives us what we need.

We also need the accessibility relation to be symmetric. Assume $\Gamma^\sharp \subseteq \Delta$; we show that $\Delta^\sharp \subseteq \Gamma$. Suppose $X \in \Delta^\sharp$. Then for some justification term t, $t{:}X \in \Delta$. Because Δ is consistent, $\neg t{:}X \notin \Delta$, and hence $\neg t{:}X \notin \Gamma^\sharp$. But then $?t{:}\neg t{:}X \notin \Gamma$ so, using the axiom $\neg t{:}X \to ?t{:}\neg t{:}X$, $\neg t{:}X \notin \Gamma$. By maximal consistency, $t{:}X \in \Gamma$, so using the reflexivity of accessibility, which we already have in LP, $X \in \Gamma$.

4.5.4 Sahlqvist Examples

(Continuing 2.7.3 and 4.3.5.)

Example One In our treatment the modal scheme $\Box(\Box X \to X)$ corresponds to the justification scheme $f(t){:}(t{:}X \to X)$, where f is a one-place function symbol. The semantic frame condition is: if $\Gamma\mathcal{R}\Delta$ then $\Delta\mathcal{R}\Delta$, Let $\mathcal{M}_{\mathsf{JSahlOne(CS)}} = \langle \mathcal{G}, \mathcal{R}, \mathcal{V}, \mathcal{E}\rangle$ be the canonical model for the justification logic extending J_0 with the scheme $f(t){:}(t{:}X \to X)$ and a constant specification CS. Assume that for some $\Gamma, \Delta \in \mathcal{G}$ we have $\Gamma\mathcal{R}\Delta$ but not $\Delta\mathcal{R}\Delta$. We derive a contradiction.

Because we do not have $\Delta\mathcal{R}\Delta$ then $\Delta^\sharp \not\subseteq \Delta$. Then for some t and X, $t{:}X \in \Delta$ but $X \notin \Delta$. Because Δ is maximally consistent, $t{:}X \to X \notin \Delta$. Because Γ is maximally consistent, $f(t){:}(t{:}X \to X) \in \Gamma$ because it is an axiom. And because $\Gamma\mathcal{R}\Delta$, $\Gamma^\sharp \subseteq \Delta$, so $t{:}X \to X \in \Delta$, and this is our contradiction.

For the evidence function we need to show $\mathcal{E}(f(t), t{:}X \to X) = \mathcal{G}$. But this is equivalent to the requirement that for every $\Gamma \in \mathcal{G}$, $f(t){:}(t{:}X \to X) \in \Gamma$. This is so because the formula is an axiom, and hence in every maximally consistent set.

Example Two The scheme $\Box X \to (X \vee \Box\neg X)$ corresponds to the frame condition $\Gamma\mathcal{R}\Delta \implies \Gamma\mathcal{R}\Gamma$. Our justification scheme, added to J_0, is $t{:}X \to (X \vee g(t){:}\neg X)$.

Let $\mathcal{M}_{\mathsf{JSahlTwo(CS)}} = \langle \mathcal{G}, \mathcal{R}, \mathcal{V}, \mathcal{E}\rangle$ be the canonical model for the justification logic extending J_0 with a constant specification CS and the scheme $t{:}X \to (X \vee g(t){:}\neg X)$. We show the frame of $\mathcal{M}_{\mathsf{JSahlTwo(CS)}}$ meets the frame condition just given.

Suppose $\Gamma\mathcal{R}\Delta$ but not $\Gamma\mathcal{R}\Gamma$; we derive a contradiction. We have $\Gamma^\sharp \subseteq \Delta$ and $\Gamma^\sharp \not\subseteq \Gamma$. By the second of these, $t{:}X \in \Gamma$ but $X \notin \Gamma$ for some $t{:}X$. Because Γ is maximally consistent, $t{:}X \to (X \vee g(t){:}\neg X) \in \Gamma$, and hence $(X \vee g(t){:}\neg X) \in \Gamma$.

And because $X \notin \Gamma$, $g(t){:}\neg X \in \Gamma$. Because $\Gamma^\sharp \subseteq \Delta$, both $X \in \Delta$ and $\neg X \in \Delta$, violating consistency.

The evidence function requirement is

$$\Gamma \in \mathcal{E}(t, X) \Rightarrow \Gamma \Vdash X \text{ or } \Gamma \in \mathcal{E}(g(t), \neg X).$$

To verify this let Γ be a possible world of the canonical model $\mathcal{M}_{\mathsf{JSahlTwo(CS)}}$ and suppose $\Gamma \in \mathcal{E}(t, X)$ but $\Gamma \nVdash X$; we show $\Gamma \in \mathcal{E}(g(t), \neg X)$. Because $\Gamma \in \mathcal{E}(t, X)$ we have $t{:}X \in \Gamma$. Because $\Gamma \nVdash X$ we have $X \notin \Gamma$ and so by maximal consistency, $\neg X \in \Gamma$. Because $t{:}X \to (X \vee g(t){:}\neg X)$ is an axiom it is in Γ, and then because maximally consistent sets are closed under consequence, it follows that $g(t){:}\neg X \in \Gamma$. From this we have $\Gamma \in \mathcal{E}(g(t), \neg X)$.

4.5.5 S4.2 and JT4.2

(Continuing 2.7.4 and 4.3.6.)

The logics S4.2 and J4.2 are especially important in this book because, as we will see in Chapter 8, they are representative cases of very general families. Because our full arguments will be generalizations of those that work for S4.2 and J4.2, we present these cases in detail, beginning with a completeness proof for the modal logic S4.2. If this work is understood, the generalizations should be much easier to follow.

We start with the modal argument, showing completeness for S4.2. Let $\mathcal{M}_{\mathsf{S4.2}} = \langle \mathcal{G}, \mathcal{R}, \mathcal{V} \rangle$ be the canonical model for the modal logic S4.2. The intended Kripke semantics for the modal logic S4.2 has a *convergent* condition, $\Gamma_1 \mathcal{R} \Gamma_2$ and $\Gamma_1 \mathcal{R} \Gamma_3$ together imply the existence of Γ_4 with $\Gamma_2 \mathcal{R} \Gamma_4$ and $\Gamma_3 \mathcal{R} \Gamma_4$. We show the canonical model $\mathcal{M}_{\mathsf{S4.2}}$ meets this condition.

Suppose, in the canonical model $\mathcal{M}_{\mathsf{S4.2}}$ for S4.2, we have $\Gamma_1 \mathcal{R} \Gamma_2$ and $\Gamma_1 \mathcal{R} \Gamma_3$. If we could show that $\Gamma_2^\sharp \cup \Gamma_3^\sharp$ was consistent, then it could be extended to a maximally consistent Γ_4, and we would have $\Gamma_2 \mathcal{R} \Gamma_4$ and $\Gamma_3 \mathcal{R} \Gamma_4$, as illustrated in Figure 4.1. So it is enough to derive a contradiction from the assumption that $\Gamma_2^\sharp \cup \Gamma_3^\sharp$ is not S4.2 consistent.

To simplify things we begin with the following, which tells us that we can replace arguments about finite subsets of $\Gamma^\sharp$ with arguments involving single members of $\Gamma^\sharp$.

Lemma 4.14 *Let $\mathcal{M} = \langle \mathcal{G}, \mathcal{R}, \mathcal{V} \rangle$ be the canonical model for any normal modal logic. Suppose $\Gamma \in \mathcal{G}$. Then $A_1, \ldots, A_n \in \Gamma^\sharp$ if and only if $A_1 \wedge \ldots \wedge A_n \in \Gamma^\sharp$ (parenthesized however).*

Proof Suppose $A_1, \ldots, A_n \in \Gamma^\sharp$. Then $\Box A_1, \ldots, \Box A_n \in \Gamma$. In any normal

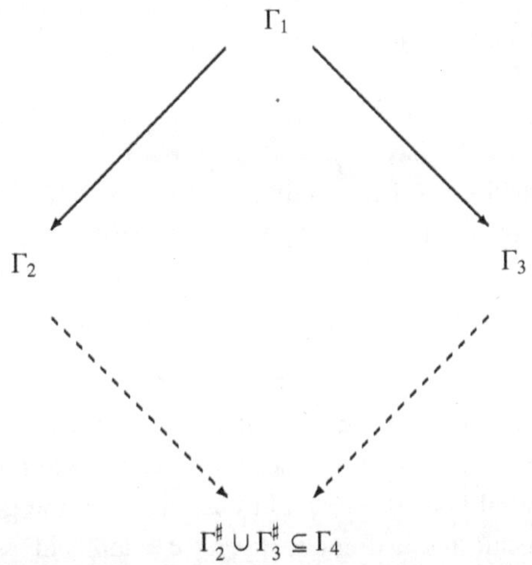

Figure 4.1 Obtaining the Convergent Condition for S4.2

modal logic, $\Box A_1 \to (\Box A_2 \to \ldots \to (\Box A_n \to \Box(A_1 \wedge \ldots \wedge A_n))\ldots)$ is valid. Because worlds in the canonical model are maximally consistent, it follows that $\Box(A_1 \wedge \ldots \wedge A_n) \in \Gamma$, and hence $A_1 \wedge \ldots \wedge A_n \in \Gamma^\#$. The converse is similar, using the validities $(\Box A_1 \wedge \ldots \wedge \Box A_i \wedge \ldots \wedge \Box A_n) \to \Box A_i$, for each i. □

Now, suppose $\Gamma_2^\# \cup \Gamma_3^\#$ is not S4.2 consistent. Then (using Lemma 4.14) there are $A \in \Gamma_2^\#$ and $B \in \Gamma_3^\#$ so that $B \to \neg A$ is provable in S4.2, and hence so is $\Box B \to \Box\neg A$. Because $A \in \Gamma_2^\#$, $\Box A \in \Gamma_2$. Similarly, $\Box B \in \Gamma_3$. Because $\Gamma_1 \mathcal{R} \Gamma_2$ then $\Gamma_1^\# \subseteq \Gamma_2$, so $\Box\neg\Box A \notin \Gamma_1$ by consistency of Γ_2, and so $\neg\Box\neg\Box A \in \Gamma_1$, by maximality of Γ_1. That is, $\Diamond\Box A \in \Gamma_1$. Then, using $\Diamond\Box A \to \Box\Diamond A$ and maximal consistency, $\Box\Diamond A \in \Gamma_1$. But $\Gamma_1 \mathcal{R} \Gamma_3$ so $\Gamma_1^\# \subseteq \Gamma_3$, and hence $\Diamond A \in \Gamma_3$, that is, $\neg\Box\neg A \in \Gamma_3$. But $\Box B \in \Gamma_3$, and $\Box B \to \Box\neg A$ is provable, so $\Box\neg A \in \Gamma_3$. This shows inconsistency of Γ_3 and is our desired contradiction.

Having shown the modal case, we now turn to completeness for the justification logic J4.2. For this we need the special assumption that we have internalization, which we have if there is an axiomatically appropriate constant specification. Such an assumption was not needed for the justification logics we discussed previously, but it comes in when we prove an analog of Lemma 4.14.

Lemma 4.15 *Suppose Γ is a possible world in the canonical Fitting justifi-*

cation model for some justification logic JL(CS) *that satisfies internalization. Then* $A_1, \ldots, A_n \in \Gamma^\sharp$ *if and only if* $A_1 \wedge \ldots \wedge A_n \in \Gamma^\sharp$.

Proof Suppose $A_1, \ldots, A_n \in \Gamma^\sharp$. Then $t_1{:}A_1, \ldots, t_n{:}A_n \in \Gamma$ for some justification terms $t_1, \ldots, t_n$. We have $A_1, \ldots, A_n \vdash_{\mathsf{JL(CS)}} (A_1 \wedge \ldots \wedge A_n)$ because it holds classically. Then the Lifting Lemma, Corollary 2.16, tells us there is a justification term u so that $t_1{:}A_1, \ldots, t_n{:}A_n \vdash_{\mathsf{JL(CS)}} u{:}(A_1 \wedge \ldots \wedge A_n)$. Because each $t_i{:}A_i \in \Gamma$, which is maximally consistent, $u{:}(A_1 \wedge \ldots \wedge A_n) \in \Gamma$, and hence $A_1 \wedge \ldots \wedge A_n \in \Gamma^\sharp$. The converse direction is simpler, and is omitted. □

Now let $\mathcal{M}_{\mathsf{J4.2(CS)}} = \langle \mathcal{M}, \mathcal{R}, \mathcal{V}, \mathcal{E} \rangle$ be the canonical J4.2 model with respect to constant specification CS, which is axiomatically appropriate. We show this model meets the semantic conditions given in Section 4.3.6. And we give this in some detail because it should make reading the generalization in Chapter 8 easier. Recall that, axiomatically, J4.2 extends LP with the scheme $\neg f(t,u){:}\neg t{:}X \rightarrow g(t,u){:}\neg u{:}\neg X$. We omit checking the LP semantic conditions. We already know the canonical model has a strong evidence function. What remains to be shown is that the evidence function satisfies $\mathcal{E}(f(t,u), \neg t{:}X) \cup \mathcal{E}(g(t,u), \neg u{:}\neg X) = \mathcal{G}$, and that the frame is convergent.

For the evidence condition, showing we have $\mathcal{E}(f(t,u), \neg t{:}X) \cup \mathcal{E}(g(t,u), \neg u{:}\neg X) = \mathcal{G}$ is equivalent to showing that for each Γ in the canonical model, $f(t,u){:}\neg t{:}X \in \Gamma$ or $g(t,u){:}\neg u{:}\neg X \in \Gamma$. This is an immediate consequence of the axiom $\neg f(t,u){:}\neg t{:}X \rightarrow g(t,u){:}\neg u{:}\neg X$

We still need convergence for $\mathcal{R}$. Suppose in $\mathcal{M}_{\mathsf{J4.2(CS)}}$ we have $\Gamma_1 \mathcal{R} \Gamma_2$ and $\Gamma_1 \mathcal{R} \Gamma_3$. Exactly as with the modal proof earlier, it is enough to show $\Gamma_2^\sharp \cup \Gamma_3^\sharp$ is consistent (though the sharp operation and the logic are different now). Suppose otherwise. Then, using Lemma 4.15 this time, there are $A \in \Gamma_2^\sharp$ and $B \in \Gamma_3^\sharp$ so that $B \rightarrow \neg A$ is provable. Because $A \in \Gamma_2^\sharp$, for some justification term t, $t{:}A \in \Gamma_2$. Similarly, because $B \in \Gamma_3^\sharp$, $u{:}B \in \Gamma_3$ for some u. Because we have internalization, for some justification term v, $v{:}(B \rightarrow \neg A)$ is provable, and hence so is $u{:}B \rightarrow (v \cdot u){:}\neg A$. Because $\Gamma_1 \mathcal{R} \Gamma_2$ then $\Gamma_1^\sharp \subseteq \Gamma_2$, so $f(t, v \cdot u){:}\neg t{:}A \notin \Gamma_1$ because $t{:}A \in \Gamma_2$ and Γ_2 is consistent. Then $\neg f(t, v \cdot u){:}\neg t{:}A \in \Gamma_1$. Using the special J4.2 axiom $\neg f(t, v \cdot u){:}\neg t{:}A \rightarrow g(t, v \cdot u){:}\neg v \cdot u{:}\neg A$ along with maximal consistency of Γ_1, we have $g(t, v \cdot u){:}\neg v \cdot u{:}\neg A \in \Gamma_1$. Because $\Gamma_1^\sharp \subseteq \Gamma_3$, $\neg v \cdot u{:}\neg A \in \Gamma_3$. Because $u{:}B \rightarrow (v \cdot u){:}\neg A$. $\neg u{:}B \in \Gamma_3$, a contradiction because $u{:}B \in \Gamma_3$.

4.5.6 KX4 and JX4

(Continuing 2.7.5 and 4.3.7.)

Recall, the modal logic KX4 is S4 with $\Box X \rightarrow X$, factivity dropped, and $\Box\Box X \rightarrow \Box X$ added in its place. Likewise the justification logic JX4 is LP with

$t{:}X \to X$, also called factivity, dropped and $t{:}u{:}X \to [t \mathbin{\mathsf{c}} u]{:}X$ added. Semantically, for both modal and justification logics a density condition is imposed: in a model, if $\Gamma\mathcal{R}\Delta$, then there is some Ω so that $\Gamma\mathcal{R}\Omega$ and $\Omega\mathcal{R}\Delta$. In addition, for JX4 an evidence condition is added: $\mathcal{E}(s, t{:}X) \subseteq \mathcal{E}(s \mathbin{\mathsf{c}} t, X)$.

Unlike with the other justification logics considered earlier, we *do not* prove completeness of JX4 now. It is an instance of a very general family that will be investigated in Chapter 8, as are several other justification logics we have looked at. Unlike those other justification logics, however, a direct proof of completeness for JX4 requires us to bring in much of the machinery of the general argument of Chapter 8, meaning that many things would be presented twice. What we have chosen to do instead is give a completeness proof for the modal logic KX4. This is uncluttered enough to be considered at this point, yet the proof has the benefit that it will serve to motivate some of the features of the general argument later on. So, for the rest of this section we work with axiomatic KX4, for which the modal semantics requires transitivity and denseness.

Recall from Definition 4.6 that for S a set of modal formulas, $S^\sharp = \{X \mid \Box X \in S\}$. We now add a kind of dual version: let $S^\flat = \{\neg\Box\neg X \mid X \in S\}$. The canonical model $\mathcal{M}_{\mathsf{KX4}} = \langle \mathcal{G}, \mathcal{R}, \mathcal{V}\rangle$ for KX4 is constructed as usual. The proof that $\mathcal{R}$ is transitive is standard. We concentrate on showing it meets the denseness condition.

Lemma 4.16 *Assume Γ and Δ are sets of modal formulas that are maximally consistent in* KX4.

(1) $\Gamma^\sharp \subseteq \Delta$ *if and only if* $\Delta^\flat \subseteq \Gamma$.
(2) *If* $\Gamma^\sharp \subseteq \Delta$, *then* $\Gamma^{\sharp\sharp} \cup \Delta$ *is consistent.*
(3) *If* $\Gamma^\sharp \subseteq \Delta$ *then* $\Gamma^{\sharp\sharp} \subseteq \Delta$.
(4) *If* $\Gamma^\sharp \subseteq \Delta$ *then* $\Gamma^\sharp \cup \Delta^\flat$ *is consistent.*

Proof Throughout we assume Γ and Δ are maximally consistent.

(1) Assume $\Gamma^\sharp \subseteq \Delta$; we show $\Delta^\flat \subseteq \Gamma$. Suppose $\neg\Box\neg X \in \Delta^\flat$. Then $X \in \Delta$, so $\neg X \notin \Delta$ by consistency. But then $\neg X \notin \Gamma^\sharp$, so $\Box\neg X \notin \Gamma$, and by maximality, $\neg\Box\neg X \in \Gamma$.

    Now assume instead that $\Delta^\flat \subseteq \Gamma$; we show $\Gamma^\sharp \subseteq \Delta$. Suppose $X \in \Gamma^\sharp$. Then $\Box X \in \Gamma$, and so $\Box\neg\neg X \in \Gamma$ (because $\Box X \to \Box\neg\neg X$ is a theorem, and maximally consistent sets are closed under consequence). Then $\neg\Box\neg\neg X \notin \Gamma$ by consistency, and so $\neg\Box\neg\neg X \notin \Delta^\flat$. But then $\neg X \notin \Delta$, so $X \in \Delta$ by maximal consistency.

(2) Suppose $\Gamma^\sharp \subseteq \Delta$ but $\Gamma^{\sharp\sharp} \cup \Delta$ is not consistent. We derive a contradiction.

Making use of Lemma 4.14, there are $A \in \Gamma^{\#\#}$ and $B \in \Delta$ with $\vdash_{\mathsf{KX4}} A \to \neg B$. Because $A \in \Gamma^{\#\#}$ then $\Box\Box A \in \Gamma$ and because $\Box\Box A \to \Box A$ is an axiom of KX4, $\Box A \in \Gamma$. But then $A \in \Gamma^{\#}$, so $A \in \Delta$ and hence $\neg B \in \Delta$, contrading consistency.

(3) If $\Gamma^{\#} \subseteq \Delta$ then $\Gamma^{\#\#} \cup \Delta$ is consistent. Because Δ is maximally consistent, it follows that $\Gamma^{\#\#} \subseteq \Delta$.

(4) Suppose $\Gamma^{\#} \subseteq \Delta$ but $\Gamma^{\#} \cup \Delta^{\flat}$ is inconsistent. We derive a contradiction.

If $\Gamma^{\#} \cup \Delta^{\flat}$ is inconsistent, there are formulas $G \in \Gamma^{\#}$ and $\neg\Box\neg D \in \Delta^{\flat}$ such that $\vdash_{\mathsf{KX4}} G \to \Box\neg D$. Then $\vdash_{\mathsf{KX4}} \Box G \to \Box\Box\neg D$. Now $\Box G \in \Gamma$ so $\Box\Box\neg D \in \Gamma$. But then $\neg D \in \Gamma^{\#\#}$, and because $\Gamma^{\#\#} \subseteq \Delta$, $\neg D \in \Delta$. But $\neg\Box\neg D \in \Delta^{\flat}$, so $D \in \Delta$, contradicting consistency of Δ.

□

Theorem 4.17 *In the canonical model $\mathcal{M}_{\mathsf{KX4}} = \langle \mathcal{G}, \mathcal{R}, \mathcal{V} \rangle$, for $\Gamma, \Delta \in \mathcal{G}$, if $\Gamma\mathcal{R}\Delta$ then for some Ω, $\Gamma\mathcal{R}\Omega$ and $\Omega\mathcal{R}\Delta$.*

Proof Suppose $\Gamma\mathcal{R}\Delta$. Then $\Gamma^{\#} \subseteq \Delta$. By Lemma 4.16, $\Gamma^{\#} \cup \Delta^{\flat}$ is consistent. Extend it to a maximally consistent set Ω. Then $\Gamma^{\#} \subseteq \Omega$ so $\Gamma\mathcal{R}\Omega$. Also $\Delta^{\flat} \subseteq \Omega$ so by Lemma 4.16 again, $\Omega^{\#} \subseteq \Delta$ and so $\Omega\mathcal{R}\Delta$. □

4.6 Formulating Justification Logics

We have examined a number of justification logics, each associated with a modal logic. In the next chapter it will be shown that these modal logic, justification logic pairs are in fact counterparts, as defined in Section 7.2. We have been following a general method for generating justification logics from modal logics, and it is time we explained what it is. It accounts for most, but not all, of the examples we have seen, and the presence of the exceptions is significant as well.

Suppose we have a modal axiom scheme, say $(\Box P \vee \Box Q) \to \Box\Box(P \vee Q)$ to take a concrete example. Assign to each negative occurrence of $\Box$ a unique justification variable. In this case we have the occurrences before P and Q, and we assign x_1 and x_2. Replace these $\Box$ occurrences with the variables as an intermediate step; in our example we get $(x_1{:}P \vee x_2{:}Q) \to \Box\Box(P \vee Q)$. Next, assign to each positive occurrence of $\Box$ a unique function symbol with the same number of arguments as we have variables, and replace the positive $\Box$ occurrences with these function symbols applied to the variables. For our example we use f and g, each two-place, and we get $(x_1{:}P \vee x_2{:}Q) \to f(x_1, x_2){:}g(x_1, x_2){:}(P \vee Q)$. The collection of all substitution instances of this (Definition 2.17) is a justification scheme we can use as a counterpart of the original modal formula. In a sense, it is the most general counterpart, something we will say more about later.

This does not always work. It does not work with Gödel–Löb logic, GL, for instance. It does apply to all the examples considered in this book, including the infinite family of Geach logics to be investigated in Chapter 8. In the large number of cases where it does work, it does not produce the only justification counterpart of a given modal logic, but it does produce what is, in a sense, the most general version. We illustrate this with a closer look at the most prominent and the historically first example, LP.

Consider positive introspection, $\Box X \to \Box\Box X$. Applying the method given earlier, we get as a justification counterpart all instances of $t{:}X \to f(t){:}g(t){:}X$, where f and g are one-place function symbols. Writing the customary $!t$ in place of $f(t)$ we get $t{:}X \to\ !t{:}g(t){:}X$, and this differs from the scheme actually adopted for LP, $t{:}X \to\ !t{:}t{:}X$, though it reduces to it by understanding g as the identify function. We need to make proper sense of this. In what follows, let us temporarily use LP′ for the justification logic that is like LP except that the scheme $t{:}X \to\ !t{:}g(t){:}X$ replaces $t{:}X \to\ !t{:}t{:}X$.

We have seen a semantics and a soundness proof for LP in Section 4.3.2, and a completeness proof in Section 4.5.1. We produce a Fitting semantics for LP′ in a similar way. Let $\mathcal{M} = \langle \mathcal{G}, \mathcal{R}, \mathcal{V}, \mathcal{E} \rangle$ be a Fitting model defined as in Section 4.3.2 except that the Monotonicity and ! Conditions are replaced with the following.

LP′ **Monotonicity Condition:** If $\Gamma\mathcal{R}\Delta$ and $\Gamma \in \mathcal{E}(t, X)$ then $\Delta \in \mathcal{E}(g(t), X)$.
LP′ ! **Condition:** $\mathcal{E}(t, X) \subseteq \mathcal{E}(!t, g(t){:}X)$.

We show we have the validity, in $\mathcal{M}$, of $t{:}X \to\ !t{:}g(t){:}X$. Specifically assume $\mathcal{M}, \Gamma \Vdash t{:}X$; we show $\mathcal{M}, \Gamma \Vdash\ !t{:}g(t){:}X$.

We have $\Gamma \in \mathcal{E}(t, X)$. Then by the LP′ ! Condition, $\Gamma \in \mathcal{E}(!t, g(t){:}X)$, so it is enough to show that for any Δ with $\Gamma\mathcal{R}\Delta$ we have $\mathcal{M}, \Delta \Vdash g(t){:}X$. Assume $\Gamma\mathcal{R}\Delta$. By the LP′ Monotonicity Condition we have $\Delta \in \mathcal{E}(g(t), X)$, so it is enough to show that for any Ω with $\Delta\mathcal{R}\Omega$, $\mathcal{M}, \Omega \Vdash X$. But because $\Gamma\mathcal{R}\Delta$ and $\Delta\mathcal{R}\Omega$ and $\mathcal{M}, \Gamma \Vdash t{:}X$, using transitivity, $\mathcal{M}, \Omega \Vdash X$, and we are done.

Other parts of the soundness proof, and the completeness proof are rather straightforward adaptations, and we omit the details.

Now here is the point. The usual formulation of LP is informally like that of LP′ but with $g(t) = t$. In fact, in justification logics we have function symbols, but we do not really have a function mechanism, though one could be introduced if there were some advantage to it. But we can proceed at the meta-level. If we take the formulation of LP′ and replace occurrences of $g(t)$ with t, we get the formulation of LP. Similarly for the Fitting semantics. If we take the soundness proof for LP′ and make the same replacement, we get the soundness argument for LP. Similarly for the completeness argument. In short, starting with the "most general" justification counterpart of S4, we also can handle more restricted versions, and LP is such a version.

One might naturally ask why LP was formulated as it was. It was done because the motivation came from the desire to create an arithmetic semantics for propositional intuitionistic logic, and the version selected was sufficient and natural for the purpose. The machinery of justification logics is rich and flexible, and this is one of its virtues.

We briefly discuss one more example, S4.2 and JT4.2, Sections 2.7.4, 4.3.6, and 4.5.5. Modal S4.2 adds to S4 the scheme $\Diamond\Box X \rightarrow \Box\Diamond X$, or equivalently $\neg\Box\neg\Box X \rightarrow \Box\neg\Box\neg X$. Our justification counterpart, JT4.2 added to LP the scheme $\neg f(t,u){:}\neg t{:}X \rightarrow g(t,u){:}\neg u{:}\neg X$, and a corresponding semantics was created. There are many weakenings of this, but perhaps the most plausible is $\neg h(u){:}\neg t{:}X \rightarrow k(t){:}\neg u{:}\neg X$. It is a good exercise to formulate the appropriate Fitting semantics, and check the correctness of the soundness and completeness proofs. These result, of course, by replacing $f(t,u)$ with $h(u)$ and $g(t,u)$ with $k(t)$. Other variations are possible, and the LP underpinning could also have been LP′, as earlier. Clearly justification counterparts are far from unique. The "most general" version has much to recommend it, because others can be derived from it with routine labor. But in general, extralogical considerations can be expected to play a role in making a choice among them.

5
Sequents and Tableaus

5.1 Background

Realization theorems connect modal logics and justification logics—at least, some modal logics and some justification logics. It is not known how extensive the phenomenon is, but it has turned out to be much broader than was originally thought. Realization theorems have been given a variety of proofs, but they fall into two general categories: constructive and nonconstructive. In this chapter we present the syntactic and proof-theoretic material that will be needed for both kinds of arguments.

All constructive proofs are based on some version of cut-free proof system—a realization is not extracted from a modal validity, but from a cut-free proof of that validity. The first and most common cut-free proof system is the *sequent calculus*, with modal versions deriving from the historic work of Gentzen on intuitionistic and classical systems. Such proofs have two features that are significant for realization arguments. First, all formulas that enter into sequent proofs are subformulas of the formula being proved. And second, all occurrences of the "same" formula in a proof have the same *polarity*, positive or negative. (Loosely, a positive subformula of a formula is one that appears in the scope of an even number of negation signs.) We begin the chapter with a formal presentation of a sequent calculus for the modal logic S4. It was used in the original realization theorem proof, Artemov (1995, 2001), a proof that is presented in Chapter 6.

Sequents and semantic tableaus have a close relationship. Tableaus, and their machinery, have features that make them particularly handy for realization proofs, both constructive and nonconstructive. In particular, we use *signed* tableaus, and signed formulas have an existence that breaks them free from sequents, while carrying with them the key information about where they were in the sequent—on the left or on the right. Much of the present chapter presents what we will need about tableaus, with applications in Chapter 7.

5.2 Classical Sequents

In this section we offer a concise introduction to the sequent calculus, a well-known and powerful tool in structural proof theory and applications. Out of a large variety of formats for sequent calculi we opted for one that provides a compact and yet rigorous path to classical logic, and subsequently to S4, and which plays a pivotal role in this area. From time to time we will mention where our formal choices could have been different. Ultimately, these differences make no difference.

We fix a language of propositional modal logic, for the time being without the modality $\Box$, but with $\wedge$, $\vee$, and $\rightarrow$. It is convenient here to take negation as defined, $\neg X = (X \rightarrow \bot)$.

Definition 5.1 A *sequent* is an ordered pair of finite sets of formulas Γ, Δ, traditionally written as

$$\Gamma \Rightarrow \Delta.$$

Here Γ is called the *antecedent* and Δ the *succedent* of the sequent.

We note that in alternate formulations, Γ and Δ are sometimes taken to be finite multisets, or even finite sequences of formulas.

We will use suggestive notational conventions: if A is a formula, then $\Gamma, A = A, \Gamma = \Gamma \cup \{A\}$. Γ or Δ or both could be empty, and then a sequent is written as "$\Rightarrow \Delta$," "$\Gamma \Rightarrow$" and "$\Rightarrow$" correspondingly. By $\bigwedge \Gamma$ and $\bigvee \Gamma$ we understand conjunction and disjunction of all formulas from Γ correspondingly. We note that semantically $\bigwedge \emptyset$ is true and $\bigvee \emptyset$ is false.

Intuitively, a sequent $\Gamma \Rightarrow \Delta$ denotes a derivation of Δ from the set of assumptions Γ. In such reading, Δ is understood to be the disjunction of all its formulas. This intuition is formalized in the following definition.

Definition 5.2 By the *formula translation of a sequent* $\Gamma \Rightarrow \Delta$ we mean formula

$$\bigwedge \Gamma \rightarrow \bigvee \Delta.$$

Definition 5.3 [Classical Sequent Calculus] We now define a sequent calculus Gcl with the following axioms and rules of inference.

Axioms: $A \Rightarrow A$ $\qquad$ $\bot \Rightarrow$

Structural rules: weakening

$$\frac{\Gamma \Rightarrow \Delta}{A, \Gamma \Rightarrow \Delta}(W \Rightarrow) \qquad \frac{\Gamma \Rightarrow \Delta}{\Gamma \Rightarrow \Delta, A}(\Rightarrow W).$$

Propositional rules:

$$\frac{A, \Gamma \Rightarrow \Delta}{A \wedge B, \Gamma \Rightarrow \Delta}(\wedge \Rightarrow), \qquad \frac{B, \Gamma \Rightarrow \Delta}{A \wedge B, \Gamma \Rightarrow \Delta}(\wedge \Rightarrow) \qquad \frac{\Gamma \Rightarrow \Delta, A \quad \Gamma \Rightarrow \Delta, B}{\Gamma \Rightarrow \Delta, A \wedge B}(\Rightarrow \wedge)$$

$$\frac{A, \Gamma \Rightarrow \Delta \quad B, \Gamma \Rightarrow \Delta}{A \vee B, \Gamma \Rightarrow \Delta}(\vee \Rightarrow) \qquad \frac{\Gamma \Rightarrow \Delta, A}{\Gamma \Rightarrow \Delta, A \vee B}(\Rightarrow \vee) \qquad \frac{\Gamma \Rightarrow \Delta, B}{\Gamma \Rightarrow \Delta, A \vee B}(\Rightarrow \vee)$$

$$\frac{\Gamma \Rightarrow \Delta, A \quad B, \Gamma \Rightarrow \Delta}{A \rightarrow B, \Gamma \Rightarrow \Delta}(\rightarrow \Rightarrow) \qquad \frac{A, \Gamma \Rightarrow \Delta, B}{\Gamma \Rightarrow \Delta, A \rightarrow B}(\Rightarrow \rightarrow).$$

Cut:

$$\frac{\Gamma \Rightarrow \Delta, A \quad A, \Gamma \Rightarrow \Delta}{\Gamma \Rightarrow \Delta}\, Cut.$$

By Gcl^- we mean the same system but without the Cut rule. A formula X is a theorem (of Gcl or of Gcl^-) if $\Rightarrow X$ is provable.

We note that in some formulations Weakening is not a rule, but rather the effect is built into the axioms. For instance, $A \Rightarrow A$ becomes $\Gamma, A \Rightarrow \Delta, A$. The two versions are easily seen to be equivalent.

The reader is probably aware of the term "cut-elimination" reflecting the desire to look for derivations that do not involve the Cut rule. What is wrong with the Cut and why should we try to avoid it? After all, the Cut appears to be a sequent counterpart of the syllogism logic rule

$$\frac{A \rightarrow B \qquad B \rightarrow C}{A \rightarrow C}.$$

The Cut rule is the only rule in Gcl that violates the so-called subformula property: every formula occurring in the premise sequents of a rule is a subformula in the conclusion sequent. Consequently, a cut-free derivation of a sequent

$\Gamma \Rightarrow \Delta$ contains only subformulas of Γ, Δ. Because this is a finite set, it makes it theoretically possible to do a bottom-up proof search, which forms the basis for semantic tableaus. Conceptually, in cut-free derivations, the aggregated complexity of formulas grows from premises to conclusions. (We note that some derivation steps could be void, i.e., do not change sequents.) This allows for the valuable possibility of complexity-based induction along derivation trees.

Perhaps, the easiest consequence of cut-freeness begins with the observation that the empty sequent "$\Rightarrow$" cannot not have a cut-free derivation (which is immediate from the subformula property). On the other hand, an inconsistent sequent-based system with Cut derives the empty sequent "$\Rightarrow$". Indeed, suppose both $\Rightarrow A$ and $\Rightarrow \neg A$ are derivable. (Recall that here we read $\neg A$ as $A \to \bot$.) Then

$$\dfrac{\Rightarrow A \to \bot \qquad \dfrac{\Rightarrow A \qquad \bot \Rightarrow}{A \to \bot \Rightarrow}(\to\Rightarrow)}{\Rightarrow}(Cut).$$

Conclusion: *cut-free derivation systems are consistent*, and in the areas of proof theory that study consistency questions, proving a cut-elimination theorem is a valuable tool for establishing consistency.

In the justification logic project, cut-free derivations were used in the early proofs of the Realization Theorem, cf. Chapter 6.

Example 5.4 Here is a proof in Gcl of $(A \to (B \to C)) \to (B \to (A \to C))$. We have combined multiple applications of Weakening into one abbreviated version called W, for simplicity.

$$\dfrac{\dfrac{\dfrac{\dfrac{\dfrac{A \Rightarrow A}{B, A \Rightarrow C, A}W \qquad \dfrac{\dfrac{B \Rightarrow B}{B, A \Rightarrow C, B}W \qquad \dfrac{C \Rightarrow C}{C, B, A \Rightarrow C}W}{B \to C, B, A \Rightarrow C}\to\Rightarrow}{A \to (B \to C), B, A \Rightarrow C}\to\Rightarrow}{A \to (B \to C), B \Rightarrow A \to C}\Rightarrow\to}{A \to (B \to C) \Rightarrow B \to (A \to C)}\Rightarrow\to}{\Rightarrow (A \to (B \to C)) \to (B \to (A \to C))}\Rightarrow\to$$

For brevity, unless the opposite is explicitly mentioned, we consider the language with Boolean connectives $\to$ and $\bot$ and regard other connectives $\wedge$, $\vee$, $\neg$ as definable.

Theorem 5.5 (Atomic Axiomatization) *Each theorem of* Gcl$^-$ *can be derived in* Gcl$^-$ *with atomic axioms only*

$$p \Rightarrow p \qquad\qquad \bot \Rightarrow$$

where p is any propositional letter.

Proof It suffices to provide a cut-free derivation of each "conventional" axiom $A \Rightarrow A$, by induction on the complexity of A. The atomic case is covered. If A is $X \to Y$ then $X \Rightarrow X$ and $Y \Rightarrow Y$ are derivable by the Induction Hypothesis. Then do as follows:

$$\dfrac{\dfrac{X \Rightarrow X \qquad Y \Rightarrow Y}{X \to Y, X \Rightarrow Y}(\to \Rightarrow)}{X \to Y \Rightarrow X \to Y}(\Rightarrow \to).$$

□

5.3 Sequents for S4

Many of the best-known modal logics have sequent proof systems, but far from all. Devising natural extensions of the machinery is common, and we now have hypersequents, nested sequents, and labeled sequents, among others. Fortunately, S4 is a logic that does have a sequent calculus, and a simple one at that. We now add $\Box$ to the formal language of Section 5.1, with the usual formation rule.

Definition 5.6 (S4 Sequent Calculus GS4) First some notation.

$$\Box\Gamma = \{\Box F \mid F \in \Gamma\}$$
$$\Gamma^{\Box} = \{\Box F \mid \Box F \in \Gamma\}$$

We add to the axioms and rules of Definition 5.3 the following modal rules:

$$\dfrac{A, \Gamma \Rightarrow \Delta}{\Box A, \Gamma \Rightarrow \Delta}(\Box \Rightarrow) \qquad \dfrac{\Box\Gamma \Rightarrow A}{\Box\Gamma \Rightarrow \Box A}(\Rightarrow \Box).$$

A GS4 proof uses the axioms and rules of Definition 5.3 and the Modal rules listed earlier. A GS4$^-$ proof is defined in the same way, but Cut is not allowed.

Example 5.7 Derivations of the S4 axioms in $\mathsf{GS4}^-$

$$
\dfrac{\dfrac{\dfrac{\dfrac{\dfrac{\dfrac{A \Rightarrow A \quad B \Rightarrow B}{A \to B, A \Rightarrow B}(\to\Rightarrow)}{\Box(A \to B), A \Rightarrow B}(\Box\Rightarrow)}{\Box(A \to B), \Box A \Rightarrow B}(\Box\Rightarrow)}{\Box(A \to B), \Box A \Rightarrow \Box B}(\Rightarrow\Box)}{\Box(A \to B) \Rightarrow \Box A \to \Box B}(\Rightarrow\to)}{\Rightarrow \Box(A \to B) \to (\Box A \to \Box B)}(\Rightarrow\to)
$$

$$
\dfrac{\dfrac{A \Rightarrow A}{\Box A \Rightarrow A}(\Box\Rightarrow)}{\Rightarrow \Box A \to A}(\Rightarrow\to)
$$

$$
\dfrac{\dfrac{\Box A \Rightarrow \Box A}{\Box A \Rightarrow \Box\Box A}(\Rightarrow\Box)}{\Rightarrow \Box A \to \Box\Box A}(\Rightarrow\to)
$$

Theorem 5.5 extends to S4 quite easily.

Theorem 5.8 *Each theorem of* $\mathsf{GS4}^-$ *can be derived in* $\mathsf{GS4}^-$ *with atomic axioms only.*

Proof The proof of Theorem 5.5 needs one more case added to it. If A is $\Box X$ and $X \Rightarrow X$ is derived, then

$$
\dfrac{\dfrac{X \Rightarrow X}{\Box X \Rightarrow X}(\Box\Rightarrow)}{\Box X \Rightarrow \Box X}(\Rightarrow\Box).
$$

□

5.4 Sequent Soundness, Completeness, and More

We are about to prove several principal facts about GS4: semantical completeness, finite model property, equivalency of sequent GS4 and Hilbert-style S4 axiomatizations, and cut-elimination in GS4, in one theorem.

Modal models are discussed briefly in Section 4.1. For convenience, we recall that an S4-model is $\langle \mathcal{G}, \mathcal{R}, \Vdash \rangle$ where $\mathcal{G}$ is a nonempty set of states or worlds, $\mathcal{R}$ is a reflexive and transitive accessibility relation on $\mathcal{G}$, and $\Vdash$ is an assignment of truth values to propositional letters at each world in $\mathcal{G}$. Furthermore, $\Vdash$ extends to all formulas by stipulating that $\Vdash$ respects Boolean connectives at each world and for every $\Gamma \in \mathcal{G}$,

$$\Gamma \Vdash \Box F \quad \textit{iff} \quad \mathcal{R}(\Gamma) \Vdash F$$

where $\mathcal{R}(\Gamma) \Vdash F$ means $\Delta \Vdash F$ for every $\Delta \in \mathcal{G}$ such that $\Gamma \mathcal{R} \Delta$.

Theorem 5.9 *The following are equivalent*

(1) $\mathsf{GS4}^- \vdash \Gamma \Rightarrow \Delta$;
(2) $\mathsf{GS4} \vdash \Gamma \Rightarrow \Delta$;
(3) $\mathsf{S4} \vdash \bigwedge \Gamma \to \bigvee \Delta$;
(4) $\bigwedge \Gamma \to \bigvee \Delta$ *holds in all* S4-*models;*
(5) $\bigwedge \Gamma \to \bigvee \Delta$ *holds in all finite* S4-*models.*

Proof Items "(1) yields (2)" and "(4) yields (5)" hold vacuously. Item "(3) yields (4)" is the well-known fact of soundness of axiomatic S4 with respect to S4-models.

Item "(2) yields (3)" states the soundness of sequent rules with respect to S4 and its proof consists of checking that axioms and rules of GS4 are S4 compliant. The cases of axioms, propositional, and structural rules are straightforward, and we only check the modal rules.

($\Box \Rightarrow$). Suppose

$$\mathsf{S4} \vdash A \wedge \bigwedge \Gamma \to \bigvee \Delta.$$

By the reflection axiom $\Box A \to A$ and easy propositional reasoning,

$$\mathsf{S4} \vdash \Box A \wedge \bigwedge \Gamma \to \bigvee \Delta.$$

($\Rightarrow \Box$). Suppose

$$\mathsf{S4} \vdash \bigwedge \Box \Gamma \to A.$$

Because $\Box$ and $\wedge$ commute in S4,

$$\mathsf{S4} \vdash \Box \bigwedge \Gamma \to A.$$

By necessitation,

$$\mathsf{S4} \vdash \Box\left[\Box \bigwedge \Gamma \to A\right],$$

by normality,

$$\mathsf{S4} \vdash \Box\Box \bigwedge \Gamma \to \Box A,$$

and, by positive introspection $\Box F \to \Box\Box F$,

$$\mathsf{S4} \vdash \Box \bigwedge \Gamma \to \Box A,$$

hence

$$\mathsf{S4} \vdash \bigwedge \Box\Gamma \to \Box A.$$

It remains to establish that (5) yields (1), which we show contrapositively as "not (1)" yields "not (5)." Specifically, we show that from a sequent $\Gamma \Rightarrow \Delta$ that is not derivable in $\mathsf{GS4}^-$, we can build a finite S4-model $\langle \mathcal{G}, \mathcal{R}, \Vdash \rangle$ and a world $\Gamma \in \mathcal{G}$ such that all formulas from Γ are true and all formulas from Δ are false at Γ.

For the rest of this argument for Theorem 5.9, we assume $\Gamma \Rightarrow \Delta$ *is not provable in* $\mathsf{GS4}^-$, *the negation of item* (1) *of the Theorem.*

Definition 5.10 A sequent $\Theta \Rightarrow \Phi$ is *admissible* iff each formula from Θ, Φ is a subformula of a member of $\Gamma \cup \Delta$. Note that there are only finitely many admissible sequents. A sequent $\Theta \Rightarrow \Phi$ is *consistent* iff it is not provable in $\mathsf{GS4}^-$. A sequent $\Theta \Rightarrow \Phi$ is *saturated* iff

- if $A \to B \in \Theta$, then $B \in \Theta$ or $A \in \Phi$;
- if $A \to B \in \Phi$, then $A \in \Theta$ and $B \in \Phi$;
- if $\Box A \in \Theta$, then $A \in \Theta$ as well.

Lemma 5.11 *Each admissible consistent sequent can be extended to a saturated admissible consistent sequent, i.e., for each admissible consistent sequent* $\Theta \Rightarrow \Phi$ *there is a saturated admissible consistent sequent* $\Theta' \Rightarrow \Phi'$ *such that* $\Theta \subseteq \Theta'$ *and* $\Phi \subseteq \Phi'$.

Proof We eliminate all violations of the saturation property one by one without repetitions, in the following way.

If $A \to B \in \Theta$, then at least one of $B, \Theta \Rightarrow \Phi$ or $\Theta \Rightarrow \Phi, A$ is consistent. Otherwise, by $(\to\Rightarrow)$, $\mathsf{GS4}^- \vdash A \to B, \Theta \Rightarrow \Phi$. Because $A \to B \in \Theta$, this would yield inconsistency of $\Theta \Rightarrow \Phi$. As the next sequent, pick a consistent one out of $B, \Theta \Rightarrow \Phi$ or $\Theta \Rightarrow \Phi, A$.

If $A \to B \in \Phi$, then add A to Θ and B to Φ. The resulting sequent $A, \Theta \Rightarrow \Phi, B$ is consistent because otherwise, by $(\Rightarrow\to)$, $\Theta \Rightarrow \Phi$ would be inconsistent.

If $\Box A \in \Theta$, then add A to Θ. The resulting sequent $A, \Theta \Rightarrow \Phi$ is consistent because otherwise, by $(\Box \Rightarrow)$, $\Theta \Rightarrow \Phi$ would be inconsistent.

This procedure respects admissibility and terminates with a saturated extension $\Theta' \Rightarrow \Phi'$ of $\Theta \Rightarrow \Phi$. □

We now define our desired S4 model $\langle \mathcal{G}, \mathcal{R}, \Vdash \rangle$:

- $\mathcal{G}$ is the set of all admissible saturated consistent sequents (a finite set).
- $(\Theta \Rightarrow \Phi)\mathcal{R}(\Theta' \Rightarrow \Phi')$ iff $\Theta^{\Box} \subseteq \Theta'$. Obviously, $\mathcal{R}$ is vacuously reflexive and transitive.
- $(\Theta \Rightarrow \Phi) \Vdash p$ iff $p \in \Theta$ for p a propositional letter.
- $(\Theta \Rightarrow \Phi) \nVdash \bot$

Lemma 5.12 (Truth Lemma) *For any formula A,*

- *if $A \in \Theta$, then $(\Theta \Rightarrow \Phi) \Vdash A$;*
- *if $A \in \Phi$, then $(\Theta \Rightarrow \Phi) \nVdash A$.*

Proof By induction on A.

A **is a propositional letter** p. If $p \in \Theta$, then, by the definition of the model, $(\Theta \Rightarrow \Phi) \Vdash p$. If $p \in \Phi$, then $p \notin \Theta$ (because otherwise $\Theta \Rightarrow \Phi$ would be inconsistent), hence by the definition of the model, $(\Theta \Rightarrow \Phi) \nVdash p$.

A **is a propositional constant** $\bot$. $(\Theta \Rightarrow \Phi) \nVdash \bot$ by definition. Also $\bot \in \Theta$ is impossible because $\Theta \Rightarrow \Phi$ is consistent. Then the Truth Lemma holds vacuously in this case.

A **is** $X \to Y$. If $X \to Y \in \Theta$ then, by saturation, $Y \in \Theta$ or $X \in \Phi$. By the Induction Hypothesis, either $(\Theta \Rightarrow \Phi) \Vdash Y$ or $(\Theta \Rightarrow \Phi) \nVdash X$. In either case, $(\Theta \Rightarrow \Phi) \Vdash (X \to Y)$.

If $X \to Y \in \Phi$ then, by saturation, $X \in \Theta$ and $Y \in \Phi$. By the Induction Hypothesis, $(\Theta \Rightarrow \Phi) \Vdash X$ and $(\Theta \Rightarrow \Phi) \nVdash Y$, hence $(\Theta \Rightarrow \Phi) \nVdash (X \to Y)$.

A **is** $\Box X$. Assume $\Box X \in \Theta$ and $(\Theta \Rightarrow \Phi)\mathcal{R}(\Theta' \Rightarrow \Phi')$. By the definition of the model, $\Box X \in \Theta'$. By saturation, $X \in \Theta'$ and, by the Induction Hypothesis, $(\Theta' \Rightarrow \Phi') \Vdash X$. Because $(\Theta' \Rightarrow \Phi')$ was arbitrary, by the definition of $\Box$ it follows that,

$$(\Theta \Rightarrow \Phi) \Vdash \Box X.$$

We come to the most complex case. Assume now that $\Box X \in \Phi$. We first claim that the sequent $\Theta^{\Box} \Rightarrow X$ is consistent. Indeed, otherwise,

$$\mathsf{GS4}^{-} \vdash \Theta^{\Box} \Rightarrow X,$$

and, by $(\Rightarrow \Box)$,

$$\mathsf{GS4}^{-} \vdash \Theta^{\Box} \Rightarrow \Box X.$$

Because $\Theta^{\Box} \subseteq \Theta$ and $\Box X \in \Phi$, by weakenings, we would get

$$\mathsf{GS4}^{-} \vdash \Theta \Rightarrow \Phi,$$

which is impossible because $\Theta \Rightarrow \Phi$ is consistent.

Pick a consistent saturation $\Theta' \Rightarrow \Phi'$ of $\Theta^{\Box} \Rightarrow X$. Because $\Theta^{\Box} \subseteq \Theta'$,

$$(\Theta \Rightarrow \Phi)\mathcal{R}(\Theta' \Rightarrow \Phi').$$

Because $X \in \Phi'$, by the Induction Hypothesis,

$$(\Theta' \Rightarrow \Phi') \not\Vdash X$$

hence

$$(\Theta \Rightarrow \Phi) \not\Vdash \Box X.$$

This completes the proof of the Truth Lemma. □

To finish the proof of Theorem 5.9, recall that $(\Gamma \Rightarrow \Delta)$ is not provable in $\mathsf{GS4}^{-}$, and hence is consistent. It is trivially admissible. By Lemma 5.11, it extends to a saturated admissible consistent sequent, $(\Gamma' \Rightarrow \Delta')$. This is a possible world in the model $\langle \mathcal{G}, \mathcal{R}, \Vdash \rangle$ that we just constructed. Because this possible world extends $(\Gamma \Rightarrow \Delta)$, all members of Γ are true at it, and all members of Δ are false at it. □

Corollary 5.13 *All the following hold.*

(1) *Equivalence of axiomatic* S4 *and* GS4*:*

$$\mathsf{GS4} \vdash \Gamma \Rightarrow \Delta \quad \textit{iff} \quad \mathsf{S4} \vdash \bigwedge \Gamma \rightarrow \bigvee \Delta.$$

In particular,

$$\mathsf{S4} \vdash F \quad \textit{iff} \quad \mathsf{GS4} \vdash \Rightarrow F.$$

(2) S4 *is complete with respect to the class of transitive reflexive models.*

(3) *Finite Model Property: if* S4 $\nvdash F$*, then* F *fails in some finite transitive reflexive model.*

(4) *Cut Elimination in* GS4*: each theorem* $\Gamma \Rightarrow \Delta$ *of* GS4 *has a cut-free proof in* GS4.

5.5 Classical Semantic Tableaus

We now move to a discussion of semantic tableaus. Sequent proofs start with axioms and work forward to the formula being proved. They are often used for proof discovery, in which case they are run backwards, starting with the formula to be proved and working upward to axioms. Tableaus take that feature and make it primary. As the name suggests, their origins were semantic—they can be thought of as searches for countermodels. Soundness and completeness have direct proofs, but because tableaus have a strong connection with sequents, we will simply use this connection to transfer these results that were already proved for sequents in Section 5.4.

Tableaus are *refutation* systems; we will say what this means later. For those logics having tableau systems, machinery can vary quite a bit. At the heart of these systems, most commonly, is classical logic, and we sketch the ideas for that here.

For starters, let us assume formulas are built up from propositional letters using just $\rightarrow$ and $\perp$—other connectives will be added later. We introduce two special symbols, T and F, and specify that $T\,X$ and $F\,X$ are *signed* formulas if X is a formula. The intended reading is that X is true, or false, respectively, but this might mean classically, or constructively, or at some possible world, depending on the logic in question. Because this is classical logic for now, Boolean truth and falsehood are the motivating intuitions behind T and F. For intuitionistic logic, signs are essential; for classical logic they could be dispensed with, but for our purposes it is much more convenient to have them.

A tableau proof is a labeled binary tree, where the labels are signed formulas meeting special conditions. A proof of X begins with a tree having only a root node, labeled $F\,X$. We noted earlier that tableaus are refutation systems—we want to show $F\,X$ is impossible. To this end a tree is "grown" using *branch extension* rules, one for each propositional connective and each sign (though we only discuss $\rightarrow$ for the time being). The initial tree and each subsequent tree is a tableau. The goal is to produce a tableau in which each branch contains an obvious contradiction. If we achieve this goal we conclude that $F\,X$ cannot happen, X cannot be false, and so must be a tautology.

Figure 5.1 shows the very simple branch extension rules for implication—the rule for $T \rightarrow$ is branching, the $F \rightarrow$ rule is not.

$$\frac{T\,X \rightarrow Y}{F\,X \mid T\,Y} \qquad \frac{F\,X \rightarrow Y}{\begin{array}{c} T\,X \\ F\,Y \end{array}}$$

Figure 5.1 Tableau Rules for $\rightarrow$

Tableaus are displayed as downward branching trees. Think of a tree as representing the disjunction of its branches, and a branch as representing the conjunction of the signed formulas on it. Because a node may be common to several branches, a formula labeling it, in effect, can occur as a constituent of several conjunctions, while being written only once. This amounts to a kind of structure sharing.

A tableau expansion is often discussed temporally—one talks about the *stages* of constructing a tableau, meaning the stages of growing a tree. The rules in Figure 5.1 are thought of as branch-lengthening rules. Thus, a branch containing $F\,X \to Y$ can be lengthened with two new nodes at the end, labeled $T\,X$ and $F\,Y$ (take the node with $F\,Y$ as the child of the one labeled $T\,X$). A branch containing $T\,X \to Y$ can be split—its leaf is given a new left and a new right child, with one labeled $F\,X$, the other $T\,Y$.

Finally we have the conditions for ending a proof. A branch is *closed* if it contains $T\,A$ and $F\,A$ for some formula A, or if it contains $T\,\bot$. (A branch is *open* if it is not closed.) If each branch is closed, the tableau is *closed*. A closed tableau for $F\,X$ is a tableau proof of X.

Figure 5.2 shows the final stage of a tableau construction beginning with (or as we will say, *for*) the signed formula $F\,(X \to (Y \to Z)) \to (Y \to (X \to Z))$. Numbers are shown for reference purposes. The order of rule application followed here is top-down, on the leftmost open branch. So, 2 and 3 are from 1 by $F\ \to$, 4 and 5 are from 2 by $T\ \to$, 6 and 7 are from 3 by $F\ \to$, 8 and 9 are from 7 by $F\ \to$, and then we move to the second branch from the left. And so on. The tableau displayed in Figure 5.2 is closed, so the formula $(X \to (Y \to Z)) \to (Y \to (X \to Z))$ has a tableau proof.

1. $F\,(X \to (Y \to Z)) \to (Y \to (X \to Z))$
2. $T\,X \to (Y \to Z)$
3. $F\,Y \to (X \to Z)$

4. $F\,X$	5. $T\,Y \to Z$	
6. $T\,Y$	10. $T\,Y$	
7. $F\,X \to Z$	11. $F\,X \to Z$	
8. $T\,X$		
9. $F\,Z$	12. $F\,Y$	13. $T\,Z$
		14. $T\,X$
		15. $F\,Z$

Figure 5.2 A Classical Tableau Example

Tableau rules are understood to be *nondeterministic*. At each stage we are allowed to choose a signed formula occurrence on a branch and apply a rule to it. Because the order of choice is arbitrary, there can be many tableaus for a single signed formula. Sometimes a prescribed order of rule application is imposed, but this is not basic to a tableau system. For example, Figure 5.3

shows another tableau proof of $(X \to (Y \to Z)) \to (Y \to (X \to Z))$ that is shorter than the one in Figure 5.2. The basic idea followed is that we do a nonbranching rule before a branching rule whenever we can, because this avoids duplication of work. Items 2 and 3 are from 1 by $F \to$; 4 and 5 are from 3 by $F \to$; 6 and 7 are from 5 by $F \to$; and then 8 and 9 are from 2 by $T \to$; and so on.

1. $F\,(X \to (Y \to Z)) \to (Y \to (X \to Z))$
2. $T\,X \to (Y \to Z)$
3. $F\,Y \to (X \to Z)$
4. $T\,Y$
5. $F\,X \to Z$
6. $T\,X$
7. $F\,Z$

8. $F\,X$ 9. $T\,Y \to Z$

10. $F\,Y$ 11. $T\,Z$

Figure 5.3 A Classical Tableau Example Revisited

For all logics we consider, closure can be taken to be *atomic*. That is, a branch is closed if it contains $T\,P$ and $F\,P$ where P is atomic, or it contains $T\,\bot$. This corresponds to *Atomic Axiomatization* for sequents, see Theorem 5.5. We will always require atomic closure.

Branch extension rules for classical connectives can be restricted to *single use*. That is, a tableau rule need never be applied to a signed formula occurrence on a branch more than once. The tableau rules allow for multiple usage, but it is never necessary. (This is not true for all logics, however, and modal logics are an example.) The tableau examples in Figures 5.2 and 5.3, in fact, follows a single use convention. All *classical* tableau rule applications here will follow this single use convention.

We now have the basics of classical tableaus. Other connectives can easily be added, either as primitive or as defined connectives. Figure 5.4 shows the rules for negation, conjunction, and disjunction. If the connectives are taken as defined, these rules are derivable rules. It is possible, and often very useful, to combine rules that appear similar into general conjunctive and disjunctive classes called α and β cases respectively. This is called *uniform notation*. It can be applied here as well, but we thought it would be information overload and have decided not to follow this route. More can be found about uniform notation in Smullyan (1963) where it was introduced, and in Smullyan (1968) and Fitting (1996).

To illustrate the full set of connectives, a classical tableau proof of $\neg(X \wedge$

$$\frac{T\,\neg X}{F\,X} \qquad \frac{F\,\neg X}{T\,X}$$

$$\frac{T\,X \wedge Y}{\begin{array}{c}T\,X\\ T\,Y\end{array}} \qquad \frac{F\,X \wedge Y}{F\,X \mid F\,Y}$$

$$\frac{T\,X \vee Y}{T\,X \mid T\,Y} \qquad \frac{F\,X \vee Y}{\begin{array}{c}F\,X\\ F\,Y\end{array}}$$

Figure 5.4 Tableau Rules for $\neg$, $\wedge$, and $\vee$

$\neg Y) \rightarrow (\neg X \vee Y)$ can be found in Figure 5.5. In it, 2 and 3 are from 1 by $F \rightarrow$, 4 is from 2 by $T\,\neg$, 5 and 6 are from 3 by $F\,\vee$, 7 and 8 are from 4 by $F\,\wedge$, 9 is from 5 by $F\,\neg$, and 10 is from 8 by $F\,\neg$.

1. $F\,\neg(X \wedge \neg Y) \rightarrow (\neg X \vee Y)$
2. $T\,\neg(X \wedge \neg Y)$
3. $F\,\neg X \vee Y$
4. $F\,X \wedge \neg Y$
5. $F\,\neg X$
6. $F\,Y$

7. $F\,X$	8. $F\,\neg Y$
9. $T\,X$	10. $T\,Y$

Figure 5.5 A Classical Tableau Example with Multiple Connectives

We have said nothing about how soundness and completeness are proved for tableaus. This is an extensive topic in itself and is not the subject of this book. However, there is room for a few illustrative remarks.

Soundness of tableau systems is usually given a direct proof, making use of the semantics for the logic. It also follows from the soundness of the sequent calculus, once a translation is defined in Section 5.9. There is also a proof-theoretic way of understanding tableaus that has relevance to the tableau quasi-realization algorithm that will be presented in Chapter 7. It has already been seen, for the sequent calculus, in Definition 5.2. The following is for classical logic, but the same ideas apply directly to the modal tableaus that will be discussed starting in the next section.

Definition 5.14 Let $\mathcal{B}$ be a branch of a tableau. By the *formula translation* for $\mathcal{B}$ we mean $\bigwedge \mathcal{B}_T \to \bigvee \mathcal{B}_F$ where $\mathcal{B}_T = \{X \mid T\,X \text{ is on } \mathcal{B}\}$ and $\mathcal{B}_F = \{X \mid F\,X \text{ is on } \mathcal{B}\}$.

Pick your favorite axiomatic presentation of classical logic—details don't matter. If a tableau branch is closed, or can be continued to closure, the formula translation for that branch will be provable in that axiom system. Here is how to show this. If a branch is closed it is obvious; because both $T\,P$ and $F\,P$ will be on the branch the formula translation for the branch will be $(\ldots \wedge P \wedge \ldots) \to (\ldots \vee P \vee \ldots)$, which is obviously a tautology and hence axiomatically provable. With a modest amount of work, one can show the following. If branch $\mathcal{B}$ extends in one step to a branch or branches having axiomatically provable formula translations, then $\mathcal{B}$ itself has an axiomatically provable formula translation. Then a kind of backward induction establishes the result for all closable branches.

If X has a tableau proof, a tableau for $F\,X$ closes, so the formula translation for the only branch of the initial tableau is axiomatically provable. But this branch contains only $F\,X$, so the formula translation is $\bigwedge \emptyset \to \bigvee X$, or $\top \to X$, or equivalently, X, and hence X is axiomatically provable. The idea will turn up again in Section 7.7.

As to completeness, we briefly sketch the ideas that were used for the sequent calculus, but applied to tableaus. If a tableau is constructed so that every signed formula has a rule applied to it on every branch on which it appears, the result will either be a proof or will yield a counterexample. This makes concrete the idea that tableaus are a search for a countermodel, and if none is found, we have a proof. It is one of the things that people find attractive about tableaus as a proof method. Completeness is an immediate consequence. We illustrate this in Figure 5.6. An attempt is made to prove $(X \to Y) \to (\neg Y \to X)$. 2 and 3 are from 1 by $F \to$; 4 and 5 are from 3 by $F \to$; 6 is from 4 by $T\,\neg$; 7 and 8 are from 2 by $T \to$. The right branch is closed because of 6 and 7. The left branch is not closed—this is not a proof. Furthermore, every nonatomic formula on the left branch has had a rule applied to it on that branch. Given our single use restriction, there is nothing more to do. And in fact, the open branch tells us why we failed to find a proof. If we set X to be false (5 or 7) and Y to be false (6), every formula on the branch will have the truth value corresponding its sign, so $(X \to Y) \to (\neg Y \to X)$ will be false, and hence is not a tautology.

1. $F\,(X \to Y) \to (\neg Y \to X)$
2. $T\,X \to Y$
3. $F\,\neg Y \to X$
4. $T\,\neg Y$
5. $F\,X$
6. $F\,Y$

7. $F\,X$ | 8. $T\,Y$

Figure 5.6 An Unclosed Classical Tableau

5.6 Modal Tableaus for K

The classical propositional language we have been using is now enhanced with $\Box$. The $\Diamond$ could easily be added too, but it plays no special role when studying justification logics, so we omit it. There are several varieties of modal tableaus. Here we will use what are called a *destructive* version. The name comes from the fact that certain rules cause branch information to disappear. Such tableaus exist for K, T, D, D4, K4, and S4, among other familiar logics, but not for S5. We will sketch rules for these systems in the next section, and for K here. We begin with a signed analog of Definition 4.6.

Definition 5.15 (Modal Sharp Operation, Signed Version) For a set S of signed modal formulas, let $S^\sharp = \{T\,X \mid T\,\Box X \in S\}$.

There is only one modal rule for K, given in Figure 5.7, in addition to the classical propositional rules given earlier. This rule is applicable to a branch containing $F\,\Box X$, with S being the set of other formulas on the branch. *The entire branch is replaced with a new branch consisting of the members of* $S^\sharp$, *and* $F\,X$. Note that information is lost passing from S to $S^\sharp$—the reason why such tableaus are called *destructive*.

$$\frac{S, F\,\Box X}{S^\sharp, F\,X}$$

Figure 5.7 Modal K Rule

Figure 5.8 shows a tableau proof using the K-rules, of $(\Box X \wedge \Box Y) \to \Box(X \wedge$

Y). It begins with several classical rule applications. After 5 has been added, we apply the modal K rule on 3. We add $F\,X \wedge Y$, but we must replace the remaining set S of signed formulas with $S^\sharp$. This modifies the branch to consist of signed formulas 6, 7, and 8. Here 1, 2, and 3 have been deleted, 4 and 5 replaced by 6 and 7, then 8 added to in place of 3. We have drawn a line to indicate the replacement of the old branch contents by the new ones. Now 8 causes branching and yields a closed tableau.

1. $F\,(\Box X \wedge \Box Y) \to \Box(X \wedge Y)$
2. $T\,\Box X \wedge \Box Y$
3. $F\,\Box(X \wedge Y)$
4. $T\,\Box X$
5. $T\,\Box Y$

---

6. $T\,X$
7. $T\,Y$
8. $F\,X \wedge Y$

Figure 5.8 K Proof of $(\Box X \wedge \Box Y) \to \Box(X \wedge Y)$

With classical propositional tableaus, *any* order of rule application must produce a proof if one exists, as long as eventually every applicable rule has been applied. This is not the case for K. If both $F\,\Box X$ and $F\,\Box Y$ are present on the same branch, applying a rule to one eliminates the other, and it may be that only one of the two possibilities will lead to a proof. *Backtracking* is critical to proof search using modal tableaus. Because the number of backtracking possibilities is always bounded, a systematic search for a K tableau proof allowing backtracking will provide a decision procedure for that logic.

5.7 Other Modal Tableau Systems

Several other standard modal logics have destructive tableau systems. Sometimes the definition of $S^\sharp$ needs some modification, sometimes variations on modal rules are introduced. The $S^\sharp$ operations for several logics are given in Figure 5.9 and the modal rules are given in Figure 5.10.

Figure 5.11 shows a proof, using the S4 rules, of $\Box X \to \Box(\Box X \vee Y)$. Lines 2 and 3 are from 1 by $F\ \to$. Next an S4 modal rule is applied to $F\,\Box(\Box X \vee Y)$ adding 4, while replacing S by $S^\sharp$ eliminates 1 and 3. Now an $F\,\vee$-rule application to 4 adds 5 and 6 and produces a closed tableau, though not an atomically closed one. Continuing, we apply an S4 modal rule again, to 5,

Logic	Definition of $S^\sharp$
K, T, D	$\{T\,X \mid T\,\Box X \in S\}$
K4, D4	$\{T\,X, T\,\Box X \mid T\,\Box X \in S\}$
S4	$\{T\,\Box X \mid T\,\Box X \in S\}$

Figure 5.9 Definitions for $S^\sharp$

Logic	Rule
K, K4, T, S4, D, D4	$\dfrac{S, F\,\Box X}{S^\sharp, F\,X}$
K, K4	no additional rules
T, S4	$\dfrac{T\,\Box X}{T\,X}$
D, D4	$\dfrac{S}{S^\sharp}$

Figure 5.10 Modal Rules

adding 7 while eliminating 4, 5, and 6. Finally, applying the second S4 modal rule to 2 adds 8, and we have atomic closure.

1. $F\,\Box X \to \Box(\Box X \vee Y)$
2. $T\,\Box X$
3. $F\,\Box(\Box X \vee Y)$

---

2. $T\,\Box X$
4. $F\,\Box X \vee Y$
5. $F\,\Box X$
6. $F\,Y$

---

2. $T\,\Box X$
7. $F\,X$
8. $T\,X$

Figure 5.11 S4 Proof of $\Box X \to \Box(\Box X \vee Y)$

The K rule $S, F\,\Box X \Rightarrow S^\sharp, F\,X$ is single usage by default. Applying it with $F\,\Box X$ eliminates the formula, so a rule cannot be applied to it a second time. The rule $T\,\Box X \Rightarrow T\,X$ for T can also be single usage, but for S4 things are trickier. For S4, if $T\,\Box X \Rightarrow T\,X$ is applied to a signed formula occurrence it need not be applied again, *until the rule* $S, F\,\Box X \Rightarrow S^\sharp, F\,X$ *has been applied.* The intuition is simple: the destructive rule might eliminate the consequent of $T\,\Box X \Rightarrow T\,X$ but for S4 it will not eliminate the premise, so a new application may be useful.

5.8 Tableaus and Annotated Formulas

When we come to Realization Theorems, starting in Chapter 6, we will need to know about the *polarity* of subformulas, that is, which subformulas occur *positively* and which *negatively*. We will also need to keep track of *occurrences* of necessity operators as formulas are decomposed. For many purposes this does not need anything elaborate, but as things get more complicated, good machinery will prove useful. Signed formulas themselves provide an appropriate device for the first of these problems. Annotated tableaus help to address the second issue. We discuss this machinery now, though it will not be employed until Chapter 7.

If a formula contains only $\wedge$ and $\neg$ as propositional connectives, a subformula occurrence is positive if it is in the scope of an even number of negation signs, and it is negative if the number is odd. If other connectives are present, translate them into $\wedge$, $\neg$ and use the previous definition. This is the idea, though we will make it more formal. (Actually, we will only be interested in the polarity of occurrences of *necessitated* formulas.) In Fitting (2009) *annotated formulas* were introduced to track the occurrences of necessitated formulas manipulated in the course of proofs, and in Fitting (2013a, b) a simpler version was used. It is the simpler version that we present now.

Definition 5.16 (Annotated Formula) An *annotated modal formula* is like a standard modal formula except for the following.

(1) Instead of a single necessity symbol $\Box$ there is an infinite family, $\Box_1$, $\Box_2$, ..., called *annotated* modal operators. Formulas are built up as usual, but using annotated modal operators instead of $\Box$. The important condition is that in an annotated formula, *no annotation may occur twice*.

(2) If A is an annotated formula, and A' is the result of replacing all annotated modal operators, $\Box_n$, with $\Box$, regardless of annotation, then A' is a conventional modal formula. We say A is an *annotated version* of A', and A' is an *unannotated version* of A.

Annotations are purely for bookkeeping purposes. Semantically they are ignored. Thus $\Box_n$ and $\Box$ are understood to behave alike in Kripke models, so that a modal formula and an annotated version of it evaluate the same at each possible world. In tableaus the propositional rules function the same way whether or not annotations are present. Thus for instance, if $T\,\Box_1 P \wedge \Box_2 Q$ occurs on a branch we can add $T\,\Box_1 P$ and $T\,\Box_2 Q$ to the branch end, and similarly for the other rules. The definition of $S^\sharp$ for S4 becomes the following. $S^\sharp = \{T\,\Box_i X \mid T\,\Box_i X \in S\}$, and similarly for other modal logics. And because we require atomic closure, closure conditions are not affected by annotations.

Given all this, Figure 5.12 shows an annotated version of the S4 proof given in Figure 5.11. Every modal tableau proof can be turned into an annotated proof simply by annotating the modal operators appearing in the root, and then propagating these annotations downward through the tree. And similarly for other modal logics as well, of course.

$$
\begin{array}{ll}
1. & F\,\Box_1 X \rightarrow \Box_2(\Box_3 X \vee Y) \\
2. & T\,\Box_1 X \\
3. & F\,\Box_2(\Box_3 X \vee Y) \\
\hline
2. & T\,\Box_1 X \\
4. & F\,\Box_3 X \vee Y \\
5. & F\,\Box_3 X \\
6. & F\,Y \\
\hline
2. & T\,\Box_1 X \\
7. & F\,X \\
8. & T\,X
\end{array}
$$

Figure 5.12 Annotated S4 Proof of $\Box_1 X \rightarrow \Box_2(\Box_3 X \vee Y)$

Signed formulas are useful for keeping track of the polarity of subformulas. We do need to make a somewhat arbitrary decision about whether, in $F\,X$, say, the occurrence of X should be counted as positive or as negative. Because a tableau proof of X will start with $F\,X$, it is most convenient for us to consider such an occurrence as a positive one, and an occurrence of X in $T\,X$ as negative, though this may be thought backwards for some other purposes. Once this decision has been, the key observation is that the conjunctive and disjunctive classical tableau rules from Section 5.5 respect polarity. For instance, suppose we use the $F \rightarrow$ tableau rule to conclude $T\,X$ and $F\,Y$ from $F\,X \rightarrow Y$. Well, if $X \rightarrow Y$ is positive according to our earlier informal characterization, the occurrence of X will be negative and the occurrence of Y will be positive, matching the way we said we would be understanding the signs.

In fact, it is only necessitated formulas whose polarity we will be interested in, and annotated versions at that. Below is a recursive characterization of polarity for this case. It will be used extensively in Chapter 7. In the definition we talk about a formula X being a subformula of a signed formula, say $T\,Z$ or $F\,Z$, by which we mean it is a subformula of Z. Recall that an annotation can occur at most once in an annotated formula so, for instance, if $\Box_n X$ occurs as a subformula of a conjunction, it must occur in exactly one of the conjuncts, and similarly for disjunctions and implications.

Definition 5.17 (Polarity) We define the polarity of necessitated subformulas of an *annotated signed* formula.

(1) The occurrence of $\Box_n Z$ in $T\,\Box_n Z$ is negative; and in $F\,\Box_n Z$ is positive.

(2)

if $\Box_n Z$ occurs as a subformula of	$\Box_n Z$ has the same polarity that it has in
$T\,\neg X$	$F\,X$
$F\,\neg X$	$T\,X$

(3)

if $\Box_n Z$ occurs as a subformula of	$\Box_n Z$ has the same polarity that it has in
$T\,X \to Y$	$F\,X$ or $T\,Y$
$F\,X \to Y$	$T\,X$ or $F\,Y$
$T\,X \wedge Y$	$T\,X$ or $T\,Y$
$F\,X \wedge Y$	$F\,X$ or $F\,Y$
$T\,X \vee Y$	$T\,X$ or $T\,Y$
$F\,X \vee Y$	$F\,X$ or $F\,Y$

(4)

if $n \neq k$ and $\Box_n Z$ occurs as a subformula of	$\Box_n Z$ has the same polarity that it has in
$T\,\Box_k X$	$T\,X$
$F\,\Box_k X$	$T\,X$

If $\Box_n Z$ is a subformula of an unsigned annotated formula X, its occurrence has the same polarity that it does in $F\,X$.

Example 5.18 We compute the polarity of $\Box_2 X$ in $(A \to \Box_2 X) \to B$. To do this we evaluate the polarity that $\Box_2 X$ has in $F\,(A \to \Box_2 X) \to B$. This is the same as the polarity it has in $T\,A \to \Box_2 X$; and this is the same polarity that it has in $T\,\Box_2 X$. Hence the polarity is negative.

It is easy to show that if $\Box_n X$ is a subformula of Z, the polarity it has in $T\,Z$ and in $F\,Z$ will be opposites.

5.9 Changing the Tableau Representation

So far tableaus have been trees with nodes labeled by signed formulas, and these formula occurrences could be common to multiple branches. While this has advantages for some purposes, it does not when our quasi-realization construction is introduced in Chapter 7. We will be associating a set of quasi-realizers with each signed formula occurrence in a tableau. How that is done will depend on the history of the branch below a given occurrence. If an occurrence is common to more than one branch it has more than one history, and things become ambiguous. Our simple solution is to change the way tableaus are represented so that a certain amount of that history is explicitly present. The result is, essentially, the *block tableaus* of (Smullyan, 1968, chapter XI)—our version differs only in that we impose a single-use restriction where Smullyan did not. This is of no deep significance for present purposes.

Definition 5.19 (Block Tableau) A *block tableau* is similar to a tableau except that each node in the tree is labeled with a finite set of signed formulas (such a set is called a block). A block tableau proof of X begins with a root node labeled with $\{F\,X\}$. For classical logic, branches are grown using the rules in Figure 5.13, and *these rules are only applied to the lowest blocks on branches.* A block is closed if it contains $T\,P$ and $F\,P$ for some atomic P, or if it contains $T\,\bot$ (we require atomic closure). A block tableau branch is closed if it contains a closed block, and a block tableau is closed if each of its branches is closed. A classical *block tableau proof* of X is a closed block tableau beginning with $\{F\,X\}$.

Notationally we often (but not always) display a block without using the enclosing curly brackets customary with sets. Also, when we write a block as $\mathcal{B}, Z$ we mean it is the set whose members are those of $\mathcal{B}$, together with signed formula Z. *This notation generally assumes that Z is not part of $\mathcal{B}$.* Single use conditions are ensured by our deleting from a block the signed formula to which we have applied a rule. For example, the $T \rightarrow$ rule in Figure 5.13 is to be read as follows. If a block tableau has block $\mathcal{B}, T\,X \rightarrow Y$ at a node, then we can branch and add two new blocks, $\mathcal{B}, F\,X$ and $\mathcal{B}, T\,Y$, and these do not contain the occurrence of $T\,X \rightarrow Y$. Similarly for the other rules.

There is one misleading aspect to the notation in Figure 5.13. In the rule for $T\,\neg$, for instance, it may happen that $F\,X$ already occurs in $\mathcal{B}$, in which case the display of $\mathcal{B}, F\,X$ below the line is not correct—it should be simply $\mathcal{B}$. We allow this mild abuse, rather than complicating notation.

$$\frac{\mathcal{B}, T\,\neg X}{\mathcal{B}, F\,X} \qquad \frac{\mathcal{B}, F\,\neg X}{\mathcal{B}, T\,X} \qquad \frac{\mathcal{B}, T\,X \rightarrow Y}{\mathcal{B}, F\,X \mid \mathcal{B}, T\,Y} \qquad \frac{\mathcal{B}, F\,X \rightarrow Y}{\mathcal{B}, T\,X, F\,Y}$$

$$\frac{\mathcal{B}, T\,X \wedge Y}{\mathcal{B}, T\,X, T\,Y} \qquad \frac{\mathcal{B}, F\,X \wedge Y}{\mathcal{B}, F\,X \mid \mathcal{B}, F\,Y} \qquad \frac{\mathcal{B}, T\,X \vee Y}{\mathcal{B}, T\,X \mid \mathcal{B}, T\,Y} \qquad \frac{\mathcal{B}, F\,X \vee Y}{\mathcal{B}, F\,X, F\,Y}$$

Figure 5.13 Classical Block Branch Extension Rules

An ordinary classical tableau can easily be converted into a block tableau.

Simply replace each signed formula occurrence Z on a branch by the set of all signed formulas on the branch that have not had a rule applied to them at the point when Z was added to the branch. Informally, a block is a snapshot of a tableau branch at a particular stage of the construction. For instance, Figure 5.5 shows a standard tableau proof. Figure 5.14 shows the same proof, converted to a block tableau.

$$1.\ F\,\neg(X \wedge \neg Y) \rightarrow (\neg X \vee Y)$$
$$2.\ 3.\ T\,\neg(X \wedge \neg Y), F\,\neg X \vee Y$$
$$4.\ F\,X \wedge \neg Y, F\,\neg X \vee Y$$
$$5.\ 6.\ F\,X \wedge \neg Y, F\,\neg X, F\,Y$$

Left branch:

$$7.\ F\,X, F\,\neg X, F\,Y$$
$$9.\ F\,X, T\,X, F\,Y$$

Right branch:

$$8.\ F\,\neg Y, F\,\neg X, F\,Y$$
$$10.\ T\,Y, F\,\neg X, F\,Y$$

Figure 5.14 Tableau from Figure 5.5 as a Block Tableau

Next we reformulate the S4 tableau rules in this style, building in the notion of single usage for tableau rules. Of course rules for other modal logics could also be presented in a similar fashion, but S4 will be sufficiently representative. Single usage is a bit tricky for S4 block modal rules, which are shown in Figure 5.15. Single usage for the $F\,\Box$ rule is automatic because the

$$\frac{\mathcal{B}, T\,\Box X}{\mathcal{B}, \cancel{T\,\Box X}, T\,X} \qquad \frac{\mathcal{B}, F\,\Box X}{\mathcal{B}^{\sharp}, F\,X}$$

$$\text{where } \mathcal{B}^{\sharp} = \{T\,\Box X \mid T\,\Box X \in \mathcal{B} \text{ or } \cancel{T\,\Box X} \in \mathcal{B}\}$$

Figure 5.15 S4 Modal Block Branch Extension Rules

signed formula $F\,\Box X$ is removed. But for the $T\,\Box$ rule of S4 single usage is meant to only apply until the next application of the $F\,\Box$ rule. We incorporate this visually into the rules of Figure 5.15 first by crossing off an occurrence of $T\,\Box X$ when a rule has been applied to it, second providing no rule that has a crossed off signed formula as a trigger; and third removing a cross off mark as part of the definition of $\mathcal{B}^{\sharp}$, when an $F\,\Box$ rule is applied. Figure 5.16 shows an example of an S4 block tableau, for the familiar $\Box(P \rightarrow Q) \rightarrow (\Box P \rightarrow \Box Q)$. (It is not the shortest block tableau proof for this formula.)

Block tableaus provide the easiest to describe link between tableaus and

1. $F\,\Box(P \to Q) \to (\Box P \to \Box Q)$
2. $T\,\Box(P \to Q), F\,\Box P \to \Box Q$
3. $\cancel{T\,\Box(P \to Q)}, T\,P \to Q, F\,\Box P \to \Box Q$
4. $\cancel{T\,\Box(P \to Q)}, T\,P \to Q, T\,\Box P, F\,\Box Q$
5. $T\,\Box(P \to Q), T\,\Box P, F\,Q$
6. $\cancel{T\,\Box(P \to Q)}, T\,P \to Q, T\,\Box P, F\,Q$

7. $\cancel{T\,\Box(P \to Q)}, F\,P, T\,\Box P, F\,Q$ 8. $\cancel{T\,\Box(P \to Q)}, T\,Q, T\,\Box P, F\,Q$

9. $\cancel{T\,\Box(P \to Q)}, F\,P, \cancel{T\,\Box P}, T\,P, F\,Q$

Figure 5.16 A Longer Block Tableau Proof than Necessary

sequents. Here is a brief description of a translation from block tableau proofs to sequent proofs; the other direction is similar and a description is omitted. Start with a block tableau. Replace each block $\mathcal{B}$ with the sequent $\mathcal{B}_T \Rightarrow \mathcal{B}_F$, where $\mathcal{B}_T$ and $\mathcal{B}_F$ are as in Definition 5.14 (ignore the distinction between crossed out and uncrossed out formulas). Finally, turn the tree over. Figure 5.17 shows an example of an S4 block tableau conversion to sequents.

Finally, block *annotated* tableaus will be needed. They have a similar structure to the block tableaus discussed earlier, and we can safely omit the obvious details in our presentation here.

1. $F\,\Box(X \wedge Y) \rightarrow (\Box X \wedge \Box Y)$
2. $T\,\Box(X \wedge Y), F\,\Box X \wedge \Box Y$

3. $T\,\Box(X \wedge Y), F\,\Box X$	4. $T\,\Box(X \wedge Y), F\,\Box Y$
5. $T\,\Box(X \wedge Y), F\,X$	8. $T\,\Box(X \wedge Y), F\,Y$
6. $\cancel{T\,\Box(X \wedge Y)}, F\,X, T\,X \wedge Y$	9. $\cancel{T\,\Box(X \wedge Y)}, F\,Y, T\,X \wedge Y$
7. $\cancel{T\,\Box(X \wedge Y)}, F\,X, T\,X, T\,Y$	10. $\cancel{T\,\Box(X \wedge Y)}, F\,Y, T\,X, T\,Y$

(a)

1. $\Rightarrow \Box(X \wedge Y) \rightarrow (\Box X \wedge \Box Y)$
2. $\Box(X \wedge Y) \Rightarrow \Box X \wedge \Box Y$

3. $\Box(X \wedge Y) \Rightarrow \Box X$	4. $\Box(X \wedge Y) \Rightarrow \Box Y$
5. $\Box(X \wedge Y) \Rightarrow X$	8. $\Box(X \wedge Y) \Rightarrow Y$
6. $\Box(X \wedge Y), X \wedge Y \Rightarrow X$	9. $\Box(X \wedge Y), X \wedge Y \Rightarrow Y$
7. $\Box(X \wedge Y), X, Y \Rightarrow X$	10. $\Box(X \wedge Y), X, Y \Rightarrow Y$

(b)

7. $\Box(X \wedge Y), X, Y \Rightarrow X$	10. $\Box(X \wedge Y), X, Y \Rightarrow Y$
6. $\Box(X \wedge Y), X \wedge Y \Rightarrow X$	9. $\Box(X \wedge Y), X \wedge Y \Rightarrow Y$
5. $\Box(X \wedge Y) \Rightarrow X$	8. $\Box(X \wedge Y) \Rightarrow Y$
3. $\Box(X \wedge Y) \Rightarrow \Box X$	4. $\Box(X \wedge Y) \Rightarrow \Box Y$

2. $\Box(X \wedge Y) \Rightarrow \Box X \wedge \Box Y$
1. $\Rightarrow \Box(X \wedge Y) \rightarrow (\Box X \wedge \Box Y)$

(c)

Figure 5.17 Block Tableau to Sequent Conversion: (a) An S4 Block Tableau Proof, (b) Intermediate Conversion Stage, (c) A Sequent Proof

6

Realization – How It Began

We have already seen examples showing that some modal derivations have natural justification counterparts: Examples 2.10–2.21. These are really instances of a very general phenomenon that we call *realization*. That is, for many common modal logics, each of their theorems can be read as a statement about explicit justifications, which can be found by an algorithmic procedure. In this chapter we present the original Realization Theorem and its proof, for S4 and its subsystems.

6.1 The Logic LP

We begin by recalling some of the machinery from Chapter 2, beginning with the Logic of Proofs, LP, itself. The language for the Logic of Proofs contains:

- the language of classical propositional logic, which includes propositional variables, propositional constant $\bot$, and Boolean connectives
- proof variables $x_0, x_1, \ldots$, proof constants $a_0, a_n, \ldots$
- function symbols: monadic !, binary $\cdot$ and $+$
- operator symbol of the type "term : formula."

Terms are defined by the grammar

$$t ::= x \mid a \mid !t \mid t \cdot t \mid t + t.$$

We call these terms *proof polynomials* and denote them by p, r, s .... Constants correspond to proofs of a finite fixed set of axiom schemas. We assume that $p \cdot r \cdot s \ldots$ should be read as $(\ldots((p \cdot r) \cdot s) \ldots)$, and $p+r+s \ldots$ as $(\ldots((p+r)+s) \ldots)$. Using t to stand for any proof polynomial and S for any propositional letter or the propositional constant $\bot$, formulas are defined by the grammar

$$F ::= S \mid F \to F \mid F \wedge F \mid F \vee F \mid \neg F \mid t{:}F.$$

Within this chapter we will use $A, B, C \ldots$ for the formulas in this language, and $\Gamma, \Delta, \ldots$ for finite sets of formulas unless otherwise explicitly stated. We will also use $\vec{x}, \vec{y}, \vec{z} \ldots$ and $\vec{p}, \vec{q}, \vec{r} \ldots$ for vectors of proof variables and proof polynomials, respectively. If $\vec{s} = (s_1, \ldots, s_n)$ and $\Gamma = (F_1, \ldots, F_n)$, then $\vec{s}{:}\Gamma$ denotes $(s_1{:}F_1, \ldots s_n{:}F_n)$, $\bigvee \Gamma = F_1 \vee \ldots \vee F_n$ and $\bigwedge \Gamma = F_1 \wedge \ldots \wedge F_n$.

The original intended semantics for the LP formula $p{:}F$ is "p is a proof of F", and this will be arithmetically formalized in Chapter 9. Note that proof systems that provide a semantics for $p{:}F$ are multiconclusion ones, i.e., p may be thought of as a proof of more than one F.

We first define the axiomatic system LP_0 in the language of LP.

Axiom Schemes:

(A0) Finite set of axiom schemes of classical propositional logic in the language of LP

(A1) $t{:}F \to F$ (*reflection*)

(A2) $s{:}(F \to G) \to (t{:}F \to [s \cdot t]{:}G)$ (*application*)

(A3) $t{:}F \to\ !t{:}t{:}F$ (*proof checker*)

(A4) $s{:}F \to [s + t]{:}F$, $t{:}F \to [s + t]{:}F$ (*union*)

Rule of inference:

(R1) $\vdash F \to G, \vdash F \Rightarrow \vdash G$ (modus ponens)

The system LP is LP_0 plus the rule

(R2) $\vdash c{:}A$ if A is an axiom and c a proof constant (*axiom necessitation*)

In this chapter, a constant specification, CS, is a finite set of formulas $\{c_1{:}A_1, \ldots c_n{:}A_n\}$ such that each c_i is a constant, and A_i is an axiom (A0)–(A4). CS is *single-conclusion* if for each constant c there is at most one formula $c{:}A \in$ CS (each constant denotes a proof of not more than one axiom). Each derivation in LP naturally generates a constant specification CS consisting of all formulas introduced in this derivation by the axiom necessitation rule (R2). For a constant specification CS, by LP(CS) we mean LP_0 plus formulas from CS as additional axioms.

There are no *a priori* restrictions on the choice of constant c and axiom A in (R2). In particular, (R2) can introduce formulas $c{:}A$ where A contains c itself. For instance, we might have $c{:}(c{:}P \to P)$. Or again, we might specify a constant several times as a proof of different axioms. One might restrict (R2) to single-conclusion constant specifications only without changing the ability of LP to emulate modal logic (this chapter), or the arithmetical completeness theorem for LP (Chapter 9).

As we noted in Chapter 2, both LP_0 and LP enjoy the deduction theorem

$$\Gamma, A \vdash B \quad \Leftrightarrow \quad \Gamma \vdash A \to B$$

and the substitution lemma: *if* $\Gamma(x, P) \vdash B(x, P)$, *then for any* t, F,

$$\Gamma(x/t, P/F) \vdash B(x/t, P/F).$$

Also, obviously

$$\mathsf{LP(CS)} \vdash F \quad \Leftrightarrow \quad \mathsf{LP}_0 \vdash \bigwedge \mathsf{CS} \to F.$$

A result of fundamental importance is the following, which was shown generally as Corollary 2.16. We restate the argument for LP, for convenience.

Lemma 6.1 (Lifting lemma) *Consider* LP *and let*

$$\vec{x}{:}\Gamma, \Delta \vdash F.$$

Then there is a proof polynomial $t(\vec{x}, \vec{y})$ *such that*

$$\vec{x}{:}\Gamma, \vec{y}{:}\Delta \vdash t(\vec{x}, \vec{y}){:}F.$$

Moreover, if the constant specification CS *in the original derivation is single-conclusion then the resulting constant specification is also single-conclusion and extends* CS.

Proof By induction on the derivation $\vec{x}{:}\Gamma, \Delta \vdash F$. If $F = x{:}G \in \vec{x}{:}\Gamma$, then put $t{:}={!x}$ and use (A3). If $F = D_j \in \Delta$, then put $t := y_j$. If F is an axiom (A0)–(A4), then pick a fresh proof constant c and put $t := c$; then by (R2), $\vdash c{:}F$. Let F be derived by modus ponens from $G \to F$ and G. Then, by the induction hypothesis, there are proof polynomials u and v such that $u{:}(G \to F)$ and $v{:}G$ are both derivable from $\vec{x}{:}\Gamma, \vec{y}{:}\Delta$. By (A2),

$$\vec{x}{:}\Gamma, \vec{y}{:}\Delta \vdash [u \cdot v]{:}G,$$

and we put $t := u \cdot v$. If F is derived by (R2), then $F = c{:}A$ for some constant c and axiom A. Use the same (R2) followed by (A3):$c{:}A \to {!c}{:}c{:}A$ and modus ponens to get ${!c}{:}F$ and put $t{:}={!c}$. □

It is easy to see from the proof that the lifting polynomial t is nothing but a blueprint of a given derivation of F. Thus, LP internalizes its own proofs as proof terms.

Corollary 6.2 (Internalization property for LP) *If* $G_1, \ldots, G_n \vdash F$, *then it is possible to construct a proof polynomial* $t(x_1, \ldots, x_n)$ *depending on fresh variables* $x_1, \ldots, x_n$, *such that*

$$x_1{:}G_1, \ldots, x_n{:}G_n \vdash t(x_1, \ldots, x_n){:}F.$$

It should be noticed that the Curry–Howard isomorphism covers only a simple instance of the internalization property when all of $G_1, \ldots, G_n, F$ are purely propositional formulas without proof terms.

Corollary 6.3 (Necessitation rule for LP) *If* LP $\vdash F$*, then* LP $\vdash p{:}F$ *for some ground proof polynomial p.*

6.2 Realization for LP

The formulation of LP and our earlier examples should have left an impression that LP is something like an explicit version of S4. The main idea driving the project that created the logic of proofs project was the observation that proof polynomials apparently denoted classical proof objects and a hope that LP indeed was capable of realizing derivations in S4 by recovering proof polynomials for every occurrence of modality. This would immediately deliver a realizability-style provability semantics for Gödel's provability calculus S4 and hence a *BHK*-style semantics of proofs for intuitionistic propositional calculus IPC. In fact, both were established in Artemov (1995).

The inverse operation to the realization of modalities of proof polynomials is the forgetful projection of LP-formulas to the modal formulas obtained by replacing all $t{:}X$'s by $\Box X$'s. It is easy to see that the forgetful projection of LP is S4-compliant. Let F° be the forgetful projection of F. By a straightforward induction on a derivation in LP one can show the following, though we omit explicit details.

Lemma 6.4 *If* LP $\vdash F$*, then* S4 $\vdash F^\circ$.

Our goal now is to establish the converse, that LP suffices to realize any theorem of S4.

By an LP-*realization* of a modal formula F we mean an assignment of proof polynomials to all occurrences of the modal operator in F along with a constant specification of all constants occurring in those proof polynomials. By F^r we understand the image of F under a realization r.

A sequent calculus for S4 is presented in Section 5.3. Chapter 5 also contains a discussion of positive and negative occurrences of modalities in a formula and in a sequent. Because we will be using sequent calculus language, GS4 in particular, in the realization algorithm below, for convenience we recall the polarity definition for a modal formula F within a given sequent. First, the indicated occurrence of $\Box$ in $\Box F$ is positive. Second, any occurrence of $\Box$ from F in $G \to F, G \wedge F, F \wedge G, G \vee F, F \vee G, \Box F$ and $\Gamma \Rightarrow \Delta, F$ has the same polarity as the corresponding occurrence of $\Box$ in F. And third, any occurrence of $\Box$ from F in $\neg F$, $F \to G$ and $F, \Gamma \Rightarrow \Delta$ has a polarity opposite to that of the corresponding occurrence of $\Box$ in F.

In a provability context, $\Box F$ is intuitively understood as "there exists a proof x of F." After an informal Skolemization, i.e. replacing quantifiers by functions, all negative occurrences of $\Box$ produce arguments of Skolem functions, whereas positive ones give functions of those arguments. For example, $\Box A \to \Box B$ could be read informally as

$$\exists x \text{ ``}x \textit{ is a proof of } A\text{''} \to \exists y \text{ ``}y \textit{ is a proof of } B\text{,''}$$

with the Skolem form

$$\text{``}x \textit{ is a proof of } A\text{''} \to \text{``}f(x) \textit{ is a proof of } B\text{.''}$$

The following definition captures this feature.

A realization r is called *normal* if all negative occurrences of $\Box$ are realized by proof variables, and the corresponding constant specification is single-conclusion.

Now, here is realization as it was first stated and proved.

Theorem 6.5 *Given a derivation* $\mathsf{S4} \vdash F$ *one can recover a normal realization r such that* $\mathsf{LP} \vdash F^r$.

Proof We use the cut-free sequent formulation of S4 from Section 5.3, GS4. Without loss of generality we assume that axioms are atomic, i.e., sequents of the form $S \Rightarrow S$, where S is a propositional letter, and also the sequent $\bot \Rightarrow$.

Suppose $\mathsf{S4} \vdash F$, and let $\mathcal{T}$ be a cut-free derivation of the sequent $\Rightarrow F$. As an illustrative example that we will follow through the proof, consider the following specific cut-free GS4 derivation, where A and B are propositional letters.

$$\dfrac{\dfrac{\dfrac{\dfrac{\dfrac{A \Rightarrow A}{\Box A \Rightarrow A}}{\Box A \Rightarrow \Box A}}{\Box A \Rightarrow \Box A \vee B}}{\Box A \Rightarrow \Box(\Box A \vee B)} \qquad \dfrac{\dfrac{\dfrac{B \Rightarrow B}{\Box B \Rightarrow B}}{\Box B \Rightarrow \Box A \vee B}}{\Box B \Rightarrow \Box(\Box A \vee B)}}{\Box A \vee \Box B \Rightarrow \Box(\Box A \vee B)}.$$

Having a cut-free derivation $\mathcal{T}$ in GS4 with atomic axioms, we use it to construct a normal realization r with a single-conclusion constant specification CS such that $\mathsf{LP}(\mathsf{CS}) \vdash \bigwedge \Gamma^r \to \bigvee \Delta^r$ for any sequent $\Gamma \Rightarrow \Delta$ occurring in $\mathcal{T}$.

Note that all $\Box$'s introduced by ($\Rightarrow \Box$) are positive, while all negative $\Box$'s are introduced by ($\Box \Rightarrow$) or by weakening.

For each rule applied in $\mathcal{T}$, every occurrence of $\Box$ in the conclusion sequent of the rule is either introduced by this rule (as in ($W \Rightarrow$), ($\Rightarrow W$), ($\wedge \Rightarrow$), ($\Rightarrow \vee$), ($\Box \Rightarrow$), ($\Rightarrow \Box$)) or has been transferred from clearly indicated predecessors in the premise sequences; in the latter case we call similar occurrences of $\Box$'s in the premise and conclusion sequents *related*. For example, in the following instance of the rule ($\vee \Rightarrow$), all displayed occurrences of $\Box$ are related

$$\frac{\Box A, \Box A \vee B \Rightarrow \qquad B, \Box A \vee B \Rightarrow}{\Box A \vee B \Rightarrow}\,(\vee \Rightarrow),$$

because $\Box A \vee B$ in the conclusion represents both a principal formula of the ($\vee \Rightarrow$) rule (hence $\Box A$ in the conclusion is related to the first occurrence of $\Box A$ in the premise) and a side formula occurring in both premise sequents.

We extend this relationship of occurrences of $\Box$'s by transitivity. Hence, all occurrences of $\Box$ in $\mathcal{T}$ are naturally split into disjoint families of related ones. Because cut-free derivations in S4 respect polarities, in any given family either all $\Box$'s are negative or all are positive. We call a family *essential* if it contains at least one instance of the ($\Rightarrow \Box$) rule. Clearly, essential families are all positive. In the example we are following, there are four families of $\Box$'s, indicated by subscripts in the diagram below.

$$\frac{\dfrac{\dfrac{\dfrac{\dfrac{A \Rightarrow A}{\Box_1 A \Rightarrow A}}{\Box_1 A \Rightarrow \Box_3 A}}{\Box_1 A \Rightarrow \Box_3 A \vee B}}{\Box_1 A \Rightarrow \Box_4(\Box_3 A \vee B)} \qquad \dfrac{\dfrac{\dfrac{B \Rightarrow B}{\Box_2 B \Rightarrow B}}{\Box_2 B \Rightarrow \Box_3 A \vee B}}{\Box_2 B \Rightarrow \Box_4(\Box_3 A \vee B)}}{\Box_1 A \vee \Box_2 B \Rightarrow \Box_4(\Box_3 A \vee B)}\,.$$

There are two negative families, 1 and 2, and two essential positive families, 3 and 4.

Now the desired r will be constructed by stages I–III as described next. For the construction we reserve a large enough set of proof variables as *provisional variables*.

Stage I. For each negative family or nonessential positive family, pick a fresh proof variable x and replace all occurrences of $\Box B$ by "$x{:}B$": in our running example, we get the following.

$$\dfrac{\dfrac{\dfrac{\dfrac{\dfrac{A \Rightarrow A}{x{:}A \Rightarrow A}}{x{:}A \Rightarrow \Box_3 A}}{x{:}A \Rightarrow \Box_3 A \vee B}}{x{:}A \Rightarrow \Box_4(\Box_3 A \vee B)} \qquad \dfrac{\dfrac{\dfrac{B \Rightarrow B}{y{:}B \Rightarrow B}}{y{:}B \Rightarrow \Box_3 A \vee B}}{y{:}B \Rightarrow \Box_4(\Box_3 A \vee B)}}{x{:}A \vee y{:}B \Rightarrow \Box_4(\Box_3 A \vee B)}.$$

Stage II. Pick an essential family f, enumerate all the occurrences of rules $(\Rightarrow \Box)$, which introduce boxes of this family. Let n_f be the total number of such rules for f. Replace all boxes of f by the polynomial

$$w_1 + \ldots + w_{n_f}$$

where w_i's are fresh provisional variables. The resulting tree $\mathcal{T}'$ is labeled by LP-formulas because all $\Box$'s have been replaced by proof polynomials.

$$\dfrac{\dfrac{\dfrac{\dfrac{\dfrac{A \Rightarrow A}{x{:}A \Rightarrow A}}{x{:}A \Rightarrow v_1{:}A}}{x{:}A \Rightarrow v_1{:}A \vee B}}{x{:}A \Rightarrow [w_1 + w_2]{:}(v_1{:}A \vee B)} \qquad \dfrac{\dfrac{\dfrac{B \Rightarrow B}{y{:}B \Rightarrow B}}{y{:}B \Rightarrow v_1{:}A \vee B}}{y{:}B \Rightarrow [w_1 + w_2]{:}(v_1{:}A \vee B)}}{x{:}A \vee y{:}B \Rightarrow [w_1 + w_2]{:}(v_1{:}A \vee B)}.$$

Stage III. Run a process going from the leaves of the tree to its root, which will update a constant specification CS and replace the provisional variables by proof polynomials of the usual variables from stage (I) and constants from CS as shown next. Simultaneously, by induction on the depth of a node in $\mathcal{T}'$ we establish that after the process passes a node the sequent assigned to this node becomes derivable in LP(CS) for the current CS.

At the start CS is empty. The axioms "$S \Rightarrow S$" and "$\Rightarrow \bot$" are derivable in LP_0. For every rule other than $(\Rightarrow \Box)$ we change neither the realization of formulas nor CS and just notice that the concluding sequent is provable in LP(CS) given that the premises are. It is easy to see that for every move down in the tree other than $(\Rightarrow \Box)$ the corresponding sequent $\Gamma \Rightarrow \Delta$ is provable in LP(CS), namely,

$$\Gamma \vdash \Delta.$$

Consider a rule $(\Rightarrow \Box)$ of a family f, and let this rule have number i in the numbering of all rules $(\Rightarrow \Box)$ from a given family f. The corresponding node in $\mathcal{T}'$ is labeled by

$$\frac{y_1{:}B_1,\ldots,y_k{:}B_k \Rightarrow B}{y_1{:}B_1,\ldots,y_k{:}B_k \Rightarrow [u_1+\ldots+u_{n_f}]{:}B}$$

where $y_1,\ldots,y_k$ are proof variables introduced in Stage I, $u_1,\ldots,u_{n_f}$ are proof polynomials, and u_i is a provisional variable. By the induction hypothesis and Corollary 6.2, construct a proof polynomial $t(y_1,\ldots,y_n)$ and extend the constant specification to get a new single-conclusion CS such that in LP(CS)

$$y_1{:}B_1,\ldots,y_k{:}B_k \vdash t(y_1,\ldots,y_n){:}B.$$

Because

$$\mathsf{LP}_0 \vdash t{:}B \rightarrow [u_1+\ldots+u_{i-1}+t+u_{i+1}+\ldots+u_{n_f}]{:}B,$$

in LP(CS),

$$y_1{:}B_1,\ldots,y_k{:}B_k \vdash [u_1+\ldots+u_{i-1}+t+u_{i+1}+\ldots+u_{n_f}]{:}B\,.$$

We declare the sequent

$$y_1{:}B_1,\ldots,y_k{:}B_k \Rightarrow [u_1+\ldots+u_{i-1}+t+u_{i+1}+\ldots+u_{n_f}]{:}B$$

the outcome of the current step assigned to the current node of $\mathcal{T}'$.

Furthermore, substitute $t(y_1,\ldots,y_n)$ for u_i everywhere in $\mathcal{T}'$ and in the current CS. The latter remains single-conclusion after such a substitution, though this operation may lead to self-referential constant specifications of the sort $c{:}A(c)$ where $A(c)$ contains c. Note that $t(y_1,\ldots,y_n)$ has no provisional variables, hence such a substitution is always possible. Moreover, after the substitution there is one less provisional variable (namely u_i) left, which guarantees termination.

The conclusion of the rule $(\Rightarrow \Box)$ under consideration becomes derivable in LP(CS), and the induction step is complete.

Eventually, we substitute polynomials of nonprovisional variables for all provisional variables and build a realization of the root sequent of the proof tree derivable in LP(CS). Obviously, the realization r built by this procedure is normal.

We conclude the proof of the realization theorem by applying Stage III procedure to our continuing example to realize the provisional variables. In the step

$$\frac{x{:}A \Rightarrow A}{x{:}A \Rightarrow v_1{:}A},$$

we can put $v_1 = x$ (though the general internalization lemma may offer some other, equally correct realization). In the step

$$\frac{x{:}A \Rightarrow x{:}A \vee B}{x{:}A \Rightarrow [w_1 + w_2]{:}(x{:}A \vee B)}$$

we can realize w_1 by any proof polynomial s such that $\mathsf{LP} \vdash x{:}A \rightarrow s{:}(x{:}A \vee B)$, e.g. $s = a{\cdot}!x$ for constant a such that $a{:}(x{:}A \rightarrow x{:}A \vee B)$. Likewise, in the step

$$\frac{y{:}B \Rightarrow x{:}A \vee B}{y{:}B \Rightarrow [w_1 + w_2]{:}(x{:}A \vee B)}$$

we can realize w_2 by any t such that $\mathsf{LP} \vdash y{:}B \rightarrow t{:}(x{:}A \vee B)$, e.g., $t = b \cdot y$ with constant b such that $b{:}(B \rightarrow x{:}A \vee B)$. □

Corollary 6.6 (Realization of S4)

$$S4 \vdash F \quad \Leftrightarrow \quad \mathsf{LP} \vdash F^r \textit{ for some realization } r.$$

6.3 Comments

The realization theorem, Theorem 6.5, is formulated for the logic of proofs LP with the total constant specification containing $c{:}A$ for each constant c and each axiom A. However, this requirement can be eased. What we need from CS in this proof is to support the Lifting Lemma, Lemma 6.1, and substitutions $t(y_1, \ldots, y_n)$ for u_i in CS: the former requires axiomatically appropriate CS and the latter a schematic CS. So, a natural sufficient requirement for this proof of the realization theorem is *axiomatically appropriate and schematic*.

The realization algorithm given in the proof of Theorem 6.5 is in fact exponential in the length of a given cut-free derivation of S4, mostly because of the repeating use of the Lifting Lemma. A polynomial time realization algorithm was offered in Brezhnev and Kuznets (2006).

Some S4-theorems admit essentially different realizations in LP. For example, among possible realizations of $\Box F \vee \Box F \rightarrow \Box F$ are the following.

$$x{:}F \vee y{:}F \rightarrow [x + y]{:}F \textit{ and } x{:}F \vee x{:}F \rightarrow x{:}F$$

The first of these formulas is a meaningful specification of the operation "+," the second one is a trivial tautology. Another example comes from the S4 theorem $\Box F \rightarrow \Box F$. The first formula that follows is a normal realization, where a is a constant that realizes $F \rightarrow F$. Again the second is simply a tautology.

This time the first formula involves in an essential way the operation "·".

$$x{:}F \rightarrow [a \cdot x]{:}F \textit{ and } x{:}F \rightarrow x{:}F$$

Modal formulas can be realized by some restricted classes of proof polynomials. For example, the standard realization of the S4-theorem $(\Box A \vee \Box B) \rightarrow \Box(A \vee B)$ gives $(x{:}A \vee y{:}B) \rightarrow [a{\cdot}x + b{\cdot}y]{:}(A \vee B)$ with the single-conclusion constant specification $a{:}(A \rightarrow A \vee B)$, $b{:}(B \rightarrow A \vee B)$. The same modal formula can be realized in LP as $(c{:}A \vee c{:}B) \rightarrow [c \cdot c]{:}(A \vee B)$ with the constant specification $c{:}(A \rightarrow A \vee B), c{:}(B \rightarrow A \vee B)$. The idea behind the design of LP, however, was to keep its language reasonably general because the realization of S4 was not the Logic of Proofs' only intended application.

7
Realization – Generalized

7.1 What We Do Here

In Chapter 6 we saw the first Realization Theorem, established with its first proof. Since then, not only have many realization proofs been created, but the realization phenomenon has also been found to apply across a much wider range of modal logics than had originally been contemplated. Here and in Chapter 8 we examine this. In order to discuss what we will be doing, the diagram in Figure 7.1 will be helpful. The rest of this introductory section is devoted to explaining what it depicts.

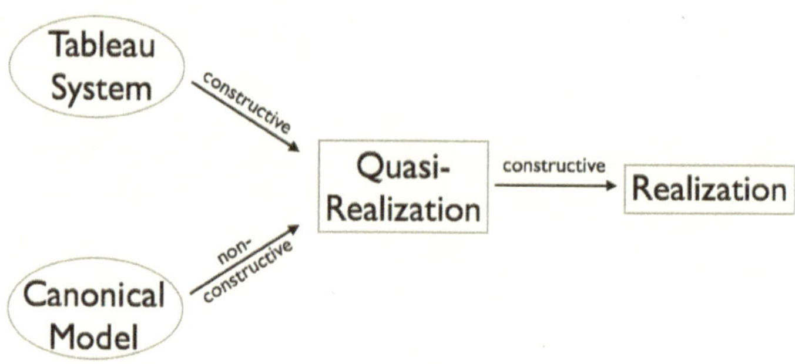

Figure 7.1 Organization of This Chapter

The original definition for realizing a formula of S4 in LP easily extends to other modal/justification logic pairs. A formal definition is in Section 7.2. The original realization proof was immediately seen to cover well-known sublogics of S4—just leave out some of the machinery. But eventually extensions of S4 were also brought into the picture and things got more complicated. In particular, the original motivation connecting intuitionistic logic with arithmetic began to fade because extensions of S4 can embody principles that don't correspond to arithmetic facts. Still while the arithmetic connection may disappear, the idea of turning modal operators into explicit justification terms retains its interest.

An important feature of the original realization proof is the role played by the substitution of complex justification terms for justification variables. The same feature appears in several subsequent proofs and, in fact, cannot be avoided. But it can be separated out, using machinery that traces back to Fitting (2005), though the significance was not understood until later. In Fitting (2013b) the name *Quasi-Realization* was introduced. Quasi-realizations are similar to realizations but they have a more complex form. Introducing them has two important benefits. First, when their existence can be established, doing so is substantially simpler than showing a similar result for realizations, insofar as it avoids any mention of substitution. In fact they still represent an embedding from a modal logic into a justification logic and can serve some of the functions that realizations were designed for, though they do not have the intuitive simplicity we would like from realization. The second important thing concerning them is that quasi-realizations can be converted into realizations—it is at this stage that substitution enters into things. Moreover this conversion can be done constructively. The algorithm for doing so proceeds by recursion on formula structure; it makes no use of the structure of cut-free proofs, or indeed, of proofs at all. And significantly, the conversion algorithm and its correctness proof can be given uniformly across all justification logics. This is shown here in Section 7.6 and is represented in Figure 7.1 by the arrow on the right.

Then establishing the existence of realizations for a modal/justification logic pair reduces to a similar problem, but for quasi-realizations. The original work on LP was constructive, and this was important because of the connection with intuitionistic logic. In Fitting (2005) a nonconstructive argument was given, and it turned out to generalize to a large number of cases for which no constructive version is currently known. All this is represented in Figure 7.1 by the two arrows on the left.

Constructive proofs of the existence of quasi-realizers all involve cut-free proof systems of some kind. Originally the sequent calculus was used. In this chapter we use tableau systems, which we covered in Chapter 5. The change is primarily one of convenience because the two kinds of proof systems can be translated into each other. Nested sequents and hypersequents have also been used. We will give some references at the end of this chapter, in Section 7.13. We will show that a quasi-realization in a justification logic can be constructed from a cut-free modal tableau proof using any of the modal tableau proof procedures from Chapter 5. We will show this for S4 and leave it to the reader to check that similar arguments work in the other cases. As is indicated in Figure 7.1, this gives us realization constructively.

The nonconstructive realization proof for LP from Fitting (2005) was eventually shown to be very general, Fitting (2014a, 2016a, b). It is semantic based and makes use of the canonical model construction for Fitting models of justification logics, from Chapter 4. This gives us a highly uniform way of showing realization, indeed for an *infinite* number of cases, with the disadvantage that the results are not constructive. This is covered in Section 7.9, with a multiplicity of examples provided in Chapter 8.

7.2 Counterparts

We have been talking about justification logics (such as LP) *corresponding to* modal logics (S4 in the case of LP). Here we say a little more about what this means, beginning with a simple mapping from the language(s) of justification logic to modal language.

Definition 7.1 (Forgetful Functor) For each justification formula X, let X° be the result of replacing every subformula $t{:}Y$ with $\Box Y$. More precisely, $P^\circ = P$ if P is atomic; $(X \rightarrow Y)^\circ = (X^\circ \rightarrow Y^\circ)$ and similarly for other connectives; and $[t{:}X]^\circ = \Box X^\circ$.

We want to know the circumstances under which the set of theorems of a justification logic is mapped by the forgetful functor to exactly the set of theorems of a modal logic, something captured formally in the following.

Definition 7.2 (Counterparts) Suppose KL is a normal modal logic (thus extending K) and JL is a justification logic. (That is, following Definition 2.6, JL extends axiomatic J_0 with additional axiom schemes and a constant specification.) We say JL is a *counterpart* of KL if the following holds.

(1) If X is provable in JL using any constant specification then X° is a theorem of KL.
(2) If Y is a theorem of KL then there is some justification formula X so that $X^\circ = Y$, where X is provable in JL using some (generally axiomatically appropriate) constant specification.

In other words, JL is a counterpart of KL if the forgetful functor is a mapping from the set of theorems of JL onto the set of theorems of KL (provided arbitrary constant specifications are allowed).

As the subject has developed so far, justification logics have been formulated so that item (1) of Definition 7.2 is simple to show. Item (2), known as a *realization* result, is not at all simple and is a central topic.

Definition 7.3 (Realization) A theorem X of justification logic JL is a *realization* of theorem Y of modal logic KL if $X^\circ = Y$ (this is condition 2 from Definition 7.2). We say X is a *normal* realization of Y if X results by the replacement of *negative* occurrences of $\Box$ in Y with *distinct justification variables* and positive occurrences by justification terms that need not be variables.

That a particular justification logic is a counterpart of a particular modal logic has always been proved by showing that *normal* realizations exist for modal theorems. For some modal logics there are algorithms for producing normal realizations, as we have seen for S4 and LP, but there are cases where no known algorithms exist, though one still has normal realizations. The extent of algorithmic realization is unclear.

For a modal logic KL and a justification logic JL to be counterparts, both items of Definition 7.2 must be satisfied. It is important to note that the two are quite independent of each other, and there are simple examples that illustrate this. We already know that S4 and LP are counterparts. Justification logic JT45 was defined in Section 2.7.2. It will be shown that S5 and JT45 are counterparts, and we make use of this in our discussion.

The pair LP and S5 shows we can have condition (1) of Definition 7.2 without condition (2). If X is a theorem of LP, X° will be a theorem of S5 because LP embeds into S4, and this is a sublogic of S5. But $\neg\Box X \rightarrow \Box\neg\Box X$ is an example of a theorem of S5 that has no realization in LP. If $\neg\Box X \rightarrow \Box\neg\Box X$ did have a realization in LP, which is a counterpart of S4, the forgetful projection of this realization would be a theorem of S4, which means $\neg\Box X \rightarrow \Box\neg\Box X$ would be an S4 theorem, and it isn't.

The pair JT45 and S4 shows we can have condition (2) of Definition 7.2 without condition (1). The forgetful functor does not map the set of theorems of JT45 into the set of theorems of S4; for example $\neg t{:}X \rightarrow ?t{:}(\neg t{:}X)$ maps outside of S4. But every theorem of S4 has a realization in JT45, because S4 and LP are counterparts, and LP is a sublogic of JT45.

7.3 Realizations

In this chapter we will be following very different realization strategies from Chapter 6, and a different, recursive, characterization of realization will be useful. We begin with a pure language construct; issues of provability will be brought in later.

We assume we have a fixed set of propositional variables; exact details don't matter. This is enough to fully determine a modal language, but more than one justification language is possible because we can still vary our choice of justification operation symbols and justification constant symbols. Of course the choice of justification variables can also be an issue, but we avoid this with the following stipulation.

Variable Convention From now on we take $v_1, v_2, \ldots$ to be an enumeration of all justification variables, with no justification variable repeated. This list is fixed once and for all.

Then only justification constants and function symbols are language dependent. Assume that some choice has been made, some set of justification terms has been specified, and a justification language is fixed.

We want to associate with each modal formula a set of *potential realizers*. As we said earlier, this is a purely linguistic issue. When, however, a specific modal logic and a specific justification logic are chosen, if a modal formula has a provable normal realization it is intended that such a realization should come from our associated set of potential realizers. The recursive Definition 7.4 is central to the algorithms and proofs that follow. But first we have some bookkeeping to take care of.

If X is a modal formula, a normal realization is sensitive to positive and negative occurrences of necessity operators. In Section 5.8 we introduced signed annotated formulas to assist with this issue of polarity, and that section should be reviewed now.

If X is a modal formula, we can turn it into an annotated formula in infinitely many ways—choose any one of them, say the result is Y. An annotated necessity operator in Y occurs positively (negatively) if it occurs positively (negatively) in the signed annotated formula $F\,Y$, as specified in Definition 5.17. So, the heart of our characterization of potential realizer is in terms of signed and annotated modal formulas. Essentially, Definition 5.17 is built directly into the definition of potential realizers.

We assume we have a full set of connectives, $\neg$, $\rightarrow$, $\wedge$, and $\vee$, though a subset would do as well.

Definition 7.4 (Potential Realizers) The mapping $[\![\cdot]\!]$ associates with each signed annotated modal formula a set of signed justification formulas called *potential realizers*. It is defined recursively, as follows. We use the enumeration, $v_1, v_2, \ldots$, of justification variables noted earlier.

(1) If A is atomic, $[\![T\,A]\!] = \{T\,A\}$ and $[\![F\,A]\!] = \{F\,A\}$

(2)
$$[\![T\,\neg X]\!] = \{T\,\neg U \mid F\,U \in [\![F\,X]\!]\}$$
$$[\![F\,\neg X]\!] = \{F\,\neg U \mid T\,U \in [\![T\,X]\!]\}$$

(3)
$$[\![T\,X \to Y]\!] = \{T\,U \to V \mid F\,U \in [\![F\,X]\!] \text{ and } T\,V \in [\![T\,Y]\!]\}$$
$$[\![F\,X \to Y]\!] = \{F\,U \to V \mid T\,U \in [\![T\,X]\!] \text{ and } F\,V \in [\![F\,Y]\!]\}$$

(4)
$$[\![T\,X \wedge Y]\!] = \{T\,U \wedge V \mid T\,U \in [\![F\,X]\!] \text{ and } T\,V \in [\![T\,Y]\!]\}$$
$$[\![F\,X \wedge Y]\!] = \{F\,U \wedge V \mid F\,U \in [\![F\,X]\!] \text{ and } F\,V \in [\![F\,Y]\!]\}$$

(5)
$$[\![T\,X \vee Y]\!] = \{T\,U \vee V \mid T\,U \in [\![F\,X]\!] \text{ and } T\,V \in [\![T\,Y]\!]\}$$
$$[\![F\,X \vee Y]\!] = \{F\,U \vee V \mid F\,U \in [\![F\,X]\!] \text{ and } F\,V \in [\![F\,Y]\!]\}$$

(6)
$$[\![T\,\Box_n X]\!] = \{T\,v_n{:}U \mid T\,U \in [\![T\,X]\!]\}$$
$$[\![F\,\Box_n X]\!] = \{F\,t{:}U \mid F\,U \in [\![F\,X]\!] \text{ and } t \text{ is any justification term}\}$$

(7) The mapping is extended to *sets* of signed annotated formulas by letting $[\![S]\!] = \cup\{[\![Z]\!] \mid Z \in S\}$.

Members of $[\![T\,Y]\!]$ are called *potential realizers* of $T\,Y$, where $T\,Y$ is a T signed, annotated modal formula; and similarly for the F-signed case. A potential realizer of an *unsigned* annotated formula Y is any justification formula Z where $F\,Z$ is a potential realizer for $F\,Y$. And finally, a potential realizer for a modal formula X that is *not annotated* is any potential realizer for Y where Y is any annotated version of X. (This involves an arbitrary choice of annotation, but different choices produce results that differ only in justification variable usage, something that is not significant here. We ignore the issue.)

Example 7.5 We compute the set of potential realizers for $\Box(\Box P \to (\Box Q \to \Box R))$. This is the set of potential realizers for any annotated version; we use $\Box_1(\Box_2 P \to (\Box_3 Q \to \Box_4 R))$. The potential realizers for this formula is the set of potential realizers for the signed formula $F\,\Box_1(\Box_2 P \to (\Box_3 Q \to \Box_4 R))$, and it is this that we now compute.

$$
\begin{aligned}
[\![T\ Q]\!] &= \{T\ Q\} \\
[\![T\ \Box_3 Q]\!] &= \{T\ v_3{:}Q\} \\
[\![F\ R]\!] &= \{F\ R\} \\
[\![F\ \Box_4 R]\!] &= \{F\ t{:}R \mid \text{any } t\} \\
[\![F\ \Box_3 Q \to \Box_4 R]\!] &= \{F\ v_3{:}Q \to t{:}R \mid \text{any } t\} \\
[\![T\ P]\!] &= \{T\ P\} \\
[\![T\ \Box_2 P]\!] &= \{T\ v_2{:}P\} \\
[\![F\ \Box_2 P \to (\Box_3 Q \to \Box_4 R)]\!] &= \{F\ v_2{:}P \to (v_3{:}Q \to t{:}R) \mid \text{any } t\} \\
[\![F\ \Box_1(\Box_2 P \to (\Box_3 Q \to \Box_4 R))]\!] &= \{F\ u{:}(v_3{:}P \to (v_2{:}Q \to t{:}R)) \mid \text{any } t, u\}
\end{aligned}
$$

So, any justification formula of the form $u{:}(v_2{:}P \to (v_3{:}Q \to t{:}R))$ is a potential realizer for $\Box(\Box P \to (\Box Q \to \Box R))$. We note that $\Box(\Box P \to (\Box Q \to \Box R))$ is not a theorem of any interesting normal modal logic. Potential realizers are about language, not about logics. (The example was actually chosen at random, but is easy to see that any normal modal logic extending T and having this as a theorem is inconsistent. If $\Box(\Box P \to (\Box Q \to \Box R))$ were a theorem of a normal modal logic, its substitution instance $\Box(\Box X \to (\Box X \to \Box\bot))$ would also be a theorem, and in an extension of T this would entail $\Box X \to \neg\Box X$.)

Potential realizers have the properties that were set out in Definition 7.3 except, perhaps, for provability. Negative occurrences of necessity are replaced by justification variables because negative subformulas will be signed with T according to Definition 5.17, and these become justification variables by case (6) of Definition 7.4. Distinct occurrences of necessity become distinct variables because of the way justification variables are assigned, and the fact that in annotated formulas no annotation can occur twice. And finally, the forgetful functor has the appropriate behavior, by the following Theorem, which has a straightforward proof by induction on formula complexity.

Theorem 7.6 *For any annotated formula X, $T\ Y \in [\![T\ X]\!]$ or $F\ Y \in [\![F\ X]\!]$ implies $Y^\circ = X$.*

7.4 Quasi-Realizations

We remind you of what we said in the introduction to this chapter. Substitution can be complicated, and while it cannot be avoided in proving realization, its complexities can be localized. *Quasi-realizers* are like realizers but with a more complicated structure, and showing their existence avoids substitution. Quasi-realizers convert into realizers, and this is where substitution comes in.

The following is exactly like Definition 7.4 except for one case, that of $F\,\Box$. We are still assuming $v_1, v_2, \ldots$ is a fixed enumeration of all justification variables with no variable repeated.

Definition 7.7 (Potential Quasi-Realizers) The mapping $\langle\!\langle \cdot \rangle\!\rangle$ is defined recursively, as follows.

(1) If A is atomic, $\langle\!\langle T\,A \rangle\!\rangle = \{T\,A\}$ and $\langle\!\langle F\,A \rangle\!\rangle = \{F\,A\}$.

(2)
$$\langle\!\langle T\,\neg X \rangle\!\rangle = \{T\,\neg U \mid F\,U \in \langle\!\langle F\,X \rangle\!\rangle\}$$
$$\langle\!\langle F\,\neg X \rangle\!\rangle = \{F\,\neg U \mid T\,U \in \langle\!\langle T\,X \rangle\!\rangle\}$$

(3)
$$\langle\!\langle T\,X \to Y \rangle\!\rangle = \{T\,U \to V \mid F\,U \in \langle\!\langle F\,X \rangle\!\rangle \text{ and } T\,V \in \langle\!\langle T\,Y \rangle\!\rangle\}$$
$$\langle\!\langle F\,X \to Y \rangle\!\rangle = \{F\,U \to V \mid T\,U \in \langle\!\langle T\,X \rangle\!\rangle \text{ and } F\,V \in \langle\!\langle F\,Y \rangle\!\rangle\}$$

(4)
$$\langle\!\langle T\,X \wedge Y \rangle\!\rangle = \{T\,U \wedge V \mid T\,U \in \langle\!\langle F\,X \rangle\!\rangle \text{ and } T\,V \in \langle\!\langle T\,Y \rangle\!\rangle\}$$
$$\langle\!\langle F\,X \wedge Y \rangle\!\rangle = \{F\,U \wedge V \mid F\,U \in \langle\!\langle F\,X \rangle\!\rangle \text{ and } F\,V \in \langle\!\langle F\,Y \rangle\!\rangle\}$$

(5)
$$\langle\!\langle T\,X \vee Y \rangle\!\rangle = \{T\,U \vee V \mid T\,U \in \langle\!\langle F\,X \rangle\!\rangle \text{ and } T\,V \in \langle\!\langle T\,Y \rangle\!\rangle\}$$
$$\langle\!\langle F\,X \vee Y \rangle\!\rangle = \{F\,U \vee V \mid F\,U \in \langle\!\langle F\,X \rangle\!\rangle \text{ and } F\,V \in \langle\!\langle F\,Y \rangle\!\rangle\}$$

(6)
$$\langle\!\langle T\,\Box_n X \rangle\!\rangle = \{T\,v_n{:}U \mid T\,U \in \langle\!\langle T\,X \rangle\!\rangle\}$$
$$\langle\!\langle F\,\Box_n X \rangle\!\rangle = \{F\,t{:}(U_1 \vee \ldots \vee U_k) \mid F\,U_1, \ldots, F\,U_k \in \langle\!\langle F\,X \rangle\!\rangle \text{ and } t \text{ is any justification term}\}$$

(7) The mapping is extended to *sets* of signed annotated formulas by letting $\langle\!\langle S \rangle\!\rangle = \cup\{\langle\!\langle Z \rangle\!\rangle \mid Z \in S\}$.

We use terminology with this mapping similar to that for potential realizers, except that of course now we talk about potential *quasi*-realizers. Thus members of $\langle\!\langle T\,Y \rangle\!\rangle$ are called *potential quasi-realizers* of $T\,Y$, where $T\,Y$ is a T signed, annotated modal formula, and so on. We omit most of the details, except to note that a *potential quasi-realizer* for a modal formula X is any *disjunction* of formulas $Z_1, \ldots, Z_n$ where $F\,Z_1, \ldots, F\,Z_n$ are members of $\langle\!\langle F\,Y \rangle\!\rangle$, where Y is any annotated version of X.

Example 7.8 In Example 7.5 we computed the set of potential realizers for the signed formula $F\,\Box_1(\Box_2 P \to (\Box_3 Q \to \Box_4 R))$. Now we compute the set of potential *quasi*-realizers.

$$
\begin{aligned}
\langle\!\langle T\ Q \rangle\!\rangle &= \{T\ Q\} \\
\langle\!\langle T\ \Box_3 Q \rangle\!\rangle &= \{T\ v_3{:}Q\} \\
\langle\!\langle F\ R \rangle\!\rangle &= \{F\ R\} \\
\langle\!\langle F\ \Box_4 R \rangle\!\rangle &= \{F\ t{:}R, F\ t{:}(R \vee R), \ldots \mid \text{any } t\} \\
\langle\!\langle F\ \Box_3 Q \to \Box_4 R \rangle\!\rangle &= \{F\ v_3{:}Q \to t{:}R, F\ v_3{:}Q \to t{:}(R \vee R), \ldots \mid \text{any } t\} \\
\langle\!\langle T\ P \rangle\!\rangle &= \{T\ P\} \\
\langle\!\langle T\ \Box_2 P \rangle\!\rangle &= \{T\ v_2{:}P\} \\
\langle\!\langle F\ \Box_2 P \to (\Box_3 Q \to \Box_4 R) \rangle\!\rangle &= \{F\ v_2{:}P \to (v_3{:}Q \to t{:}R), \\
&\qquad F\ v_2{:}P \to v_3{:}Q \to t{:}(R \vee R), \ldots \mid \text{any } t\}
\end{aligned}
$$

Finally, $\langle\!\langle F\ \Box_1(\Box_2 P \to (\Box_3 Q \to \Box_4 R)) \rangle\!\rangle$ is rather ungainly to display. Instead of a general representation we just display a few of its members. For all justification terms s, t, u

$$
\begin{aligned}
&F\ s{:}(v_2{:}P \to (v_3{:}Q \to t{:}R)) \\
&F\ s{:}((v_2{:}P \to (v_3{:}Q \to t{:}R)) \vee v_2{:}(P \to (v_3{:}Q \to u{:}R))) \\
&F\ s{:}((v_2{:}P \to (v_3{:}Q \to t{:}R)) \vee (v_2{:}P \to (v_3{:}Q \to u{:}(R \vee R)))) \\
&\text{etc.}
\end{aligned}
$$

Obviously the structure of potential quasi-realizers can be much more complex than that of potential quasi-realizers. But we still have the following analog of Theorem 7.6. The proof is a straightforward induction on formula complexity, which we omit.

Theorem 7.9 *For any annotated formula X, $T\ Y \in \langle\!\langle T\ X \rangle\!\rangle$ or $F\ Y \in \langle\!\langle F\ X \rangle\!\rangle$ implies $Y^\circ \leftrightarrow X$ in any normal modal logic.*

7.5 Substitution

We will show how potential quasi-realizers convert to potential realizers in the next section. As we have already said, substitutions play a critical role. Much of what we need was discussed in Section 2.5, and this should be reviewed now. But we also need a few more technical items concerning substitution, peculiar to the problem at hand. In the next section we give an algorithm for converting potential quasi-realizers to realizers. Using implication as representative of the propositional cases of the algorithm, for $A \to B$ we will have substitutions σ_A and σ_B associated with the antecedent and the consequent, respectively, and we must construct a substitution appropriate to the implication itself. For this, we need to know that σ_A and σ_B commute, among other things. What follows in this section are general results that have the things we need as special cases.

The following definition may be a little confusing at first look. Substitutions are defined on justification formulas, 2.17, but the definition talks about a substitution having certain properties on *annotated formulas*. However there is a simple connection. First, recall that we have introduced a fixed enumeration, $v_1, v_2, \ldots$, of all justification variables. Also recall that in the definitions of potential realizer and potential quasi-realizer, variable v_n replaces the annotated modal operator $\Box_n$ in negative positions, so annotations and variable indexes are somewhat interchangeable. (Positiveness and negativeness of occurrences are not explicitly mentioned in the definition—it is not needed because this is taken care of automatically by the recursive definition.)

Definition 7.10 Let σ be a substitution, and A be an annotated modal formula.

(1) σ *lives on* A if, for every justification variable v_k in the domain of σ, $\Box_k$ occurs in A;
(2) σ *lives away from* A if, for every justification variable v_k in the domain of σ, $\Box_k$ does not occur in A;
(3) σ meets the *no new variable* condition if, for every v_k in the domain of σ, the justification term $v_k\sigma$ contains no variables other than v_k.

Now here is the rather specialized result concerning substitution that we will need in the next section.

Theorem 7.11 *Assume A is an annotated modal formula, σ_A is a substitution that lives on A, and σ_Z is a substitution that lives away from A.*

(1) *If $T\ U \in [\![T\ A]\!]$ then $T\ U\sigma_Z \in [\![T\ A]\!]$.*
If $F\ U \in [\![F\ A]\!]$ then $F\ U\sigma_Z \in [\![F\ A]\!]$.
(2) *If both σ_A and σ_Z meet the no new variable condition, then $\sigma_A\sigma_Z = \sigma_Z\sigma_A$.*

Proof Part 1: The proof is by induction on the complexity of A. The atomic case is trivial because no justification variables are present, and the propositional cases are straightforward. This leaves the modal cases. Suppose $A = \Box_n B$, and the result is known for simpler formulas.

Assume that $T\ v_n{:}U \in [\![T\ \Box_n B]\!]$. Because σ_Z lives away from A, $v_n\sigma_Z = v_n$. By the induction hypothesis $T\ U\sigma_Z \in [\![T\ B]\!]$. Then $T\ (v_n{:}U)\sigma_Z = T\ v_n{:}(U\sigma_Z) \in [\![T\ \Box_n B]\!]$.

Assume $F\,t{:}U \in [\![F\,\Box_n B]\!]$. By the induction hypothesis, $F\,U\sigma_Z \in [\![F\,B]\!]$. Then $F\,(t{:}U)\sigma_Z = F\,t\sigma_Z{:}U\sigma_Z \in [\![F\,\Box_n B]\!]$

Part 2: Assume the hypothesis, and let v_k be an arbitrary justification variable; we show $v_k\sigma_A\sigma_Z = v_k\sigma_Z\sigma_A$.

First, suppose $\Box_k$ occurs in A. Because σ_A meets the no new variable condition, the only justification variable that can occur in $v_k\sigma_A$ is v_k. Because σ_Z lives away from A, $v_k\sigma_Z = v_k$, and so $v_k\sigma_A\sigma_Z = v_k\sigma_A$. But also, $v_k\sigma_Z\sigma_A = v_k\sigma_A$, hence $v_k\sigma_A\sigma_Z = v_k\sigma_Z\sigma_A$.

Second, suppose $\Box_k$ does not occur in A. Because σ_A lives on A, $v_k\sigma_A = v_k$. And because σ_Z meets the no new variable condition, v_k is the only variable that can occur in $v_k\sigma_Z$. Then $v_k\sigma_Z\sigma_A = v_k\sigma_Z$, and $v_k\sigma_A\sigma_Z = v_k\sigma_Z$, so $v_k\sigma_A\sigma_Z = v_k\sigma_Z\sigma_A$. □

7.6 Quasi-Realizations to Realizations

We now address the right arrow in Figure 7.1. Potential quasi-realizers can be algorithmically converted into potential realizers. It is here that substitution comes in. It is important to note that the work in this section is logic independent and can be carried out in any justification logic.

We give an algorithm that we informally say *condenses* a potential quasi-realization set to a single potential realizer. The construction and its verification trace back to Fitting (2005), with a modification and correction supplied in Fitting (2010). Throughout the section, JL is any justification logic. We make much use of Definition 2.19, and for convenience we remind you of what it says: $\vdash_{\mathsf{JL}} X$ means that there is some axiomatically appropriate constant specification CS so that $\vdash_{\mathsf{JL(CS)}} X$. We showed in Theorem 2.20 that $\vdash_{\mathsf{JL}} X$ implies $\vdash_{\mathsf{JL}} X\sigma$ for any substitution σ, and $\vdash_{\mathsf{JL}} X$ and $\vdash_{\mathsf{JL}} X \to Y$ implies $\vdash_{\mathsf{JL}} Y$. The constant specification may change in both cases, but this doesn't matter for our purposes.

We introduce some special notation to make our algorithm more easily presentable. First and simplest, if $\mathcal{A}$ is a set of annotated formulas, we write $T\,\mathcal{A}$ for $\{T\,A \mid A \in \mathcal{A}\}$, and similarly for $F\,\mathcal{A}$. And now we come to the central notation. Informally, $\mathcal{A} \xrightarrow{T\,A} (A', \sigma)$ can be read as saying that $\mathcal{A}$ is some finite set of potential quasi-realizers for $T\,A$, and it *condenses* to the single potential realizer $T\,A'$ once substitution σ is applied, and similarly for $\mathcal{A} \xrightarrow{F\,A} (A', \sigma)$. Here is the formal definition.

Definition 7.12 (Condensing in JL) Let A be an annotated modal formula,

$\mathcal{A}$ be a finite set of justification formulas, A' be a single justification formula, and σ be a substitution that lives on A and meets the no new variable condition (Definition 7.10).

Notation	Meaning
$\mathcal{A} \xrightarrow{TA} (A', \sigma)$	$\begin{cases} T\mathcal{A} \subseteq \langle\!\langle TA \rangle\!\rangle \\ TA' \in [\![TA]\!] \\ \vdash_{\mathsf{JL}} A' \to (\bigwedge \mathcal{A})\sigma \end{cases}$
$\mathcal{A} \xrightarrow{FA} (A', \sigma)$	$\begin{cases} F\mathcal{A} \subseteq \langle\!\langle FA \rangle\!\rangle \\ FA' \in [\![FA]\!] \\ \vdash_{\mathsf{JL}} (\bigvee \mathcal{A})\sigma \to A' \end{cases}$

The following is our central result about condensing. It says that any finite set of potential quasi-realizers condenses to a potential realizer, and it does so constructively.

Theorem 7.13 (Condensing Theorem) *Let A be an annotated modal formula. For each nonempty finite set $\mathcal{A}$ of justification formulas:*

(1) *If $T\mathcal{A} \subseteq \langle\!\langle TA \rangle\!\rangle$ then there are A' and σ so that $\mathcal{A} \xrightarrow{TA} (A', \sigma)$.*
(2) *If $F\mathcal{A} \subseteq \langle\!\langle FA \rangle\!\rangle$ then there are A' and σ so that $\mathcal{A} \xrightarrow{FA} (A', \sigma)$.*

Algorithm 7.14 provides a mechanism for generating A' and σ.

We give the algorithm and follow it with a discussion of how to read a few of the cases. Then we prove it is correct, thus establishing Theorem 7.13. We give cases for all of $\to$, $\wedge$, and $\vee$, though any one would suffice because the other connectives are definable.

Algorithm 7.14 (Potential Quasi-Realization to Potential Realization Condensing) The algorithm proceeds by induction on the complexity of the annotated formula A. The atomic case is direct. For the other cases, if the schemes above the line hold, so does the scheme below, as will be shown in the correctness proof. We remind the reader that we are still assuming $v_1, v_2, \ldots$ is a fixed list of all justification variables.

Atomic Cases Trivial, because if P is atomic $\langle\!\langle P \rangle\!\rangle = [\![P]\!] = \{P\}$, and we can use the empty substitution, ϵ. So we have the following.

$$\{P\} \xrightarrow{TP} (P, \epsilon) \qquad\qquad \{P\} \xrightarrow{FP} (P, \epsilon)$$

T ¬ Case

$$\frac{\{A_1, \ldots, A_k\} \xrightarrow{FA} (A', \sigma_A)}{\{\neg A_1, \ldots, \neg A_k\} \xrightarrow{T\neg A} (\neg A', \sigma_A)}$$

F ¬ Case

$$\frac{\{A_1,\ldots,A_k\} \xrightarrow{T\,A} (A',\sigma_A)}{\{\neg A_1,\ldots,\neg A_k\} \xrightarrow{F\,\neg A} (\neg A',\sigma_A)}$$

T → Case

$$\frac{\{A_1,\ldots,A_k\} \xrightarrow{F\,A} (A',\sigma_A) \qquad \{B_1,\ldots,B_k\} \xrightarrow{T\,B} (B',\sigma_B)}{\{A_1 \to B_1,\ldots,A_k \to B_k\} \xrightarrow{T\,A\to B} (A'\sigma_B \to B'\sigma_A,\sigma_A\sigma_B)}$$

F → Case

$$\frac{\{A_1,\ldots,A_k\} \xrightarrow{T\,A} (A',\sigma_A) \qquad \{B_1,\ldots,B_k\} \xrightarrow{F\,B} (B',\sigma_B)}{\{A_1 \to B_1,\ldots,A_k \to B_k\} \xrightarrow{F\,A\to B} (A'\sigma_B \to B'\sigma_A,\sigma_A\sigma_B)}$$

T ∧ Case

$$\frac{\{A_1,\ldots,A_k\} \xrightarrow{T\,A} (A',\sigma_A) \qquad \{B_1,\ldots,B_k\} \xrightarrow{T\,B} (B',\sigma_B)}{\{A_1 \wedge B_1,\ldots,A_k \wedge B_k\} \xrightarrow{T\,A\wedge B} (A'\sigma_B \wedge B'\sigma_A,\sigma_A\sigma_B)}$$

F ∧ Case

$$\frac{\{A_1,\ldots,A_k\} \xrightarrow{F\,A} (A',\sigma_A) \qquad \{B_1,\ldots,B_k\} \xrightarrow{F\,B} (B',\sigma_B)}{\{A_1 \wedge B_1,\ldots,A_k \wedge B_k\} \xrightarrow{F\,A\wedge B} (A'\sigma_B \wedge B'\sigma_A,\sigma_A\sigma_B)}$$

T ∨ Case

$$\frac{\{A_1,\ldots,A_k\} \xrightarrow{T\,A} (A',\sigma_A) \qquad \{B_1,\ldots,B_k\} \xrightarrow{T\,B} (B',\sigma_B)}{\{A_1 \vee B_1,\ldots,A_k \vee B_k\} \xrightarrow{T\,A\vee B} (A'\sigma_B \vee B'\sigma_A,\sigma_A\sigma_B)}$$

F ∨ Case

$$\frac{\{A_1,\ldots,A_k\} \xrightarrow{F\,A} (A',\sigma_A) \qquad \{B_1,\ldots,B_k\} \xrightarrow{F\,B} (B',\sigma_B)}{\{A_1 \vee B_1,\ldots,A_k \vee B_k\} \xrightarrow{F\,A\vee B} (A'\sigma_B \vee B'\sigma_A,\sigma_A\sigma_B)}$$

T □ Case

$$\frac{\{A_1,\ldots,A_k\} \xrightarrow{T\,A} (A',\sigma_A)}{\{v_n{:}A_1,\ldots,v_n{:}A_k\} \xrightarrow{T\,\Box_n A} (v_n{:}A'\sigma,\sigma_A\sigma)}$$

where $\vdash_{\mathsf{JL}} t_i{:}(A' \to A_i\sigma_A)$
for $i = 1,\ldots,k$, and
$\sigma = \{v_n/(s \cdot v_n)\}$
where $s = t_1 + \ldots + t_k$

F □ Case

$$\frac{\mathcal{A}_1 \cup \ldots \cup \mathcal{A}_k \xrightarrow{F\,A} (A',\sigma_A)}{\{t_1{:}\bigvee \mathcal{A}_1,\ldots,t_k{:}\bigvee \mathcal{A}_k\} \xrightarrow{F\,\Box_n A} (t\sigma_A{:}A',\sigma_A)}$$

where
$\vdash_{\mathsf{JL}} u_i{:}(\bigvee \mathcal{A}_i\sigma_A \to A')$
for $i = 1,\ldots,k$, and
$t = u_1 \cdot t_1 + \ldots + u_k \cdot t_k$

Before we prove correctness of Algorithm 7.14 we discuss how to read two representative cases, to make sure the content is properly understood. The first is the $T \rightarrow$ case, which is similar to all the cases involving propositional connectives. Let us suppose $T\,A_1 \rightarrow B_1, \ldots, T\,A_k \rightarrow B_k$ are all members of $\langle\!\langle T\,A \rightarrow B \rangle\!\rangle$, and condensing is known to hold for signed, annotated formulas simpler than $A \rightarrow B$. From Definition 7.7, $\{F\,A_1, \ldots, F\,A_k\} \subseteq \langle\!\langle F\,A \rangle\!\rangle$ and $\{T\,B_1, \ldots, T\,B_k\} \subseteq \langle\!\langle T\,B \rangle\!\rangle$. (Some of the A_i may be the same, and similarly for the B_i.) By our induction assumption that condensing holds in simpler cases, there are a justification formula A' and a substitution σ_A so that $\{A_1, \ldots, A_k\} \xrightarrow{F\,A} (A', \sigma_A)$, and a B' and σ_B so that $\{B_1, \ldots, B_k\} \xrightarrow{T\,B} (B', \sigma_B)$. The Algorithm asserts that $\{T\,A_1 \rightarrow B_1, \ldots, T\,A_k \rightarrow B_k\}$ condenses to the justification formula $A'\sigma_B \rightarrow B'\sigma_A$ using the substitution $\sigma_A\sigma_B$.

The second case we discuss is $F\,\Box$, which has a somewhat more complex appearance. Suppose $t_1{:} \bigvee \mathcal{A}_1, \ldots, t_k{:} \bigvee \mathcal{A}_k$ are all members of $\langle\!\langle F\,\Box_n A \rangle\!\rangle$, and condensing is known to hold for signed, annotated formulas simpler than $\Box_n A$. Using part (6) of Definition 7.7, each $\mathcal{A}_i$ is a finite subset of $\langle\!\langle F\,A \rangle\!\rangle$, so of course $\mathcal{A}_1 \cup \ldots \cup \mathcal{A}_k \subseteq \langle\!\langle F\,A \rangle\!\rangle$. By the induction assumption, this set condenses to a justification formula A' using substitution σ_A. As part of this, we have $\vdash_{\mathsf{JL}} \bigvee(\mathcal{A}_1 \cup \ldots \cup \mathcal{A}_k)\sigma_A \rightarrow A'$. Then of course for each i we have $\vdash_{\mathsf{JL}} \bigvee \mathcal{A}_i\sigma_A \rightarrow A'$. Because JL has an axiomatically appropriate constant specification then, by Internalization, for each i there is a justification term u_i so that $\vdash_{\mathsf{JL}} u_i{:}(\bigvee \mathcal{A}_i\sigma_A \rightarrow A)$. Then the algorithm says that $\{t_1{:} \bigvee \mathcal{A}_1, \ldots, t_k{:} \bigvee \mathcal{A}_k\}$ condenses to $t\sigma_A{:}A'$ using substitution σ_A, where $t = u_1 \cdot t_1 + \ldots + u_k \cdot t_k$.

We hope this discussion clarified any confusions about how to read the schematic cases of the algorithm. Now the formal details. Showing correctness of the Algorithm serves to establish Theorem 7.13.

Proof of Correctness for Algorithm 7.14 Each case of the algorithm must be justified. We give two propositional cases in considerable detail and omit the rest. The modal cases are fully presented.

Atomic Cases These cases are immediate.

$T \rightarrow$ **Case** Assume we are given $\{T\,A_1 \rightarrow B_1, \ldots, T\,A_k \rightarrow B_k\} \subseteq \langle\!\langle T\,A \rightarrow B \rangle\!\rangle$, and also we have both of the following.

$$\{A_1, \ldots, A_k\} \xrightarrow{F\,A} (A', \sigma_A) \qquad\qquad \{B_1, \ldots, B_k\} \xrightarrow{T\,B} (B', \sigma_B)$$

Then we also have that $\{F\,A_1, \ldots, F\,A_k\} \subseteq \langle\!\langle F\,A \rangle\!\rangle$ and $\{T\,B_1, \ldots, T\,B_k\} \subseteq \langle\!\langle T\,B \rangle\!\rangle$. Because $\{A_1, \ldots, A_k\} \xrightarrow{F\,A} (A', \sigma_A)$ we have A' and σ_A so that σ_A lives on A and meets the no new variable condition, $F\,A' \in [\![F\,A]\!]$, and $(A_1 \vee \ldots \vee A_k)\sigma_A \rightarrow A'$ is provable in JL. Similarly, because $\{B_1, \ldots, B_k\} \xrightarrow{T\,B}$

(B', σ_B) we have B' and σ_B so that σ_B lives on B and meets the no new variable condition, $T\,B' \in [\![T\,B]\!]$, and $B' \to (B_1 \wedge \ldots \wedge B_k)\sigma_B$ is provable in JL.

Because $A \to B$ is an annotated modal formula then A and B have no indexes in common, because indexes can appear only once in a formula. Then σ_A and σ_B have disjoint domains. In particular, σ_A lives on A and so lives away from B, while σ_B lives on B and so lives away from A. Then $\sigma_A\sigma_B = \sigma_B\sigma_A$ by Proposition 7.11. It is easy to see that $\sigma_A\sigma_B$ lives on $A \to B$ and meets the no new variable condition.

Again by Proposition 7.11, $F\,A'\sigma_B \in [\![F\,A]\!]$ because $F\,A' \in [\![F\,A]\!]$ and σ_B lives away from A. Likewise $T\,B'\sigma_A \in [\![T\,B]\!]$. Then $T\,A'\sigma_B \to B'\sigma_A \in [\![T\,A \to B]\!]$.

Finally, because $(A_1 \vee \ldots \vee A_k)\sigma_A \to A'$ is provable in JL, so is $[(A_1 \vee \ldots \vee A_k)\sigma_A \to A']\sigma_B = (A_1 \vee \ldots \vee A_k)\sigma_A\sigma_B \to A'\sigma_B$. Similarly $B'\sigma_A \to (B_1 \wedge \ldots \wedge B_k)\sigma_B\sigma_A$ is provable, or equivalently, $B'\sigma_A \to (B_1 \wedge \ldots \wedge B_k)\sigma_A\sigma_B$. Then by classical logic, the following is JL provable.

$$(A'\sigma_B \to B'\sigma_A) \to [(A_1 \to B_1) \vee \ldots \vee (A_k \to B_k)]\sigma_A\sigma_B$$

We have now established the following, completing the case.

$$\{A_1 \to B_1, \ldots, A_k \to B_k\} \xrightarrow{T\,A \to B} (A'\sigma_B \to B'\sigma_A, \sigma_A\sigma_B)$$

F → Case Assume we are given $\{F\,A_1 \to B_1, \ldots, F\,A_k \to B_k\} \subseteq \langle\!\langle F\,A \to B \rangle\!\rangle$, and we also have the following.

$$\{A_1, \ldots, A_k\} \xrightarrow{T\,A} (A', \sigma_A) \qquad \{B_1, \ldots, B_k\} \xrightarrow{F\,B} (B', \sigma_B)$$

As in the previous case, $\sigma_A\sigma_B = \sigma_B\sigma_A$. Also $T\,A'\sigma_B \in [\![T\,A]\!]$ and $F\,B'\sigma_A \in [\![F\,B]\!]$, so $F\,A'\sigma_B \to B'\sigma_A \in [\![F\,A \to B]\!]$. Likewise the following is provable in JL.

$$(A'\sigma_B \to B'\sigma_A) \to [(A_1 \to B_1) \wedge \ldots \wedge (A_k \to B_k)]\sigma_A\sigma_B$$

All this establishes the following.

$$\{A_1 \to B_1, \ldots, A_k \to B_k\} \xrightarrow{F\,A \to B} (A'\sigma_B \to B'\sigma_A, \sigma_A\sigma_B)$$

T ∧, F ∧, T ∨, F ∨, T ¬, F ¬ Cases These are similar to the implication cases and are left to the reader.

T □ Case Assume we are given $\{T\,v_n{:}A_1, \ldots, T\,v_n{:}A_k\} \subseteq \langle\!\langle T\,\Box_n A \rangle\!\rangle$, and also the following.

$$\{A_1, \ldots, A_k\} \xrightarrow{T\,A} (A', \sigma_A).$$

From the first assumption, $\{T\,A_1, \ldots, T\,A_k\} \subseteq \langle\!\langle T\,A \rangle\!\rangle$. Then by the second assumption, there are σ_A and $T\,A' \in [\![T\,A]\!]$ such that $\vdash_{\mathsf{JL}} A' \rightarrow (A_1 \wedge \ldots \wedge A_k)\sigma_A$, where σ_A lives on A and meets the no new variable condition.

For each $i = 1, \ldots, k$, the formula $A' \rightarrow A_i\sigma_A$ is provable, so by Internalization, 2.13, there is a justification term t_i (with no justification variables) such that $t_i{:}(A' \rightarrow A_i\sigma_A)$ is JL provable. Let s be the justification term $t_1 + \ldots + t_k$. Then $s{:}(A' \rightarrow A_i\sigma_A)$ is provable, for each i.

Let σ be the substitution $\{v_n/(s \cdot v_n)\}$. For each $i = 1, \ldots, k$, $s{:}(A' \rightarrow A_i\sigma_A)$ is provable, hence so is $[s{:}(A' \rightarrow A_i\sigma_A)]\sigma$. Because s is a justification term with no justification variables, $s{:}(A'\sigma \rightarrow A_i\sigma_A\sigma)$ is provable. Then for each i, $v_n{:}A'\sigma \rightarrow (s \cdot v_n){:}A_i(\sigma_A\sigma)$ is provable. Because $\Box_n A$ is an annotated modal formula indexes cannot occur more than once, hence index n cannot occur in A. Substitution σ_A lives on A, hence v_n is not in its domain. It follows that $v_n(\sigma_A\sigma) = v_n\sigma = (s \cdot v_n)$, and so $[v_n{:}A_i](\sigma_A\sigma) = (s \cdot v_n){:}A_i(\sigma_A\sigma)$. Then for each i, $v_n{:}A'\sigma \rightarrow [v_n{:}A_i](\sigma_A\sigma)$ is provable, and so $v_n{:}A'\sigma \rightarrow [v_n{:}A_1 \wedge \ldots \wedge v_n{:}A_k](\sigma_A\sigma)$ is provable.

The substitution σ lives away from A so, because $T\,A' \in [\![T\,A]\!]$ then also $T\,A'\sigma \in [\![T\,A]\!]$ by Proposition 7.11. Then $T\,v_n{:}A'\sigma \in [\![T\,\Box_n A]\!]$.

Finally, it is easy to check that $\sigma_A\sigma$ lives on $\Box_n A$ and meets the no new variable condition.

This is enough to establish the following.

$$\{v_n{:}A_1, \ldots, v_n{:}A_k\} \xrightarrow{T\,\Box_n A} (v_n{:}A'\sigma, \sigma_A\sigma)$$

F $\Box$ Case Assume we are given $\{F\,t_1{:}\bigvee \mathcal{A}_1, \ldots, F\,t_k{:}\bigvee \mathcal{A}_k\} \subseteq \langle\!\langle F\,\Box_n A \rangle\!\rangle$, and also the following.

$$(\mathcal{A}_1 \cup \ldots \cup \mathcal{A}_k) \xrightarrow{F\,A} (A', \sigma_A)$$

From the first assumption, $F\,\mathcal{A}_1 \cup \ldots \cup \mathcal{A}_k \subseteq \langle\!\langle F\,A \rangle\!\rangle$. Then by the second assumption there are σ_A and $F\,A' \in [\![F\,A]\!]$ such that $\vdash_{\mathsf{JL}} \bigvee\{\mathcal{A}_1 \cup \ldots \cup \mathcal{A}_k\}\sigma_A \rightarrow A'$, where σ_A lives on A and meets the no new variable condition.

For each i, $\bigvee \mathcal{A}_i\sigma_A \rightarrow A'$ is provable, so by the Lifting Lemma, 2.16, there is a justification term u_i (with no justification variables) such that $\vdash_{\mathsf{JL}} u_i{:}(\bigvee \mathcal{A}_i\sigma_A \rightarrow A')$. Then $(t_i\sigma_A){:}\bigvee \mathcal{A}_i\sigma_A \rightarrow (u_i \cdot (t_i\sigma_A)){:}A'$ is also provable, or equivalently, $(t_i{:}\bigvee \mathcal{A}_i)\sigma_A \rightarrow (u_i \cdot (t_i\sigma_A)){:}A'$. Because u_i contains no justification variables, this in turn is equivalent to $(t_i{:}\bigvee \mathcal{A}_i)\sigma_A \rightarrow ((u_i \cdot t_i)\sigma_A){:}A'$. Now let $t = u_1 \cdot t_1 + \ldots + u_k \cdot t_k$. Then for each i, $\vdash_{\mathsf{JL}} (t_i{:}\bigvee \mathcal{A}_i)\sigma_A \rightarrow (t\sigma_A){:}A'$. This gives us JL provability of the following.

$$\left[t_1{:}\bigvee \mathcal{A}_1 \vee \ldots \vee t_k{:}\bigvee \mathcal{A}_k\right]\sigma_A \rightarrow (t\sigma_A){:}A'$$

It is immediate that $F\,(t\sigma_A){:}A' \in [\![F\,\Box_n A]\!]$, and that σ_A lives on $\Box_n A$. We already know it meets the no new variable condition. We have thus verified the following.

$$\{t_1{:}\bigvee \mathcal{A}_1, \ldots, t_k{:}\bigvee \mathcal{A}_k\} \xrightarrow{F\,\Box_n A} (t\sigma_A{:}A', \sigma_A)$$

□

With this somewhat elaborate algorithm verified, we easily get the corollary that we have been after. Loosely it says that to show there is a provable realization it is enough to show there is a provable *quasi*-realization.

Corollary 7.15 *Suppose the modal formula X has a potential quasi-realizer* $Z_1 \vee \ldots \vee Z_n$ *such that* $\vdash_{\mathsf{JL}} Z_1 \vee \ldots \vee Z_n$*. Then X has a potential realizer Z such that* $\vdash_{\mathsf{JL}} Z$.

Proof By Definition 7.7, a potential quasi-realizer for X is a disjunction $Z_1 \vee \ldots \vee Z_n$ where $\{Z_1, \ldots, Z_n\} \subseteq \langle\!\langle F\,X' \rangle\!\rangle$, for some annotated version X' of X. By Theorem 7.13 we can construct a justification formula Z and a substitution σ so that $\{F\,Z_1, \ldots, F\,Z_n\} \xrightarrow{F\,X'} (F\,Z, \sigma)$. Then $F\,Z \in [\![F\,X']\!]$, so Z is a potential realizer of X. We also have that $\vdash_{\mathsf{JL}} (Z_1 \vee \ldots \vee Z_n)\sigma \rightarrow Z$. If the potential quasi-realizer is provable, that is if $\vdash_{\mathsf{JL}} Z_1 \vee \ldots \vee Z_n$, then also $\vdash_{\mathsf{JL}} (Z_1 \vee \ldots \vee Z_n)\sigma$, (2.18), and hence $\vdash_{\mathsf{JL}} Z$, and we have a provable realizer. □

7.7 Proving Realization Constructively

We have shown that establishing the existence of a quasi-realizer is sufficient for a realization result because these convert to realizers. We now start filling in the details of the upper left arrow in Figure 7.1, constructively converting a modal tableau proof into a quasi-realization. Unlike the original proof given in Chapter 6, the construction used now proceeds one proof step at a time. Details vary from logic to logic. We present things for S4 and LP; small and obvious changes adapt the work to any of the logics discussed in Section 5.7.

If there is an S4 tableau proof, it can be annotated, see Section 5.8. The result can then be turned into an annotated block tableau, see Section 5.9. All this is preprocessing. Now we describe how a closed S4 annotated block tableau $\mathcal{T}$ can be converted into what we call a justification sound quasi-realization Tree, something that will let us read off a potential quasi-realizer that is provable in LP, corresponding to the formula that the S4 tableau proves.

Definition 7.16 (Quasi-Realization Tree) Suppose $\mathcal{T}$ is an annotated block

S4 tableau. A *quasi-realization Tree* $\mathcal{T}^q$ corresponding to $\mathcal{T}$ has the same tree structure as $\mathcal{T}$, but with each signed annotated formula Z in $\mathcal{T}$ replaced with a finite, nonempty subset of $\langle\!\langle Z \rangle\!\rangle$, potential quasi-realizers of Z.

Thus each node in $\mathcal{T}$, labeled with a block $\mathcal{B}$ of signed annotated modal formulas, becomes a node in $\mathcal{T}^q$ labeled with a set $\mathcal{B}^q$ whose members are sets of potential quasi-realizers for the members of $\mathcal{B}$.

Definition 7.17 (Justification Sound) Let $\mathcal{T}^q$ be a quasi-realization tree corresponding to annotated block S4 tableau $\mathcal{T}$, and let $\mathcal{B}^q$ label a node in $\mathcal{T}^q$. By $\mathcal{B}^q_T$ we mean the set of justification formulas X such that $T\,X$ occurs in any member of $\mathcal{B}^q$, and likewise $\mathcal{B}^q_F$ is the set of X such that $F\,X$ occurs in any member of $\mathcal{B}^q$. By the *justification formula translation* for $\mathcal{B}^q$ we mean $\bigwedge \mathcal{B}^q_T \rightarrow \bigvee \mathcal{B}^q_F$. We say $\mathcal{B}^q$ is *justification sound* if its formula translation is provable in LP, that is, if $\vdash_{\mathsf{LP}} \bigwedge \mathcal{B}^q_T \rightarrow \bigvee \mathcal{B}^q_F$. We say $\mathcal{T}^q$ is justification sound if each node is labeled with a justification sound set.

The definition of justification formula translation is, intentionally, similar to that of formula translation, given in Definition 5.14. It might be helpful to reread the discussion following that definition.

Our goal is to show the following.

Theorem 7.18 *Let $\mathcal{T}$ be an annotated* S4 *block tableau that is closed. Then, $\mathcal{T}$ has a corresponding quasi-realization tree $\mathcal{T}^q$ that is justification sound.*

Our constructive proof of this is somewhat spread out. Algorithm 7.21 in Section 7.8 displays our method to construct justification sound mixed tableaus from closed S4 tableaus. In Section 7.9, a proof of the correctness of the algorithm is given. All this establishes Theorem 7.18, which, in the following Corollary, provides our desired result. An example of the Algorithm at work is given in Section 7.10.

Corollary 7.19 *Let X be an annotated modal formula. Given a block tableau proof of X in* S4, *a finite set $\{F\,Q_1, \ldots, F\,Q_k\}$ of quasi-realizers for $F\,X$ can be constructed so that $Q_1 \vee \ldots \vee Q_k$ is a theorem of* LP.

Proof Suppose X is an annotated modal formula, and we have a closed S4 block tableau proof $\mathcal{T}$ for X. Let $\mathcal{T}^q$ be a corresponding quasi-realization tree that is justification sound—the existence is guaranteed by Theorem 7.18. The root node of $\mathcal{T}$ is labeled with $\{F\,X\}$. Then the root node of $\mathcal{T}^q$ will be labeled with $\{\{F\,Q_1, \ldots, F\,Q_k\}\}$, where $\{F\,Q_1, \ldots, F\,Q_k\} \subseteq \langle\!\langle F\,X \rangle\!\rangle$. Because this set must be justification sound, we have $\vdash_{\mathsf{LP}} \bigwedge \emptyset \rightarrow \bigvee \{Q_1, \ldots, Q_k\}$. That is, $Q_1 \vee \ldots \vee Q_k$ is provable, where $F\,Q_1, \ldots, F\,Q_k$ are quasi-realizers for $F\,X$. □

7.8 Tableau to Quasi-Realization Algorithm

The construction in our algorithm involves a kind of backward induction. Suppose $\mathcal{T}$ is a closed annotated S4 block tableau. It is constructed using the tableau rules from Figures 5.13 and 5.15. Each branch terminates with a closed block. To construct a quasi-realization tree counterpart $\mathcal{T}^q$, we first say explicitly what the counterpart of a closed block is. And then, for each tableau rule application, if we know a counterpart for the conclusion (or conclusions when there is branching), we use this to compute the counterpart of the premise. Thus the construction of tableau $\mathcal{T}$ proceeds from root node downward to leaves, while the construction of a corresponding quasi-realization tree $\mathcal{T}^q$ proceeds from leaves upward to the root.

A branch extension rule application extends only one branch—all others remain unchanged. Consequently the algorithm is stated in terms of branch extension rules applied to single branches. The rest of the quasi-realization tree being constructed does not change, so unaffected branches are not explicitly displayed.

In a few of the algorithm cases we need what can be called a *trivial counterpart*. This is for those cases where choices don't matter. For instance, if we have a modal tableau block that is closed because it contains $T\, P$ and $F\, P$, the rest of the block doesn't matter, but any modal operators present still need to be turned into justification terms. To be definite, we take the trivial counterpart of a signed annotated formula Z to be the set whose only member is the result of replacing each $\Box_i$ with justification variable v_i. (Recall that v_1, v_2, ... is a fixed enumeration of all justification variables of LP with no variable repeated.) Likewise the trivial counterpart of a block of modal signed annotated formulas replaces each member of the set with its trivial counterpart. We denote the trivial counterpart of $\mathcal{B}$ by $\mathcal{B}^t$.

We introduce some notation that will make stating our algorithm easier. Let $\mathcal{T}$ be an annotated S4 block tableau, and let $\mathcal{T}^q$ be a corresponding quasi-realization tree. If $\mathcal{B}$ is a block labeling a node in $\mathcal{T}$, the corresponding node in $\mathcal{T}^q$ will generally be denoted by $\mathcal{B}^q$. It may happen that the same block labels more than one node in $\mathcal{T}$, and so may correspond to different collections of potential quasi-realization sets in $\mathcal{T}^q$. In this case, rather than just $\mathcal{B}$, we denote the collections by $\mathcal{B}^{q_1}$, $\mathcal{B}^{q_2}$, ... We use similar notational conventions for individual signed formula occurrences as well. That is, if Z is a signed formula in $\mathcal{B}$ then Z^b is the set of potential quasi-realizers for Z in $\mathcal{B}^q$, and similarly for $\mathcal{B}^{q_1}$, $\mathcal{B}^{q_2}$, ... Finally, by $\mathcal{B}^{q_1} \dot{\cup} \mathcal{B}^{q_2}$ we mean $\{Z^{q_1} \cup Z^{q_2} \mid Z \in \mathcal{B}\}$.

In one case, $F\,\Box$, we use the Lifting Lemma, though not exactly as stated in

Lifting Lemma 2.16. Rather, we need an easy consequence of it that is applicable to LP.

Theorem 7.20 (Lifting Lemma Specialized) *Let* JL *be a justification logic that has the internalization property relative to* CS *(in particular, if* CS *is axiomatically appropriate), and such that* $\vdash_{\mathsf{JL(CS)}} t{:}Z \rightarrow\ !t{:}t{:}Z$ *for every justification term* t *and justification formula* Z*. If* $X_1, \ldots, X_n, t_{n+1}{:}X_{n+1}, \ldots, t_{n+k}{:}X_{n+k} \vdash_{\mathsf{JL(CS)}} Y$ *then for any justification terms* $t_1, \ldots, t_n$ *there is a justification term* u *so that* $t_1{:}X_1, \ldots, t_n{:}X_n, t_{n+1}{:}X_{n+1}, \ldots, t_{n+k}{:}X_{n+k} \vdash_{\mathsf{JL(CS)}} u{:}Y$.

Proof Suppose $X_1, \ldots, X_n, t_{n+1}{:}X_{n+1}, \ldots, t_{n+k}{:}X_{n+k} \vdash_{\mathsf{JL(CS)}} Y$. By Lifting Lemma 2.16, using the justification terms $t_1, \ldots, t_n, !t_{n+1}, \ldots, !t_{n+k}$, there is a justification term u such that

$$t_1{:}X_1, \ldots, t_n{:}X_n, !t_{n+1}{:}t_{n+1}{:}X_{n+1}, \ldots, !t_{n+k}{:}t_{n+k}{:}X_{n+k} \vdash_{\mathsf{JL(CS)}} u{:}Y.$$

The theorem now follows because $\vdash_{\mathsf{JL(CS)}} t_{n+i}{:}X_{n+i} \rightarrow\ !t_{n+i}{:}t_{n+i}{:}X_{n+i}$ for each $i = 1, \ldots, k$. □

Algorithm 7.21 gives a schematic presentation of our construction. In order to make it clear how it is to be read, we follow it by describing the $T \rightarrow$ and the $F\,\Box$ cases in some detail.

Algorithm 7.21 (S4 Block Tableau To LP Quasi-Realization Tree)

Case	Modal Rule	Justification Counterpart
Atomic	$\mathcal{B}, T\,P, F\,P$ $\mathcal{B}, T\,\bot$	$\mathcal{B}^t, \{T\,P\}, \{F\,P\}$ $\mathcal{B}^t, \{T\,\bot\}$
$T\,\neg$	$\dfrac{\mathcal{B}, T\,\neg X}{\mathcal{B}, F\,X}$	$\dfrac{\mathcal{B}^q, \{T\,\neg A \mid F\,A \in (F\,X)^q\}}{\mathcal{B}^q, (F\,X)^q}$
$F\,\neg$	$\dfrac{\mathcal{B}, F\,\neg X}{\mathcal{B}, T\,X}$	$\dfrac{\mathcal{B}^q, \{F\,\neg A \mid T\,A \in (T\,X)^q\}}{\mathcal{B}^q, (T\,X)^q}$
$T \rightarrow$	$\dfrac{\mathcal{B}, T\,X \rightarrow Y}{\mathcal{B}, F\,X \mid \mathcal{B}, T\,Y}$	$\dfrac{\mathcal{B}^{q_1} \dot{\cup} \mathcal{B}^{q_2}, \{T\,A \rightarrow B \mid F\,A \in (F\,X)^q \text{ and } T\,B \in (T\,Y)^q\}}{\mathcal{B}^{q_1}, (F\,X)^q \mid \mathcal{B}^{q_2}, (T\,Y)^q}$
$F \rightarrow$	$\dfrac{\mathcal{B}, F\,X \rightarrow Y}{\mathcal{B}, T\,X, F\,Y}$	$\dfrac{\mathcal{B}^q, \{F\,A \rightarrow B \mid T\,A \in (T\,X)^q \text{ and } F\,B \in (F\,Y)^q\}}{\mathcal{B}^q, (T\,X)^q, (F\,Y)^q}$

$T\wedge$	$\dfrac{\mathcal{B}, T\,X \wedge Y}{\mathcal{B}, T\,X, T\,Y}$	$\dfrac{\mathcal{B}^q, \{T\,A \wedge B \mid T\,A \in (T\,X)^q \text{ and } T\,B \in (T\,Y)^q\}}{\mathcal{B}^q, (T\,X)^q, (T\,Y)^q}$
$F\wedge$	$\dfrac{\mathcal{B}, F\,X \wedge Y}{\mathcal{B}, F\,X \mid \mathcal{B}, F\,Y}$	$\dfrac{\mathcal{B}^{q_1} \dot{\cup} \mathcal{B}^{q_2}, \{F\,A \wedge B \mid F\,A \in (F\,X)^q \text{ and } F\,B \in (F\,Y)^q\}}{\mathcal{B}^{q_1}, (F\,X)^q \mid \mathcal{B}^{q_2}, (F\,Y)^q}$
$T\vee$	$\dfrac{\mathcal{B}, T\,X \vee Y}{\mathcal{B}, T\,X \mid \mathcal{B}, T\,Y}$	$\dfrac{\mathcal{B}^{q_1} \dot{\cup} \mathcal{B}^{q_2}, \{T\,A \vee B \mid T\,A \in (T\,X)^q \text{ and } T\,B \in (T\,Y)^q\}}{\mathcal{B}^{q_1}, (T\,X)^q \mid \mathcal{B}^{q_2}, (T\,Y)^q}$
$F\vee$	$\dfrac{\mathcal{B}, F\,X \vee Y}{\mathcal{B}, F\,X, F\,Y}$	$\dfrac{\mathcal{B}^q, \{F\,A \vee B \mid F\,A \in (F\,X)^q \text{ and } F\,B \in (F\,Y)^q\}}{\mathcal{B}^q, (F\,X)^q, (F\,Y)^q}$
$T\Box$	$\dfrac{\mathcal{B}, T\,\Box_k X}{\mathcal{B}, \cancel{T\,\Box_k X}, T\,X}$	$\dfrac{\mathcal{B}^q, (T\,\Box_k X)^q \cup \{T\,v_k{:}A \mid T\,A \in (T\,X)^q\}}{\mathcal{B}^q, (\cancel{T\,\Box_k X})^q, (T\,X)^q}$
$F\Box$	$\dfrac{\mathcal{B}, F\,\Box_k X}{\mathcal{B}^\sharp, F\,X}$	$\dfrac{(\mathcal{B}^\sharp)^q, (\mathcal{B} - \mathcal{B}^\sharp)^t, F\,u{:}\bigvee\{A \mid F\,A \in (F\,X)^q\}}{(\mathcal{B}^\sharp)^q, (F\,X)^q}$

In the $F\Box$ case the justification term u is determined as follows. We will have $\vdash_{\mathsf{LP}} \bigwedge(\mathcal{B}^\sharp)^q \rightarrow \bigvee(F\,X)^q$ and so by Theorem 7.20 there is a justification term u such that $\vdash_{\mathsf{LP}} \bigwedge(\mathcal{B}^\sharp)^q \rightarrow u{:}\bigvee(F\,X)^q$

We examine two cases from the algorithm, to illustrate how the schematics are to be read.

$T \rightarrow$. Suppose that in the modal S4 proof a branching step occurs, and we split a branch ending with $\mathcal{B}, T\,X \rightarrow Y$ into two branches beginning with $\mathcal{B}, F\,X$ and $\mathcal{B}, T\,Y$, respectively. This is shown in the Modal Rule column. Informally, for modal tableaus our work proceeds from root node downward to leaves, and the tableau rule for $T\,X \rightarrow Y$ is read from top to bottom; but for the justification part, work is in the other direction. Suppose we have constructed counterparts for each of $\mathcal{B}, F\,X$ and $\mathcal{B}, T\,Y$. The counterparts for $\mathcal{B}$ may not be the same in the two cases, so let us say we have counterparts $\mathcal{B}^{q_1}, (F\,X)^q$ and $\mathcal{B}^{q_2}, (T\,Y)^q$. We use these to construct a counterpart for $\mathcal{B}, T\,X \rightarrow Y$. For $\mathcal{B}$ we use $\mathcal{B}^{q_1} \dot{\cup} \mathcal{B}^{q_2}$. And for $T\,X \rightarrow Y$ we use the set of all $F\,A \rightarrow B$ where $T\,A \in (T\,X)^q$ and $F\,B \in (T\,Y)^q$. This is shown in the Justification Counterpart column, and the display showing the construction is read from bottom upward.

$F\Box$. This time suppose that in the modal proof we apply a rule to the block $\mathcal{B}, F\Box_k X$, producing the new block $\mathcal{B}^\sharp, F X$. Suppose we have a counterpart for the consequent, and it is $(\mathcal{B}^\sharp)^q, (F X)^q$. Using the S4 definition, all members of $\mathcal{B}^\sharp$ are of the form $T\Box Z$ (possibly crossed out), so all members of $(\mathcal{B}^\sharp)^q$ are of the form $T\, t{:}W$ for some justification term t and justification formula W. Then the justification formula translation for $\mathcal{B}^\sharp, F X$ is $\bigwedge(\mathcal{B}^\sharp)^q \to \bigvee(F X)^q$, and it will be established in our algorithm correctness proof that this formula is provable in LP. Using Lifting Lemma Specialized 7.20, there is some u so that $\vdash_{\mathsf{LP}} \bigwedge(\mathcal{B}^\sharp)^q \to u{:}\bigvee(F X)^q$. Then, reading the $F\Box$ entry in the Justification Counterpart column from bottom to top, our counterpart of $\mathcal{B}, F\Box_k X$ will consist of $(\mathcal{B}^\sharp)^q$, $(\mathcal{B} - \mathcal{B}^\sharp)^t$ (using the trivial counterpart), and $\{F\, u{:}\bigvee Z \mid F Z \in (F X)^q\}$.

7.9 Tableau to Quasi-Realization Algorithm Correctness

This section is devoted to showing the correctness of Algorithm 7.21, hence proving Theorem 7.18.

Correctness for Tableau to Quasi-Realization Algorithm We need an argument for each case of Algorithm 7.21.

Atomic Cases Consider the first of the two atomic cases—the second is similar. The block $\mathcal{B}, T P, F P$ has as its counterpart $\mathcal{B}^t, \{T P\}, \{F P\}$. Of course $\{T P\} \subseteq \langle\!\langle T P \rangle\!\rangle$ and $\{F P\} \subseteq \langle\!\langle F P \rangle\!\rangle$, and $\mathcal{B}^t$ is the trivial counterpart of $\mathcal{B}$, so we have appropriate potential quasi-realizers, Definition 7.16. The justification formula translation for this is $[\bigwedge \mathcal{B}^t_T \wedge P] \to [\bigvee \mathcal{B}^t_F \vee P]$, and this is trivially an LP theorem, so $\mathcal{B}^t, \{T P\}, \{T P\}$ is justification sound.

$T\neg$, $F\neg$ Cases These are similar to the propositional connective cases, but are simpler and are left to the reader.

$T\to$ Case Assume that $\mathcal{B}^{q_1}, (F X)^q$ and $\mathcal{B}^{q_2}, (T Y)^q$ consist of appropriate potential quasi-realizers for the blocks $\mathcal{B}, F X$ and $\mathcal{B}, T Y$ respectively, and both are justification sound. We show this also to be the case for the block $\mathcal{B}, T X \to Y$ and its counterpart $\mathcal{B}^{q_1} \dot\cup \mathcal{B}^{q_2}, \{T A \to B \mid F A \in (F X)^q \text{ and } T B \in (T Y)^q\}$.

Because $(F X)^q \subseteq \langle\!\langle F X \rangle\!\rangle$ and $(T Y)^q \subseteq \langle\!\langle T Y \rangle\!\rangle$, Definition 7.7 tells us that $\{T A \to B \mid F A \in (F X)^q \text{ and } T B \in (T Y)^q\} \subseteq \langle\!\langle T X \to Y \rangle\!\rangle$. We also need a similar result for the members of $\mathcal{B}^{q_1} \dot\cup \mathcal{B}^{q_2}$. That is, we must show that for each $Z \in \mathcal{B}$, $Z^{q_1} \cup Z^{q_2} \subseteq \langle\!\langle Z \rangle\!\rangle$. But this follows because, by our induction assumption, $Z^{q_1} \subseteq \langle\!\langle Z \rangle\!\rangle$ and $Z^{q_2} \subseteq \langle\!\langle Z \rangle\!\rangle$. After this case we omit such arguments about having appropriate potential quasi-realizers..

Next we show justification soundness. Suppose $(F\,X)^q = \{F\,A_1, \ldots, F\,A_m\}$ and $(T\,Y)^q = \{T\,B_1, \ldots, T\,B_n\}$. Then the provable justification formula translations for $\mathcal{B}^{q_1}, (F\,X)^q$ and $\mathcal{B}^{q_2}, (T\,Y)^{q_2}$ are the following.

$$\bigwedge(\mathcal{B}^{q_1})_T \to \left[\bigvee(\mathcal{B}^{q_1})_F \vee \bigvee\{A_1, \ldots, A_m\}\right]$$

$$\left[\bigwedge(\mathcal{B}^{q_2})_T \wedge \bigwedge\{B_1, \ldots, B_n\}\right] \to \bigvee(\mathcal{B}^{q_2})_F$$

It is easy to check that $(\mathcal{B}^{q_1} \dot\cup \mathcal{B}^{q_2})_T = (\mathcal{B}^{q_1})_T \cup (\mathcal{B}^{q_2})_T$ and $(\mathcal{B}^{q_1} \dot\cup \mathcal{B}^{q_2})_F = (\mathcal{B}^{q_1})_F \cup (\mathcal{B}^{q_2})_F$, so we easily have provability of the following.

$$\bigwedge(\mathcal{B}^{q_1} \dot\cup \mathcal{B}^{q_2})_T \to \left[\bigvee(\mathcal{B}^{q_1} \dot\cup \mathcal{B}^{q_2})_F \vee \bigvee\{A_1, \ldots, A_m\}\right]$$

$$\left[\bigwedge(\mathcal{B}^{q_1} \dot\cup \mathcal{B}^{q_2})_T \wedge \bigwedge\{B_1, \ldots, B_n\}\right] \to \bigvee(\mathcal{B}^{q_1} \dot\cup \mathcal{B}^{q_2})_F$$

By classical logic this gives provability of the following, where i ranges over $1, 2, \ldots, m$ and j ranges over $1, 2, \ldots, n$.

$$\left[\bigwedge(\mathcal{B}^{q_1} \dot\cup \mathcal{B}^{q_2})_T \wedge \bigwedge_{i,j}(A_i \to B_j)\right] \to \bigvee(\mathcal{B}^{q_1} \dot\cup \mathcal{B}^{q_2})_F$$

Thus the justification formula translation for $\mathcal{B}^{q_1}\dot\cup\mathcal{B}^{q_2}, \{T\,A \to B \mid F\,A \in (F\,X)^q \text{ and } T\,B \in (T\,Y)^q\}$ is provable.

$F \to$ **Case** Assume that $\mathcal{B}^q, (T\,X)^q, (F\,Y)^q$ are made up of appropriate potential quasi-realizers, and this is justification sound. We must show a similar result for $\mathcal{B}^q, \{F\,A \to B \mid T\,A \in (T\,X)^q \text{ and } F\,B \in (F\,Y)^q\}$. We omit the argument that we have appropriate potential quasi-realizers and only show justification soundness.

Let us say $(T\,X)^q = \{T\,A_1, \ldots, T\,A_m\}$ and $(F\,Y)^q = \{F\,B_1, \ldots, F\,B_n\}$. Then the provable justification formula translation for $\mathcal{B}^q, (T\,X)^q, (F\,Y)^q$ is the following.

$$\left[\bigwedge(\mathcal{B}^q)_T \wedge \bigwedge\{A_1, \ldots, A_m\}\right] \to \left[\bigvee(\mathcal{B}^q)_F \vee \bigvee\{B_1, \ldots, B_n\}\right]$$

By classical logic we also have provability of the following, where i ranges over $1, \ldots, m$ and j ranges over $1, \ldots, n$.

$$\bigwedge(\mathcal{B}^q)_T \to \left[\bigvee(\mathcal{B}^q)_F \vee \bigvee_{i,j}(A_i \to B_j)\right]$$

Thus the justification formula translation for $\mathcal{B}^q, \{F\,A \to B \mid T\,A \in (T\,X)^q \text{ and } F\,B \in (F\,Y)^q\}$ is provable.

$T\wedge, F\wedge, T\vee, F\vee$ **Cases** These are similar to the implication cases and are omitted.

$T\,\Box$ **Case** Assume that $\mathcal{B}^q, (\cancel{T\,\Box_k X})^q, (T\,X)^q$ contains appropriate potential quasi-realizers and is justification sound. We show that $\mathcal{B}^q, (T\,\Box_k X)^q \cup \{T\,v_k{:}A \mid T\,A \in (T\,X)^q\}$ is justification sound.

Suppose $(T\,\Box_k X)^q = \{T\,v_k{:}A_1, \ldots, T\,v_k{:}A_m\}$ and $(T\,X)^q = \{T\,B_1, \ldots, T\,B_h\}$. Then for $\mathcal{B}^q, (\cancel{T\,\Box_k X}, S_0), (T\,X)^q$ the provable justification formula translation is the following.

$$\Big[\bigwedge(\mathcal{B}^q)_T \wedge \bigwedge\{v_k{:}A_1, \ldots, v_k{:}A_m\} \wedge \bigwedge\{B_1, \ldots, B_h\}\Big] \rightarrow \bigvee(\mathcal{B}^q)_F$$

Using Factivity we have LP provability of the following.

$$\Big[\bigwedge(\mathcal{B}^q)_T \wedge \bigwedge\{v_k{:}A_1, \ldots, v_k{:}A_m, v_k{:}B_1, \ldots, v_k{:}B_h\}\Big] \rightarrow \bigvee(\mathcal{B}^q)_F$$

This is the justification formula translation for $\mathcal{B}^q, (T\,\Box_k X)^q \cup \{T\,v_k{:}A \mid T\,A \in (T\,X)^q\}$.

$F\,\Box$ **Case** Assume $(\mathcal{B}^\sharp)^q, (F\,X)^q$ contains appropriate potential quasi-realizers and is justification sound. Because all members of $\mathcal{B}^\sharp$ are T-signed, the LP-provable justification formula translation for this is simply $\bigwedge(\mathcal{B}^\sharp)^q_T \rightarrow \bigvee\{A \mid F\,A \in (F\,X)^q\}$. Also members of $\mathcal{B}^\sharp$ are necessitated, so by the Lifting Lemma Specialized 7.20, for some justification term u, $\vdash_{\mathsf{LP}} \bigwedge(\mathcal{B}^\sharp)^q_T \rightarrow u{:}\bigvee\{A \mid F\,A \in (F\,X)^q\}$. Then, trivially, the following is also LP-provable

$$\Big[\bigwedge(\mathcal{B}^\sharp)^q_T \wedge \bigwedge(\mathcal{B} - \mathcal{B}^\sharp)^t_T\Big] \rightarrow \Big[u{:}\bigvee\{A \mid F\,A \in (F\,X)^q\} \vee \bigvee(\mathcal{B} - \mathcal{B}^\sharp)^t_F\Big]$$

and this is the justification formula translation for $(\mathcal{B}^\sharp)^q, (\mathcal{B} - \mathcal{B}^\sharp)^t, F\,u{:}\bigvee\{A \mid F\,A \in (F\,X)^q\}$, as specified by the algorithm.

$\Box$

7.10 An Illustrative Example

We give an example to illustrate how our algorithms work. We begin with an annotated S4 block tableau proof of $(\Box_1 A \vee \Box_2 B) \rightarrow \Box_3(A \vee B)$ where A and B are atomic. We generate a quasi-realization tree using Algorithm 7.21. We then convert this to a realization using Algorithm 7.14.

Figure 7.2 shows an S4 block tableau proof of the annotated formula $(\Box_1 A \vee \Box_2 B) \rightarrow \Box_3(A \vee B)$ where A and B are atomic, as described in Section 5.9. In detail: 1 is the initial block; 2 follows from 1 by $F \rightarrow$; a $T \vee$ rule application gives 3 and 4; $F\,\Box$ rule applications in 3 and 4 produce 5 and 6; $T\,\Box$ rule applications in these give 7 and 8; and finally, $F \vee$ rule applications give 9 and 10, both of which are atomically closed. This is not the only tableau proof

1. $F\,(\Box_1 A \vee \Box_2 B) \to \Box_3(A \vee B)$
2. $T\,\Box_1 A \vee \Box_2 B, F\,\Box_3(A \vee B)$

3. $T\,\Box_1 A, F\,\Box_3(A \vee B)$	4. $T\,\Box_2 B, F\,\Box_3(A \vee B)$
5. $T\,\Box_1 A, F\,A \vee B$	6. $T\,\Box_2 B, F\,A \vee B$
7. ~~$T\,\Box_1 A$~~, $F\,A \vee B, T\,A$	8. ~~$T\,\Box_2 B$~~, $F\,A \vee B, T\,B$
9. ~~$T\,\Box_1 A$~~, $T\,A, F\,A, F\,B$	10. ~~$T\,\Box_2 B$~~, $T\,B, F\,A, F\,B$

Figure 7.2 S4 Annotated Block Tableau Proof

for the formula; it is a good exercise to construct a different one and extract a quasi-realization from it.

The block tableau shown in Figure 7.2 is converted to a quasi-realization tree using Algorithm 7.21, with the result shown in Figure 7.3. The work proceeds from bottom up.

1. $\{F\,(v_1{:}A \vee v_2{:}B) \to t{:}(A \vee B), F\,(v_1{:}A \vee v_2{:}B) \to u{:}(A \vee B)\}$
2. $\{T\,v_1{:}A \vee v_2{:}B\}, \{F\,t{:}(A \vee B), F\,u{:}(A \vee B)\}$

3. $\{T\,v_1{:}A\}, \{F\,t{:}(A \vee B)\}$	4. $\{T\,v_2{:}B\}, \{F\,u{:}(A \vee B)\}$
5. $\{T\,v_1{:}A\}, \{F\,A \vee B\}$	6. $\{T\,v_2{:}B\}, \{F\,A \vee B\}$
7. $\{T\,v_1{:}A\}, \{F\,A \vee B\}, \{T\,A\}$	8. $\{T\,v_2{:}B\}, \{F\,A \vee B\}, \{T\,B\}$
9. $\{T\,v_1{:}A\}, \{T\,A\}, \{F\,A\}, \{F\,B\}$	10. $\{T\,v_2{:}B\}, \{T\,B\}, \{F\,A\}, \{F\,B\}$

Figure 7.3 Quasi-Realization Tree

In Figure 7.2, 9 is an atomically closed block. The Atomic Cases of the algorithm convert $T\,A$ and $F\,A$ to $\{T\,A\}$ and $\{F\,A\}$. The remaining signed formulas on the branch are $F\,B$ and $T\,\Box_1 A$ (which is crossed out). For these the algorithm uses the *trivial* expansion, giving us $\{F\,B\}$ and $\{T\,v_1{:}A\}$. Then block 9 in Figure 7.2 becomes 9 in Figure 7.3. A similar discussion applies to block 10.

Block 7 in Figure 7.2 yields block 9 by the $F\vee$ rule. Because 9 in Figure 7.2 converts to 9 in Figure 7.3, 7 of Figure 7.2 converts to 7 of Figure 7.3 by the $F\vee$ case of the Algorithm. Similarly for 8 and 10.

Then block 5 of Figure 7.2 converts to 5 of Figure 7.3 because of the 7 conversion, and the $T\,\Box$ case of the Algorithm, and similarly for 6 and 8.

Block 3 of Figure 7.2 yields block 5 by the $F\Box$ rule. The justification formula translation for 5 in Figure 7.3 is $v_1{:}A \to (A \vee B)$. Justification term t, in 3, is such that $v_1{:}A \to t{:}(A \vee B)$ is provable in LP. Existence is guaranteed by

Theorem 7.20. Similarly u in 4 of Figure 7.3 is such that $v_2{:}B \to u{:}(A \vee B)$ is provable in LP.

Block 2 of Figure 7.2 yields blocks 3 and 4 using the $T\vee$ rule. Note that in 2 of Figure 7.3, the set associated with $F\,\Box_3(A \vee B)$ is the union of the corresponding sets from 3 and 4.

Finally 1 is a straightforward application of the $F \to$ rule.

Then, according to the algorithm, $\{F\,(v_1{:}A \vee v_2{:}B) \to t{:}(A \vee B), F\,(v_1{:}A \vee v_2{:}B) \to u{:}(A \vee B)\}$ is a set of quasi-realizers for $F\,(\Box_1 A \vee \Box_2 B) \to \Box_3(A \vee B)$, and is justification sound. Then the following is provable in LP.

$$\bigvee\{(v_1{:}A \vee v_2{:}B) \to t{:}(A \vee B), (v_1{:}A \vee v_2{:}B) \to u{:}(A \vee B)\}$$

where $\vdash_{\mathsf{LP}} v_1{:}A \to t{:}(A \vee B)$ and $\vdash_{\mathsf{LP}} v_2{:}B \to u{:}(A \vee B)$.

Now we can apply Algorithm 7.14, Quasi-Realization to Realization Condensing. This is rather simple, and we leave the details to you—the final stage is the following.

$$\begin{array}{l}\{(v_1{:}A \vee v_2{:}B) \to t{:}(A \vee B),\\ (v_1{:}A \vee v_2{:}B) \to u{:}(A \vee B)\}\\ \xrightarrow{F\,(\Box_1 A \vee \Box_2 B) \to \Box_3(A \vee B)}\\ \qquad\qquad ((v_1{:}A \vee v_2{:}B) \to ((c \cdot t + c \cdot u){:}(A \vee B))\sigma_0, \sigma_0)\end{array}$$

In this a internalizes a proof of $A \to A$, b internalizes a proof of $B \to B$, c internalizes a proof of $(A \vee B) \to (A \vee B)$, and σ_0 is the substitution $\{v_1/a \cdot v_1, v_2/b \cdot v_2\}$. Because A and B are atomic, and c contains no justification variables, this can be rewritten as the following.

$$\begin{array}{l}\{(v_1{:}A \vee v_2{:}B) \to t{:}(A \vee B),\\ (v_1{:}A \vee v_2{:}B) \to u{:}(A \vee B)\}\\ \xrightarrow{F\,(\Box_1 A \vee \Box_2 B) \to \Box_3(A \vee B)}\\ \qquad\qquad ((v_1{:}A \vee v_2{:}B) \to (c \cdot t\sigma_0 + c \cdot u\sigma_0){:}(A \vee B), \sigma_0)\end{array}$$

In fact, $((v_1{:}A \vee v_2{:}B) \to (c \cdot t\sigma_0 + c \cdot u\sigma_0){:}(A \vee B), \sigma_0)$ is an LP provable normal realization of $(\Box_1 A \vee \Box_2 B) \to \Box_3(A \vee B)$.

7.11 Realizations, Nonconstructively

We now fill in the details of the last part of Figure 7.1, the lower left. Constructive proofs of the existence of realizations have all been tied to cut-free proof systems, and thus they are limited to modal logics for which such proof systems are known to exist. There is also a nonconstructive approach that is very general in its applicability. The argument was implicit in Fitting (2005), and clarified in Fitting (2014a, 2016a). In what was something of a surprise, the family of modal logics having justification counterparts turns out to be infinite. We give the central theorem here, and the entire of the next chapter is devoted to examples illustrating its use. Because of Corollary 7.15, it is enough to show the existence of quasi-realizations.

Theorem 7.22 (Nonconstructive Quasi-Realization Theorem) *Let* KL *be a normal modal logic characterized by a class of frames* $\mathcal{F}(\mathsf{KL})$. *Let* JL *be a justification logic,* CS *be a constant specification for it, and assume* JL(CS) *has the internalization property. If the canonical Fitting model (Section 4.4.2) for* JL(CS) *is based on a frame in* $\mathcal{F}(\mathsf{KL})$, *then every validity of* KL *has a quasi-realization provable in* JL(CS).

The Theorem is an easy consequence of the following Lemma. Here are some notational conventions we use. For a Fitting model $\mathcal{M}_{\mathsf{JL(CS)}}$ and an annotated formula X, $\mathcal{M}_{\mathsf{JL(CS)}}, \Gamma \Vdash \langle\!\langle T\, X\rangle\!\rangle$ means $\mathcal{M}_{\mathsf{JL(CS)}}, \Gamma \Vdash Y$ for every justification formula Y such that $T\, Y \in \langle\!\langle T\, X\rangle\!\rangle$, and $\mathcal{M}_{\mathsf{JL(CS)}}, \Gamma \Vdash \langle\!\langle F\, X\rangle\!\rangle$ means $\mathcal{M}_{\mathsf{JL(CS)}}, \Gamma \nVdash Y$ for every justification formula Y such that $F\, Y \in \langle\!\langle F\, X\rangle\!\rangle$. (It may be that neither $\mathcal{M}_{\mathsf{JL(CS)}}, \Gamma \Vdash \langle\!\langle T\, X\rangle\!\rangle$ nor $\mathcal{M}_{\mathsf{JL(CS)}}, \Gamma \Vdash \langle\!\langle F\, X\rangle\!\rangle$ holds.)

Lemma 7.23 *Let* $\mathcal{M}_{\mathsf{JL(CS)}} = \langle \mathcal{G}, \mathcal{R}, \mathcal{V}, \mathcal{E}\rangle$ *be the canonical Fitting model for the axiomatic justification logic with the internalization property,* JL(CS). *Let* $\mathcal{N} = \langle \mathcal{G}, \mathcal{R}, \mathcal{V}\rangle$ *be the corresponding modal model that is formed by dropping the evidence function from* $\mathcal{M}$. *Then for every annotated modal formula* X *we have the following.*

(1) $\mathcal{M}_{\mathsf{JL(CS)}}, \Gamma \Vdash \langle\!\langle T\, X\rangle\!\rangle \Longrightarrow \mathcal{N}, \Gamma \Vdash X$
(2) $\mathcal{M}_{\mathsf{JL(CS)}}, \Gamma \Vdash \langle\!\langle F\, X\rangle\!\rangle \Longrightarrow \mathcal{N}, \Gamma \nVdash X$

Proof By induction on the degree of the annotated modal formula X.

Atomic Cases Suppose $X = A$ is atomic. The only member of $\langle\!\langle T\, A\rangle\!\rangle$ is $T\, A$, so $\mathcal{M}_{\mathsf{JL(CS)}}, \Gamma \Vdash \langle\!\langle T\, A\rangle\!\rangle$ iff $\mathcal{M}_{\mathsf{JL(CS)}}, \Gamma \Vdash A$ iff $\Gamma \in \mathcal{V}(A)$ iff $\mathcal{N}, \Gamma \Vdash A$. The F case is similar.

Negation Cases Suppose $X = \neg A$ and the result is known for A. By definition, $T\, \neg U \in \langle\!\langle T\, \neg A\rangle\!\rangle$ if and only if $F\, U \in \langle\!\langle F\, A\rangle\!\rangle$. Now assume $\mathcal{M}_{\mathsf{JL(CS)}}, \Gamma \Vdash \langle\!\langle T\, \neg A\rangle\!\rangle$. Then $\mathcal{M}_{\mathsf{JL(CS)}}, \Gamma \Vdash \neg U$ for all $T\, \neg U \in \langle\!\langle T\, \neg A\rangle\!\rangle$, so $\mathcal{M}_{\mathsf{JL(CS)}}, \Gamma \nVdash U$ for all $F\, U \in \langle\!\langle F\, A\rangle\!\rangle$, and so $\mathcal{M}_{\mathsf{JL(CS)}}, \Gamma \Vdash \langle\!\langle F\, A\rangle\!\rangle$. Then by the induction hypothesis $\mathcal{N}, \Gamma \nVdash A$, and hence $\mathcal{N}, \Gamma \Vdash \neg A$.

This is case *(1)*; case *(2)* is similar.

Implication Cases Suppose $X = A \to B$ and the results are known for A and for B.

(i) Assume $\mathcal{M}_{\mathsf{JL(CS)}}, \Gamma \Vdash \langle\!\langle T\, A \to B \rangle\!\rangle$. We divide the argument into two parts.
Suppose first that $\mathcal{M}_{\mathsf{JL(CS)}}, \Gamma \Vdash \langle\!\langle F\, A \rangle\!\rangle$. By the induction hypothesis, $\mathcal{N}, \Gamma \not\Vdash A$, so $\mathcal{N}, \Gamma \Vdash A \to B$.
Suppose next that $\mathcal{M}_{\mathsf{JL(CS)}}, \Gamma \not\Vdash \langle\!\langle F\, A \rangle\!\rangle$. Then for some U with $F\, U \in \langle\!\langle F\, A \rangle\!\rangle$, $\mathcal{M}_{\mathsf{JL(CS)}}, \Gamma \Vdash U$. Let $T\, V$ be an arbitrary member of $\langle\!\langle T\, B \rangle\!\rangle$. Then $T\, U \to V \in \langle\!\langle T\, A \to B \rangle\!\rangle$. So by the assumption, $\mathcal{M}_{\mathsf{JL(CS)}}, \Gamma \Vdash U \to V$, and hence $\mathcal{M}_{\mathsf{JL(CS)}}, \Gamma \Vdash V$. Because $T\, V$ was arbitrary, $\mathcal{M}_{\mathsf{JL(CS)}}, \Gamma \Vdash \langle\!\langle T\, B \rangle\!\rangle$. Then by the induction hypothesis, $\mathcal{N}, \Gamma \Vdash B$ and hence $\mathcal{N}, \Gamma \Vdash A \to B$.

(ii) Assume $\mathcal{M}_{\mathsf{JL(CS)}}, \Gamma \Vdash \langle\!\langle F\, A \to B \rangle\!\rangle$. Let $T\, U \in \langle\!\langle T\, A \rangle\!\rangle$ and $F\, V \in \langle\!\langle F\, B \rangle\!\rangle$ both be arbitrary. Then $F\, U \to V \in \langle\!\langle F\, A \to B \rangle\!\rangle$ so by the assumption, $\mathcal{M}_{\mathsf{JL(CS)}}, \Gamma \not\Vdash U \to V$. Then $\mathcal{M}_{\mathsf{JL(CS)}}, \Gamma \Vdash U$ and $\mathcal{M}_{\mathsf{JL(CS)}}, \Gamma \not\Vdash V$. Because $T\, U$ and $F\, V$ were arbitrary, $\mathcal{M}_{\mathsf{JL(CS)}}, \Gamma \Vdash \langle\!\langle T\, A \rangle\!\rangle$ and $\mathcal{M}_{\mathsf{JL(CS)}}, \Gamma \Vdash \langle\!\langle F\, B \rangle\!\rangle$, so by the induction hypothesis, $\mathcal{N}, \Gamma \Vdash A$ and $\mathcal{N}, \Gamma \not\Vdash B$. Hence $\mathcal{N}, \Gamma \not\Vdash A \to B$.

Conjunction and Disjunction Cases These cases are similar to Implication, and are left to the reader.

Modal Cases Suppose $X = \Box_n A$ and the results are known for A.

(i) Assume $\mathcal{M}_{\mathsf{JL(CS)}}, \Gamma \Vdash \langle\!\langle T\, \Box_n A \rangle\!\rangle$. Let $T\, U \in \langle\!\langle T\, A \rangle\!\rangle$ be arbitrary. By the assumption, $\mathcal{M}_{\mathsf{JL(CS)}}, \Gamma \Vdash v_n{:}U$. Let $\Delta \in \mathcal{G}$ be arbitrary, with $\Gamma \mathcal{R} \Delta$. Then $\mathcal{M}_{\mathsf{JL(CS)}}, \Delta \Vdash U$, and because $T\, U$ was arbitrary, $\mathcal{M}_{\mathsf{JL(CS)}}, \Delta \Vdash \langle\!\langle T\, A \rangle\!\rangle$. By the induction hypothesis, $\mathcal{N}, \Delta \Vdash A$, and because Δ was arbitrary, $\mathcal{N}, \Gamma \Vdash \Box_n A$.

(ii) Assume $\mathcal{M}_{\mathsf{JL(CS)}}, \Gamma \Vdash \langle\!\langle F\, \Box_n A \rangle\!\rangle$. This case depends on the following Claim. We first show how the Claim is used, then we prove the Claim itself.

Claim: (Using notation from Section 4.4.2.) The set $S = \Gamma^\sharp \cup \{\neg U \mid F\, U \in \langle\!\langle F\, A \rangle\!\rangle\}$ is consistent in the justification logic $\mathsf{JL(CS)}$.

The Claim is used as follows. Because (assuming the claim) S is consistent in $\mathsf{JL(CS)}$, it can be extended to a maximally consistent set, Δ. Because $\mathcal{M}_{\mathsf{JL(CS)}}$ is a canonical model, $\Delta \in \mathcal{G}$ and $\Gamma \mathcal{R} \Delta$ because $\Gamma^\sharp \subseteq \Delta$. Also $\{\neg U \mid F\, U \in \langle\!\langle F\, A \rangle\!\rangle\} \subseteq \Delta$ so by the Truth Lemma 10.41, $\mathcal{M}_{\mathsf{JL(CS)}}, \Delta \not\Vdash U$ for all U with $F\, U \in \langle\!\langle F\, A \rangle\!\rangle$, and so $\mathcal{M}_{\mathsf{JL(CS)}}, \Delta \Vdash \langle\!\langle F\, A \rangle\!\rangle$. Then by the induction hypothesis $\mathcal{N}, \Delta \not\Vdash A$, and hence $\mathcal{N}, \Gamma \not\Vdash \Box_n A$, as desired. Now to complete things, we establish the claim itself.

Proof of Claim: Suppose S is not consistent in $\mathsf{JL(CS)}$. Then there are formulas $G_1, \ldots, G_m \in \Gamma^\sharp$, and signed formulas $F\, U_1, \ldots, F\, U_k \in \langle\!\langle F\, A \rangle\!\rangle$ such that $\{G_1, \ldots, G_m, \neg U_1, \ldots, \neg U_k\}$ is not consistent, and so $G_1, G_2, \ldots, G_m \vdash_{\mathsf{JL(CS)}}$

$U_1 \vee \ldots \vee U_k$. For each $1 \leq i \leq m$, $G_i \in \Gamma^\sharp$, and so there is some justification term g_i so that $g_i{:}G_i \in \Gamma$. Then by Theorem 2.16 there is some justification term u so that $g_1{:}G_1, g_2{:}G_2, \ldots, g_m{:}G_m \vdash_{\mathsf{JL(CS)}} u{:}(U_1 \vee \ldots \vee U_k)$. It follows from maximal consistency of Γ that $u{:}(U_1 \vee \ldots \vee U_k) \in \Gamma$, and hence by the Truth Lemma, $\mathcal{M}_{\mathsf{JL(CS)}}, \Gamma \Vdash u{:}(U_1 \vee \ldots \vee U_k)$. But this contradicts the assumption that $\mathcal{M}_{\mathsf{JL(CS)}}, \Gamma \Vdash \langle\!\langle F\,\Box_n A\rangle\!\rangle$.

□

With the proof of Lemma 7.23 out of the way, we can now establish the main result of this section.

Proof of Theorem 7.22 The proof is by contraposition. Suppose Y is a modal formula. Let X be an annotated version of Y and suppose that X has no provable quasi-realization in JL(CS). We show Y is not a validity of KL.

Because X has no provable quasi-realization, for every $U_1, \ldots, U_n$ such that $F\,U_1, \ldots, F\,U_n \in \langle\!\langle F\,X\rangle\!\rangle, \nvdash_{\mathsf{JL(CS)}} U_1 \vee \ldots \vee U_n$ (Section 7.4). It follows that $\{\neg U \mid F\,U \in \langle\!\langle F\,X\rangle\!\rangle\}$ is consistent in JL(CS). (Because otherwise, $\{\neg U \mid F\,U \in \langle\!\langle F\,X\rangle\!\rangle\} \vdash_{\mathsf{JL(CS)}} \bot$, so $\neg U_1, \ldots, \neg U_n \vdash_{\mathsf{JL(CS)}} \bot$ for some $F\,U_1, \ldots, F\,U_n \in \langle\!\langle F\,X\rangle\!\rangle$, and hence $\vdash_{\mathsf{JL(CS)}} U_1 \vee \ldots \vee U_n$, contrary to what has been established.)

Because $\{\neg U \mid F\,U \in \langle\!\langle F\,X\rangle\!\rangle\}$ is consistent, it can be extended to a maximal consistent set Γ. In the canonical justification model $\mathcal{M}_{\mathsf{JL(CS)}}$, Γ is a possible world. Using the Truth Lemma, Theorem 10.41, $\mathcal{M}_{\mathsf{JL(CS)}}, \Gamma \Vdash \langle\!\langle F\,X\rangle\!\rangle$. So by Lemma 7.23, $\mathcal{N}, \Gamma \nVdash X$. A hypothesis of Theorem 7.22 is that the canonical justification model is based on a frame in the class $\mathcal{F}(\mathsf{KL})$, which determines the modal logic KL. We thus have a *modal* model $\mathcal{N}$ for KL in which X fails, and hence so does Y, the unannotated version of X. Then Y is not a validity of KL. □

7.12 Putting Things Together

Theorem 7.22 tells us, nonconstructively, when provable quasi-realizers exist. Theorem 7.13 and its accompanying algorithm tells us how to convert a quasi-realizer to a realizer. Putting these together we get the following central result.

Theorem 7.24 (Realization, Nonconstructively) *Let* KL *be a normal modal logic that is characterized by a class of frames* $\mathcal{F}(\mathsf{KL})$. *Let* JL *be a justification logic and* CS *be an axiomatically appropriate constant specification for it. If the canonical Fitting model for* JL(CS) *is based on a frame in* $\mathcal{F}(\mathsf{KL})$ *then every theorem of* KL *has a* JL *provable normal realizer.*

Proof Assume KL is a normal modal logic, characterized by the class of frames $\mathcal{F}(\mathsf{KL})$, and that JL is a justification logic with an axiomatically appropriate constant specification CS. Finally, assume the canonical Fitting model for JL(CS) is based on a frame from $\mathcal{F}(\mathsf{KL})$.

Directly by Theorem 7.22, Y has a provable quasi-realization in JL(CS). More precisely there are $F\,Z_1, \ldots, F\,Z_n \in \langle\!\langle F\,Y \rangle\!\rangle$ so that $\vdash_{\mathsf{JL(CS)}} Z_1 \vee \ldots \vee Z_n$, and hence (making use of Definition 2.19) $\vdash_{\mathsf{JL}} Z_1 \vee \ldots \vee Z_n$.

Now using Corollary 7.15 there is a potential realization Z such that $\vdash_{\mathsf{JL}} Z$ (though the constant specification may no longer be CS). □

Examples illustrating this nonconstructive way of showing realization are many and varied—too much for this already long chapter. We have devoted the entirety of Chapter 8 to them.

7.13 A Brief Realization History

Proofs of realization for various modal logics have a history of interest, and some discussion and references are pertinent.

As we have already noted, the first realization proof that connected S4 and LP was constructive and made use of a cut-free sequent calculus for S4. It was developed by Sergei Artemov and appears in Artemov (1995, 2001). It was given here in Chapter 6. The algorithm used was modified to improve its computational complexity in Brezhnev (2000) and Brezhnev and Kuznets (2006).

The nonconstructive approach to realization originated in Fitting (2003, 2005) with a correction in Fitting (2010). It was extended to a broad range of modal logics, in Fitting (2014a), with these results discussed in the present chapter.

Semantic tableaus were used in Fitting (2013a) to provide a constructive proof of realization for S4 that proceeded step-by-step, making an implementation relatively easy. Such an implementation was given in Prolog.

S5 was a kind of exploration ground for a number of years. A somewhat nonstandard axiomatization was given in Artemov et al. (1999) and Kazakov (1999), and a realization theorem was shown constructively using hypersequents. A standard axiomatization provided the basis for a nonconstructive proof of the S5 realization theorem in Rubtsova (2006b) and Pacuit (2005, 2006). In Fitting (2011b), a somewhat peculiar sequent calculus for S5 was used to give a constructive realization proof. This sequent calculus was based on a tableau system given in Fitting (1999) and has no known generalizations to other logics.

A Merging Theorem was stated and proved in Fitting (2009). This gives a constructive algorithm for combining realizations and provides appropriate machinery for handling branching rules in tableau or sequent calculi.

Kuznets and Goetschi (2012) and Borg and Kuznets (2015) used nested sequent calculi to constructively show realization theorems for all the modal logics in the so-called modal cube.

Finally, the Model Existence Theorem for S4 was used in Fitting (2013b) to show realization for S4. Similar proofs could be give for those other modal logics known to have a Model Existence Theorem, but this is a rather small number.

8
The Range of Realization

Theorem 7.24 gives us a very general, though nonconstructive, criterion for the existence of realizations, and hence for when a modal logic and a justification logic are counterparts, see Definition 7.2. There are the obvious counterparts, S4 and LP, K and J, and so on, for which realization has a constructive proof. But we also discussed some less common modal logics, with some suggested justification counterparts. Justification soundness results are in Section 4.3, and completeness with respect to Fitting models was shown in Section 4.5 using a canonical model construction, and this is enough to invoke Theorem 7.24. Indeed, at this point realization for several logics has already been established, and we just need to point it out. We begin by wrapping up our discussion of these logics and then go on to investigate an infinite family of modal logics we call *Geach Logics*, and a corresponding family of justification logics. It takes considerable work, but we will show we have realization results for all of these logics. This covers many standard cases and also shows that the family of modal/justification counterparts is infinite. The extent of the realization phenomenon is not actually known.

8.1 Some Examples We Already Discussed

Scattered through earlier chapters were several examples of modal logic, justification logic pairs that were said to be counterparts. We are now in a position to prove this properly. The work has already been done for most of these examples, and the final details are presented in this section.

$\mathsf{K4}^3$ **and** $\mathsf{J4}^3$ These were introduced in Section 2.7.1. $\mathsf{K4}^3$ is complete with respect to frames meeting the $\Gamma\mathcal{R}^3\Delta \Longrightarrow \Gamma\mathcal{R}\Delta$ condition. Discussion continued in Section 4.3.3, and finally in Section 4.5.2 we showed the canonical model for $\mathsf{J4}^3$ has a frame meeting the $\mathsf{K4}^3$ condition. Then by Theorem 7.24, there is a realization result connecting $\mathsf{J4}^3$ and $\mathsf{K4}^3$.

S5 **and** JT45 These were introduced in Section 2.7.2, with soundness shown in Section 4.3.4 and completeness in Section 4.5.3. The canonical JT45 model is based on an S5 frame, so we have a realization connection.

Sahlqvist Examples Two examples that are in the Sahlqvist class were introduced in Section 2.7.3, with soundness shown in Section 4.3.5. The completeness arguments in Section 4.5.4, along with Theorem 7.24, establish realization.

S4.2 **and** JT4.2 These were discussed in Sections 2.7.4, 4.3.6, and 4.5.5. Modal completeness was shown by a canonical model construction. In the rest of the chapter this example will be greatly expanded, and realization will follow as a special case.

KX4 **and** JX4 These can be found in Sections 2.7.5, 4.3.7, and 4.5.6. The canonical completeness proof gives us realization in this case too.

8.2 Geach Logics

The modal axiom, $\Diamond\Box X \to \Box\Diamond X$ was introduced by Peter Geach and is commonly known as G. Added to S4 it gives the modal logic known as S4.2, which we considered in Sections 2.7.4, 4.3.6, and 4.5.5. In Lemmon and Scott (1977) the axiom was generalized to an infinite family that included many familiar axioms. It was shown that all logics axiomatized by members of the family were canonical. This was a significant forerunner of Sahlqvist's later work, which goes beyond our considerations here. We call the family of modal logics thus axiomatized *Geach logics* (they are also called *Lemmon–Scott logics*). Our major result is that *Geach logics all have justification counterparts, with realization theorems connecting them.* This tells us that realization is a phenomenon that is not rare—infinitely many modal logics have justification counterparts.

The Lemmon–Scott generalization allows iterated $\Box$ and $\Diamond$ occurrences and, semantically, needs iterated accessibility relations. Here are the formal definitions.

Definition 8.1 Syntactically:

$$\Box^0 X = X$$
$$\Box^{n+1} X = \Box\Box^n X$$
$$\Diamond^0 X = X$$
$$\Diamond^{n+1} X = \Diamond\Diamond^n X$$

And semantically:

$$\Gamma \mathcal{R}^0 \Delta \text{ if } \Gamma = \Delta$$
$$\Gamma \mathcal{R}^{n+1} \Delta \text{ if } \Gamma \mathcal{R} \Omega \text{ and } \Omega \mathcal{R}^n \Delta \text{ for some } \Omega$$

Informally $\Box^n X$ is $\Box\Box \cdots \Box X$ with n necessity symbols and similarly for $\Diamond^n$. Likewise $\Gamma \mathcal{R}^n \Delta$, means there is a chain $\Gamma \mathcal{R} \Omega_1 \mathcal{R} \Omega_2 \mathcal{R} \cdots \mathcal{R} \Omega_{n-1} \mathcal{R} \Delta$, with n accessibility instances. Of course $n = 0$ is a special case.

Definition 8.2 (Geach Modal Logics) We use notation from Chellas (1980). Formulas of the form $\Diamond^k \Box^l X \to \Box^m \Diamond^n X$ where $k, l, m, n \geq 0$ are $\mathsf{G}^{k,l,m,n}$ formulas. All such formulas are *Geach formulas. Geach modal logics* are those axiomatized over K using one or more of these Geach axiom schemes. Semantically the frame condition for $\mathsf{G}^{k,l,m,n}$ is a generalized convergence condition: if $\Gamma_1 \mathcal{R}^k \Gamma_2$ and $\Gamma_1 \mathcal{R}^m \Gamma_3$ then for some Γ_4, $\Gamma_2 \mathcal{R}^l \Gamma_4$ and $\Gamma_3 \mathcal{R}^n \Gamma_4$.

The generalized convergence condition on frames for $\mathsf{G}^{k,l,m,n}$ is illustrated in Figure 8.1. If $\Gamma_1 \mathcal{R}^k \Gamma_2$ and $\Gamma_1 \mathcal{R}^m \Gamma_3$ then for some Γ_4, $\Gamma_2 \mathcal{R}^l \Gamma_4$ and $\Gamma_3 \mathcal{R}^n \Gamma_4$. It is important to recall that the parameters k, l, m, and n are allowed to be 0, so some of the possible worlds mentioned may coincide.

Many standard modal logics are Geach logics.

D	=	$\mathsf{G}^{0,1,0,1}$	$\Box X \to \Diamond X$
T	=	$\mathsf{G}^{0,1,0,0}$	$\Box X \to X$
B	=	$\mathsf{G}^{0,0,1,1}$	$X \to \Box\Diamond X$
4	=	$\mathsf{G}^{0,1,2,0}$	$\Box X \to \Box\Box X$
5	=	$\mathsf{G}^{1,0,1,1}$	$\Diamond X \to \Box\Diamond X$
G	=	$\mathsf{G}^{1,1,1,1}$	$\Diamond\Box X \to \Box\Diamond X$

Axiomatic Geach logics are canonical with respect to the corresponding possible world semantics. A proof of this can be found in Chellas (1980). Essentially, what we do here is transfer the modal argument from Chellas (1980) to justification logics, paying careful attention to details.

Much of the background material that we need consists of rather technical results. We begin with this in Section 8.3. In Section 8.4 we show completeness for all justification counterparts of Geach logics with respect to their Fitting

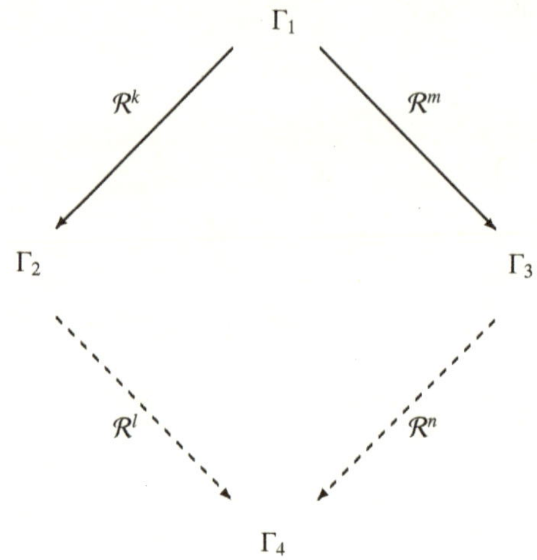

Figure 8.1 Frame Condition for $\mathsf{G}^{k,l,m,n}$

models. For the special case of JT4.2 the parameters k, l, m, and n can all be set to 1 in Section 8.4 and much of the technical material can be ignored. Thinking about this special case first might help in understanding what is going on in the general case.

8.3 Technical Results

When working with justification terms things are more complicated than in the modal case because each box of $\Box^n A$ will probably be realized by a different justification term. We introduce the following vector-style notation to help formulate this.

Definition 8.3 Let $\mathcal{T}$ be the set of justification terms of a justification logic. By $\mathcal{T}^h$ we mean $\{\langle t_1, \ldots, t_h\rangle \mid t_1, \ldots, t_h \in \mathcal{T}\}$. We allow h to be 0, where the only member of $\mathcal{T}^0$ is $\langle\,\rangle$, the 0-tuple. If $\vec{t} \in \mathcal{T}^h$, we define $\vec{t}{:}X$ as follows.

$$\langle\,\rangle{:}X = X$$
$$\langle t_{h+1}, t_h, \ldots, t_1\rangle{:}X = t_{h+1}{:}\langle t_h, \ldots, t_1\rangle{:}X$$

It follows that for $h > 0$, $\langle t_h, \ldots, t_1\rangle{:}X = t_h{:} \ldots t_1{:}X$. In Section 4.4.1 we used the notation $\Gamma^\sharp = \{X \mid \Box X \in \Gamma\}$ for a set of modal formulas Γ. In Section 4.4.2 we adapted this to justification logics, $\Gamma^\sharp = \{X \mid t{:}X \in \Gamma$ for some justification term t}. The following extends the justification version to vector notation, and also defines a dual version.

Definition 8.4 Let Γ be a set of justification formulas and $h \geq 0$.

(1) $\Gamma^{\sharp h} = \{X \mid \vec{t}{:}X \in \Gamma \text{ for some } \vec{t} \in \mathcal{T}^h\}$
(2) $\Gamma^{\flat h} = \{\neg\vec{t}{:}\neg X \mid X \in \Gamma \text{ and } \vec{t} \in \mathcal{T}^h\}$

Note that $\Gamma^{\sharp 0}$ is simply Γ. $\Gamma^{\sharp 1}$ coincides with the justification version of $\Gamma^{\sharp}$ as used earlier and will generally be written that way for simplicity.

For this section, assume JL is an axiomatically formulated justification logic with a constant specification (which is suppressed in the notation to make things more readable). We assume JL has the internalization property. Also we assume $\mathcal{T}$ is the set of justification terms of JL. All mention of consistency is with respect to JL.

Theorem 8.5 (Lifting Lemma Generalized) *Suppose $X_1, \ldots, X_n \vdash_{\mathsf{JL}} Y$. Then for any members $\vec{t_1}, \ldots, \vec{t_n}$ of $\mathcal{T}^h$ there is some $\vec{u}$ in $\mathcal{T}^h$ such that $\vec{t_1}{:}X_1, \ldots, \vec{t_n}{:}X_n \vdash_{\mathsf{JL}} \vec{u}{:}Y$.*

Proof This follows by repeated use of the Lifting Lemma, Corollary 2.16. □

Theorem 8.6 *Let Γ, Δ be maximal consistent sets of* JL. *Then*

$$\Gamma^{\sharp h} \subseteq \Delta \Longleftrightarrow \Delta^{\flat h} \subseteq \Gamma.$$

Proof We follow the modal presentation of Chellas (1980), Theorem 4.29.

Left to right. Assume $\Gamma^{\sharp h} \subseteq \Delta$. Suppose $\neg\vec{t}{:}\neg X \in \Delta^{\flat h}$. We show $\neg\vec{t}{:}\neg X \in \Gamma$. By our supposition, $X \in \Delta$. By consistency, $\neg X \notin \Delta$, hence $\neg X \notin \Gamma^{\sharp h}$. Then by definition, $\vec{t}{:}\neg X \notin \Gamma$, so by maximality, $\neg\vec{t}{:}\neg X \in \Gamma$.

Right to left. Assume $\Delta^{\flat h} \subseteq \Gamma$. Suppose $X \in \Gamma^{\sharp h}$, but also that $X \notin \Delta$. We derive a contradiction. By our supposition, $\vec{t}{:}X \in \Gamma$ for some $\vec{t} \in \mathcal{T}^h$. Because $X \vdash_{\mathsf{JL}} \neg\neg X$, by the Lifting Lemma Generalized there is some $\vec{u}$ so that $\vec{t}{:}X \vdash_{\mathsf{JL}} \vec{u}{:}\neg\neg X$. Then by maximal consistency of Γ, $\vec{u}{:}\neg\neg X \in \Gamma$. Also $X \notin \Delta$ so by maximality $\neg X \in \Delta$, so $\neg\vec{u}{:}\neg\neg X \in \Delta^{\flat h}$, and hence by our initial assumption, $\neg\vec{u}{:}\neg\neg X \in \Gamma$, contradicting consistency of Γ.

A remark. If $n = 0$ the theorem reduces to $\Gamma \subseteq \Delta$ iff $\Delta \subseteq \Gamma$, which is true because these are *maximal* consistent sets. □

Theorem 8.7 *Assume Γ is maximally consistent. Then $\Gamma^{\sharp h}$ is closed under consequence. In particular, $X_1, \ldots, X_k \in \Gamma^{\sharp h}$ if and only if $(X_1 \wedge \ldots \wedge X_k) \in \Gamma^{\sharp h}$.*

Proof It is enough to show closure under consequence. Suppose $X_1, \ldots, X_n \in \Gamma^{\sharp h}$, and $X_1, \ldots, X_n \vdash_{\mathsf{JL}} Y$. Then there are $\vec{t_1}, \ldots, \vec{t_n} \in \mathcal{T}^h$ so that $\vec{t_1}{:}X_1, \ldots \vec{t_n}{:}$

$X_n \in \Gamma$. By Theorem 8.5, for some $\vec{u} \in \mathcal{T}^h$, $\vec{t_1}{:}X_1, \ldots, \vec{t_n}{:}X_n \vdash_{\mathsf{JL}} \vec{u}{:}Y$. Because Γ is maximally consistent, it is closed under consequence, so $\vec{u}{:}Y \in \Gamma$, and hence $Y \in \Gamma^{\sharp h}$. □

Theorem 8.8 *For each $\vec{t_1}, \ldots, \vec{t_k} \in \mathcal{T}^h$ there is some $\vec{u} \in \mathcal{T}^h$ so that $\vdash (\vec{t_1}{:}X_1 \vee \ldots \vee \vec{t_k}{:}X_k) \rightarrow \vec{u}{:}(X_1 \vee \ldots \vee X_k)$.*

Proof To keep notation simple we assume $k = 2$. The proof is by induction on h.

Base case. If $h = 0$ the theorem says $\vdash (X_1 \vee X_2) \rightarrow (X_1 \vee X_2)$, which is trivial.

Induction step. Suppose the result is known for h, and we now have members $\langle c_{h+1}, c_h, \ldots, c_1\rangle$ and $\langle d_{h+1}, d_h, \ldots, d_1\rangle$ in $\mathcal{T}^{h+1}$. Let us write $\vec{c}$ for $\langle c_h, \ldots, c_1\rangle$ and $\vec{d}$ for $\langle d_h, \ldots, d_1\rangle$. By the induction hypothesis, there is $\vec{e} = \langle e_h, \ldots, e_1\rangle \in \mathcal{T}^h$ so that $\vdash_{\mathsf{JL}} (\vec{c}{:}X_1 \vee \vec{d}{:}X_2) \rightarrow \vec{e}{:}(X_1 \vee X_2)$. We then have $\vec{c}{:}X_1 \vdash_{\mathsf{JL}} \vec{e}{:}(X_1 \vee X_2)$, so using the Lifting Lemma there is some e^c_{h+1} so that $c_{h+1}{:}\vec{c}{:}X_1 \vdash_{\mathsf{JL}} e^c_{h+1}{:}\vec{e}{:}(X_1 \vee X_2)$. Similarly there is some e^d_{h+1} so that $d_{h+1}{:}\vec{d}{:}X_2 \vdash_{\mathsf{JL}} e^d_{h+1}{:}\vec{e}{:}(X_1 \vee X_2)$. Now, $e^c_{h+1}{:}\vec{e}{:}(X_1 \vee X_2) \rightarrow (e^c_{h+1} + e^d_{h+1}){:}\vec{e}{:}(X_1 \vee X_2)$ and $e^d_{h+1}{:}\vec{e}{:}(X_1 \vee X_2) \rightarrow (e^c_{h+1} + e^d_{h+1}){:}\vec{e}{:}(X_1 \vee X_2)$ hence, setting $e_{h+1} = e^c_{h+1} + e^d_{h+1}$, we have the following, establishing the $h + 1$ case.

$$\begin{aligned}
\langle c_{h+1}, c_h, \ldots, c_1\rangle{:}X_1 \vee \langle d_{h+1}, d_h, \ldots, d_1\rangle{:}X_2 &= (c_{h+1}{:}\vec{c}{:}X_1 \vee d_{h+1}{:}\vec{d}{:}X_2) \\
&\vdash_{\mathsf{JL}} (e^c_{h+1}{:}\vec{e}{:}(X_1 \vee X_2) \\
&\qquad \vee e^d_{h+1}{:}\vec{e}{:}(X_1 \vee X_2)) \\
&\vdash_{\mathsf{JL}} (e^c_{h+1} + e^d_{h+1}){:}\vec{e}{:}(X_1 \vee X_2) \\
&= e_{h+1}{:}\vec{e}{:}(X_1 \vee X_2) \\
&= \langle e_{h+1}, e_h, \ldots, e_1\rangle{:}(X_1 \vee X_2)
\end{aligned}$$

Now apply the Deduction Theorem. □

Theorem 8.9 *Assume that Γ, Δ are maximally consistent sets of justification formulas. Then*

$\Gamma^{\sharp(h+1)} \subseteq \Delta$ *iff for some maximally consistent set Ω, both $\Gamma^{\sharp} \subseteq \Omega$ and $\Omega^{\sharp h} \subseteq \Delta$.*

Proof We follow the modal version in Chellas (1980), where it is given as Theorem 4.31.

The argument from right to left is easy. Assume $\Gamma^{\sharp} \subseteq \Omega$ and $\Omega^{\sharp h} \subseteq \Delta$, where Ω is maximally consistent. Suppose $X \in \Gamma^{\sharp(h+1)}$; we show $X \in \Delta$. By the supposition, for some $\vec{t} \in \mathcal{T}^{h+1}$, $\vec{t}{:}X \in \Gamma$. Let us say $\vec{t} = \langle t_{h+1}, t_h, \ldots, t_1\rangle$. Then $\langle t_h, \ldots, t_1\rangle{:}X \in \Gamma^{\sharp}$, and so $\langle t_h, \ldots, t_1\rangle{:}X \in \Omega$. Then $X \in \Omega^{\sharp h}$, and hence $X \in \Delta$.

The left to right argument is harder. Assume $\Gamma^{\sharp(h+1)} \subseteq \Delta$. We need to show that $\Gamma^{\sharp} \cup \Delta^{\flat h}$ is consistent. Once this has been established, we can extend $\Gamma^{\sharp} \cup \Delta^{\flat h}$ to a maximal consistent set Ω. Of course $\Gamma^{\sharp} \subseteq \Omega$. But also $\Delta^{\flat h} \subseteq \Omega$ and hence $\Omega^{\sharp h} \subseteq \Delta$ by Theorem 8.6.

We now show by contradiction that $\Gamma^{\sharp} \cup \Delta^{\flat h}$ is consistent. Suppose $\Gamma^{\sharp} \cup \Delta^{\flat h} \vdash_{\mathsf{JL}} \bot$. Then there are $G_1, \ldots, G_k \in \Gamma^{\sharp}$ and $\neg\vec{t_1}{:}\neg D_1, \ldots, \neg\vec{t_m}{:}\neg D_m \in \Delta^{\flat h}$ so that $G_1, \ldots, G_k \vdash_{\mathsf{JL}} (\vec{t_1}{:}\neg D_1 \vee \ldots \vee \vec{t_m}{:}\neg D_m)$. Using Theorem 8.8, for some $\vec{u} \in \mathcal{T}^h$ we have $G_1, \ldots, G_k \vdash_{\mathsf{JL}} \vec{u}{:}(\neg D_1 \vee \ldots \vee \neg D_m)$. Because $(\neg D_1 \vee \ldots \vee \neg D_m) \vdash_{\mathsf{JL}} \neg(D_1 \wedge \ldots \wedge D_m)$, using the Lifting Lemma Generalized 8.5 we have that there is some $\vec{v} \in \mathcal{T}^h$ so that $\vec{u}{:}(\neg D_1 \vee \ldots \vee \neg D_m) \vdash_{\mathsf{JL}} \vec{v}{:}\neg(D_1 \wedge \ldots \wedge D_m)$, and hence we have $G_1, \ldots, G_k \vdash_{\mathsf{JL}} \vec{v}{:}\neg(D_1 \wedge \ldots \wedge D_m)$. Each G_i is in $\Gamma^{\sharp}$, and so $g_i{:}G_i \in \Gamma$ for some justification term g_i. Using the Lifting Lemma, for some justification term h, we have $g_1{:}G_1, \ldots, g_k{:}G_k \vdash_{\mathsf{JL}} h{:}\vec{v}{:}\neg(D_1 \wedge \ldots \wedge D_m)$. Because Γ is maximally consistent, it follows that $h{:}\vec{v}{:}\neg(D_1 \wedge \ldots \wedge D_m) \in \Gamma$. But then $\neg(D_1 \wedge \ldots \wedge D_m)$ is in $\Gamma^{\sharp(h+1)}$, and hence in Δ. But also $\neg\vec{t_1}{:}\neg D_1, \ldots, \neg\vec{t_m}{:}\neg D_m \in \Delta^{\flat h}$, so $D_1, \ldots, D_m \in \Delta$, and this implies the inconsistency of Δ. We have our contradiction. □

8.4 Geach Justification Logics Axiomatically

We have already looked at S4.2 and JT4.2 in detail, in Sections 2.7.4, 4.3.6, and 4.5.5. But for the general Geach logic, $\mathsf{G}^{k,l,m,n}$, things can be complicated. We give what amounts to a recipe for constructing justification analogs. We use $\mathsf{JG}^{k,l,m,n}$ to designate our justification analog of $\mathsf{G}^{k,l,m,n}$. There are infinitely many cases to cover now, which we want to do in the simplest way possible. Assume k, l, m, and n are fixed for the rest of this section.

For $\mathsf{JG}^{k,l,m,n}$ we enlarge the basic language of justification logic with function symbols besides the $\cdot$ and $+$ of J_0. Specifically we add function symbols $f_1, \ldots, f_k$ and $g_1, \ldots, g_m$, all distinct, each taking $l+n$ arguments. Justification terms are built up using these function symbols, as well as $\cdot$ and $+$.

In order to compactly formulate justification counterparts for all the various Geach logics it is convenient to make use of the special notation introduced in Definition 8.3. Recall that $\mathcal{T}^h$ consists of all length h vectors of justification terms. Now suppose $\vec{t} = \langle t_1, \ldots, t_l \rangle$ is a member of $\mathcal{T}^l$ and $\vec{u} = \langle u_1, \ldots, u_n \rangle$ is a member of $\mathcal{T}^n$. We write $f_i(\vec{t}, \vec{u})$ as short for $f_i(t_1, \ldots, t_l, u_1, \ldots, u_n)$. If $l = 0$, we must have $\vec{t} = \langle\,\rangle$, and we take $f_i(\vec{t}, \vec{u}) = f_i(\vec{u}) = f_i(u_1, \ldots, u_n)$. Similarly if $n = 0$. If both $l = 0$ and $n = 0$, $f_i(\vec{t}, \vec{u}) = f_i$, which can be understood as a 0 argument function, that is, a constant. Of course the same notational conventions apply to $g_i(\vec{t}, \vec{u})$.

It is convenient to make use of vector notation for function symbols too.

Suppose $\vec{f} = \langle f_1, \ldots, f_k \rangle$, $\vec{t} \in \mathcal{T}^l$ and $\vec{u} \in \mathcal{T}^n$. We write $\vec{f}(\vec{t}, \vec{u})$ as short for $\langle f_1(\vec{t}, \vec{u}), \ldots, f_k(\vec{t}, \vec{u}) \rangle$. If $k = 0$ we understand $\vec{f}(\vec{t}, \vec{u})$ to be $\langle\,\rangle$. Of course similarly for $\vec{g} = \langle g_1, \ldots, g_m \rangle$ and $\vec{g}(\vec{t}, \vec{u})$.

Finally Definition 8.3 already supplies a meaning for $\vec{f}(\vec{t}, \vec{u}){:}X$. In particular if $k = 0$ then $\vec{f}(\vec{t}, \vec{u}) = \langle\,\rangle$, in which case $\vec{f}(\vec{t}, \vec{u}){:}X$ is just X. Also if $\vec{f} = \langle f_1 \rangle$ then $\vec{f}(\vec{t}, \vec{u}){:}X$ is $f_1(\vec{t}, \vec{u}){:}X$. Similar comments apply to $\vec{g}(\vec{t}, \vec{u})$ too, of course.

Now we give an axiomatization for the justification counterpart of modal scheme $\mathsf{G}^{k,l,m,n}$, which we formulate as $\neg\Box^k\neg\Box^l X \rightarrow \Box^m\neg\Box^n\neg X$ because the possibility operator does not match well with justification logic machinery. Let $\vec{f} = \langle f_1, \ldots, f_k \rangle$ and $\vec{g} = \langle g_1, \ldots, g_m \rangle$. The axiom scheme we want is all formulas of the following form, for any formula X, and any $\vec{t} \in \mathcal{T}^l$ and $\vec{u} \in \mathcal{T}^n$. This should be compared with Section 2.7.4.

$$\neg\vec{f}(\vec{t}, \vec{u}){:}\neg\vec{t}{:}X \rightarrow \vec{g}(\vec{t}, \vec{u}){:}\neg\vec{u}{:}\neg X \tag{8.1}$$

Equivalently, and more symmetrically, we could also adopt the following.

$$\vec{f}(\vec{t}, \vec{u}){:}\neg\vec{t}{:}X \vee \vec{g}(\vec{t}, \vec{u}){:}\neg\vec{u}{:}\neg X \tag{8.2}$$

These formulas are to be added to J, with the resulting justification logic denoted $\mathsf{JG}^{k,l,m,n}$. According to our vector notation conventions, if $k = 0$ then (8.1) reduces to $\vec{t}{:}X \rightarrow \vec{g}(\vec{t}, \vec{u}){:}\neg\vec{u}{:}\neg X$; if $l = 0$ it reduces to $\neg\vec{f}(\vec{t}, \vec{u}){:}\neg X \rightarrow \vec{g}(\vec{t}, \vec{u}){:}\neg\vec{u}{:}\neg X$; and if both $k = 0$ and $l = 0$ it becomes $X \rightarrow \vec{g}(\vec{t}, \vec{u}){:}\neg\vec{u}{:}\neg X$. Similarly for m and n.

We give three examples, one of no special interest, the other two very familiar.

Example 1 $\mathsf{G}^{2,1,1,0}$, $\Diamond\Diamond\Box X \rightarrow \Box X$, or equivalently $\neg\Box\Box\neg\Box X \rightarrow \Box X$. For $\mathsf{JG}^{2,1,1,0}$, $\vec{f} = \langle f_1, f_2 \rangle$ and $\vec{g} = \langle g_1 \rangle$, where all function symbols are $1 + 0 = 1$ place. These are fixed. We have $\vec{t} = \langle t_1 \rangle$, where t_1 is arbitrary, and $\vec{u} = \langle\,\rangle$. Then (8.1) specializes to $\neg f_1(t_1){:}f_2(t_1){:}\neg t_1{:}X \rightarrow g_1(t_1){:}X$.

Example 2 $\mathsf{G}^{0,1,2,0}$, $\Box X \rightarrow \Box\Box X$, the 4 axiom scheme. Here $\vec{f} = \langle\,\rangle$ and $\vec{g} = \langle g_1, g_2 \rangle$. $\vec{t} = \langle t_1 \rangle$ while $\vec{u} = \langle\,\rangle$. (8.1) becomes $t_1{:}X \rightarrow g_1(t_1){:}g_2(t_1){:}X$. We saw this before (with different function symbols), in Section 4.6. That section also included a brief discussion of why there were differences between this counterpart of $\Box X \rightarrow \Box\Box X$ and the one chosen for LP.

Example 3 $\mathsf{G}^{0,0,1,1}$, $X \rightarrow \Box\Diamond X$, equivalently $X \rightarrow \Box\neg\Box\neg X$, known as B. For this, $\vec{f} = \langle\,\rangle$, $\vec{t} = \langle\,\rangle$, $\vec{g} = \langle g_1 \rangle$, and $\vec{u} = \langle u_1 \rangle$. Then (8.1) becomes $X \rightarrow g(u_1){:}\neg u_1{:}\neg X$ (after eliminating a double negation).

8.5 Geach Justification Logics Semantically

For justification versions of Geach logics we use Fitting models as defined in Section 4.2. Beyond the basics, we need special conditions on evidence functions and frames. The frame condition is the same as for modal $\mathsf{G}^{k,l,m,n}$. To state the evidence function conditions in a compact way we introduce yet another piece of notation.

Definition 8.10 (First and ButFirst) Suppose $\vec{t} = \langle t_1, \ldots, t_m \rangle \in \mathcal{T}^m$ where $m > 0$. Define $F(\vec{t}) = t_1$ and $B(\vec{t}) = \langle t_2, \ldots, t_m \rangle$. The operator names are meant to suggest *first* and *but-first*. The operators are undefined on $\langle\,\rangle$ so conditions for the Evidence Function fall into four cases, depending on whether or not $\vec{f}(\vec{t}, \vec{u}) = \langle\,\rangle$ or $\vec{g}(\vec{t}, \vec{u}) = \langle\,\rangle$.

Definition 8.11 (Evidence Function Condition for $\mathsf{JG}^{k,l,m,n}$) We require that $\mathcal{E}$ be a *strong* evidence function, and that the following must hold.

(1) If $k > 0$ and $m > 0$,
$\mathcal{E}(F(\vec{f}(\vec{t}, \vec{u})), B(\vec{f}(\vec{t}, \vec{u})){:}\neg\vec{t}{:}X) \cup \mathcal{E}(F(\vec{g}(\vec{t}, \vec{u})), B(\vec{g}(\vec{t}, \vec{u})){:}\neg\vec{u}{:}\neg X) = \mathcal{G}$.
(2) If $k = 0$ and $m > 0$, $\{\Gamma \mid \Gamma \Vdash \vec{t}{:}X\} \subseteq \mathcal{E}(F(\vec{g}(\vec{t}, \vec{u})), B(\vec{g}(\vec{t}, \vec{u})){:}\neg\vec{u}{:}\neg X)$.
(3) If $k > 0$ and $m = 0$, $\{\Gamma \mid \Gamma \Vdash \vec{u}{:}\neg X\} \subseteq \mathcal{E}(F(\vec{f}(\vec{t}, \vec{u})), B(\vec{f}(\vec{t}, \vec{u})){:}\neg\vec{t}{:}X)$.
(4) If $k = 0$ and $m = 0$, no conditions.

Definition 8.12 (Frame Condition for $\mathsf{JG}^{k,l,m,n}$) If $\Gamma_1 \mathcal{R}^k \Gamma_2$ and $\Gamma_1 \mathcal{R}^m \Gamma_3$ then for some Γ_4, $\Gamma_2 \mathcal{R}^l \Gamma_4$ and $\Gamma_3 \mathcal{R}^n \Gamma_4$.

We examine again the examples from Section 8.4, discussing Evidence Function Conditions. In all cases we require strong evidence functions.

Example 1 $\mathsf{G}^{2,1,1,0}$, $\neg\Box\Box\neg\Box X \to \Box X$. This falls into case (1), and the evidence function should satisfy $\mathcal{E}(f_1(t_1), f_2(t_1){:}\neg t_1{:}X) \cup \mathcal{E}(g_1(t_1), X) = \mathcal{G}$.

Example 2 $\mathsf{G}^{0,1,2,0}$, $\Box X \to \Box\Box X$. This is in case (2). The evidence function should satisfy the condition $\{\Gamma \mid \Gamma \Vdash t_1{:}X\} \subseteq \mathcal{E}(g_1(t_1), g_2(t_1){:}X)$. Note that because we have a strong evidence function, we could also express this as $\mathcal{E}(t_1, X) \subseteq \mathcal{E}(g_1(t_1), g_2(t_1){:}X)$.

Example 3 $\mathsf{G}^{0,0,1,1}$, $X \to \Box\neg\Box\neg X$. This also falls into case (2), and we have the evidence condition $\{\Gamma \mid \Gamma \Vdash X\} \subseteq \mathcal{E}(g(u_1), \neg u_1{:}\neg X)$. Unlike the previous example, this cannot be rewritten using only $\mathcal{E}$.

Case 4 for Evidence Function Conditions is vacuous. The familiar examples $\Box X \to X$ and $\Box X \to \Diamond X$ fall into this case.

8.6 Soundness, Completeness, and Realization

Axiomatic soundness is quite straightforward. If $k > 0$ and $m > 0$ soundness for $\mathsf{JG}^{k,l,m,n}$ with respect to Fitting models meeting case (1) of Definition 8.11 is just an extension of the argument for the soundness for JT4.2, which was presented in Section 4.3.6. We omit the details here. The other three cases of Definition 8.11 are simpler and are also omitted.

To show axiomatic completeness we must show that the canonical $\mathsf{JG}^{k,l,m,m}$ model is a model that meets the evidence conditions of Definition 8.11 and the frame conditions of Definition 8.12. Showing the evidence function conditions is straightforward. Showing the frame of the canonical $\mathsf{JG}^{k,l,m,n}$ model meets the convergence condition is harder, and we give the argument in full. *For what follows, internalization is assumed.* (In practice, this means assuming we have an axiomatically appropriate constant specification.) We remind the reader that the frame of the canonical model for $\mathsf{JG}^{k,l,m,n}$ is $\langle \mathcal{G}, \mathcal{R}\rangle$ where: $\mathcal{G}$ is the collection of maximally consistent sets in axiomatic $\mathsf{JG}^{k,l,m,n}$ and, for $\Gamma, \Delta \in \mathcal{G}$, $\Gamma\mathcal{R}\Delta$ if $\Gamma^{\sharp} \subseteq \Delta$, with $\Gamma^{\sharp}$ as in Definition 4.9. We also need an iterated version of $\sharp$.

$$\Gamma^{\sharp 0} = \Gamma$$
$$\Gamma^{\sharp(n+1)} = (\Gamma^{\sharp n})^{\sharp}$$

We begin with a simple looking fact that nontheless is of fundamental importance.

Lemma 8.13 *Let $\langle \mathcal{G}, \mathcal{R}\rangle$ be the frame of the canonical model for $\mathsf{JG}^{k,l,m,n}$, and let $\Gamma, \Delta \in \mathcal{G}$. For any integer $n \geq 0$, $\Gamma\mathcal{R}^{n}\Delta$ if and only if $\Gamma^{\sharp n} \subseteq \Delta$.*

Proof First, some easy observations. Suppose $\Gamma \subseteq \Delta$ and $X \in \Gamma^{\sharp}$. Then for some justification term t, $t{:}X \in \Gamma$, so $t{:}X \in \Delta$ and thus $X \in \Delta^{\sharp}$. That is, if $\Gamma \subseteq \Delta$ then $\Gamma^{\sharp} \subseteq \Delta^{\sharp}$; the $\sharp$ operation is monotonic. It follows that $\sharp n$ is also a monotonic operation.

The next easy observation is that for each $\Gamma \in \mathcal{G}$, $(\Gamma^{\sharp n})^{\sharp} = (\Gamma^{\sharp})^{\sharp n}$. This is an easy induction on n. If $n = 0$ we have $(\Gamma^{\sharp 0})^{\sharp} = \Gamma^{\sharp} = (\Gamma^{\sharp})^{\sharp 0}$. And if we assume we have this for n, we can reason as follows. $(\Gamma^{\sharp(n+1)})^{\sharp} = ((\Gamma^{\sharp n})^{\sharp})^{\sharp} = ((\Gamma^{\sharp})^{\sharp n})^{\sharp} = (\Gamma^{\sharp})^{\sharp(n+1)}$.

Now we show the implication from left to right, by induction on n. Recall Definition 8.1 here. For the $n = 0$ case, $\Gamma\mathcal{R}^{0}\Delta$ means $\Gamma = \Delta$, so trivially $\Gamma^{\sharp 0} = \Gamma = \Delta$.

For the induction case, assume we have the left–right implication for n and any members of $\mathcal{G}$. And suppose $\Gamma\mathcal{R}^{n+1}\Delta$. Then for some Ω, $\Gamma\mathcal{R}\Omega$ and $\Omega\mathcal{R}^{n}\Delta$.

By the induction hypothesis, $\Omega^{\sharp n} \subseteq \Delta$. Because we have the canonical model and $\Gamma\mathcal{R}\Omega$, $\Gamma^{\sharp} \subseteq \Omega$. Using monotonicity, $\Gamma^{\sharp(n+1)} = (\Gamma^{\sharp n})^{\sharp} = (\Gamma^{\sharp})^{\sharp n} \subseteq \Omega^{\sharp n} \subseteq \Delta$.

Next we show the implication from right to left, again by induction. First the $n = 0$ case. Suppose $\Gamma^{\sharp 0} \subseteq \Delta$. Because $\Gamma^{\sharp 0} = \Gamma$, we have $\Gamma \subseteq \Delta$. And because Γ and Δ are maximally consistent, it follows that $\Gamma = \Delta$, and hence $\Gamma\mathcal{R}^0\Delta$.

For the induction step, assume we have the result for n, and $\Gamma^{\sharp(n+1)} \subseteq \Delta$. By Theorem 8.9, for some Ω, $\Gamma^{\sharp} \subseteq \Omega$ and $\Omega^{\sharp n} \subseteq \Delta$. By the induction hypothesis, $\Omega\mathcal{R}^n\Delta$, and because we have a canonical model $\Gamma\mathcal{R}\Omega$, so $\Gamma^{n+1}\Delta$. □

We can now finish the completeness proof with respect to Fitting models.

Theorem 8.14 *Let $\langle \mathcal{G}, \mathcal{R}\rangle$ be the frame of the canonical model for* $\mathsf{JG}^{k,l,m,n}$ *and Γ_1, Γ_2, and Γ_3 be members of $\mathcal{G}$, that is, maximally consistent sets in* $\mathsf{JG}^{k,l,m,n}$. *If $\Gamma_1\mathcal{R}^k\Gamma_2$ and $\Gamma_1\mathcal{R}^m\Gamma_3$ then there is some $\Gamma_4 \in \mathcal{G}$, so that $\Gamma_2\mathcal{R}^l\Gamma_4$ and $\Gamma_3\mathcal{R}^n\Gamma_4$.*

Figure 8.2 shows a diagram that may help to keep things in mind.

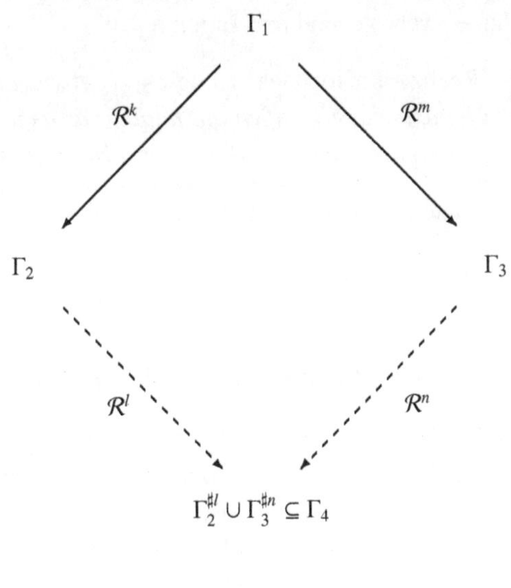

Figure 8.2 Illustration for Theorem 8.14

Proof The main thing is to show that $\Gamma_2^{\sharp l} \cup \Gamma_3^{\sharp n}$ is consistent. After that, things are short and simple. To do this, suppose $\Gamma_2^{\sharp l} \cup \Gamma_3^{\sharp n}$ is not consistent; we derive a contradiction.

To keep notation simple we write $\vdash$ for provability/derivability in justification logic $\mathsf{JG}^{k,l,m,n}$ using an axiomatically appropriate constant specification. If $\Gamma_2^{\sharp l} \cup \Gamma_3^{\sharp n}$ is not consistent, there are $X_1, \ldots, X_i \in \Gamma_2^{\sharp l}$ and $Y_1, \ldots, Y_j \in \Gamma_3^{\sharp n}$ so that $X_1, \ldots, X_i, Y_1, \ldots, Y_j \vdash \bot$. Again, to keep notation simple let $X = X_1 \wedge \ldots \wedge X_i$ and $Y = Y_1 \wedge \ldots \wedge Y_j$. Using Theorem 8.7, $X \in \Gamma_2^{\sharp l}$ and $Y \in \Gamma_3^{\sharp n}$. We have $X, Y \vdash \bot$, so $Y \vdash \neg X$.

Because $X \in \Gamma_2^{\sharp l}$, for some $\vec{t} \in \mathcal{T}^l$, $\vec{t}{:}X \in \Gamma_2$. Similarly, because $Y \in \Gamma_3^{\sharp n}$, $\vec{u}{:}Y \in \Gamma_3$ for some $\vec{u} \in \mathcal{T}^n$. And because $Y \vdash \neg X$, by the Lifting Lemma 2.16 iterated n times, for some $\vec{v} \in \mathcal{T}^n$, $\vec{u}{:}Y \vdash \vec{v}{:}\neg X$, and so $\vec{v}{:}\neg X \in \Gamma_3$. Because $\Gamma_1 \mathcal{R}^k \Gamma_2$ then $\Gamma_1^{\sharp k} \subseteq \Gamma_2$, so $\vec{f}(\vec{t}, \vec{v}){:}\neg\vec{t}{:}X \notin \Gamma_1$ because $\vec{t}{:}X \in \Gamma_2$ and Γ_2 is consistent. Then $\neg\vec{f}(\vec{t}, \vec{v}){:}\neg\vec{t}{:}X \in \Gamma_1$. Using the axiom $\vec{f}(\vec{t}, \vec{v}){:}\neg\vec{t}{:}X \vee \vec{g}(\vec{t}, \vec{v}){:}\neg\vec{v}{:}\neg X$ and maximal consistency of Γ_1, we have $\vec{g}(\vec{t}, \vec{v}){:}\neg\vec{v}{:}\neg X \in \Gamma_1$. Because $\Gamma_1 \mathcal{R}^m \Gamma_3$ then $\Gamma_1^{\sharp m} \subseteq \Gamma_3$, and so $\neg\vec{v}{:}\neg X \in \Gamma_3$. This contradicts the consistency of Γ_3.

We have now shown that that $\Gamma_2^{\sharp l} \cup \Gamma_3^{\sharp n}$ is consistent. Extend it to a maximally consistent set Γ_4, which will be a possible world in the canonical model. Because $\Gamma_2^{\sharp l} \subseteq \Gamma_4$ then $\Gamma_2 \mathcal{R}^l \Gamma_4$ by Lemma 8.13. Similarly $\Gamma_3 \mathcal{R}^n \Gamma_4$. Thus the canonical model meets the Frame Condition for $\mathsf{G}^{k,l,m,n}$. □

Finally, we have a very general realization result.

Theorem 8.15 (Realization for Geach Logics) *Let* KL *be a Geach modal logic, and* JL *be its axiomatic justification analog, as defined in Section 8.4.* KL *and* JL *are counterparts; in particular,* KL *realizes into* JL.

Proof An immediate consequence of the completeness result earlier and Theorem 7.24. □

8.7 A Concrete S4.2/JT4.2 Example

The modal logic S4.2 is the paradigm case of a Geach logic. It extends S4 with $\Diamond\Box X \to \Box\Diamond X$ and thus is axiomatized over K with Geach schemes $\mathsf{G}^{1,1,1,1}$, $\mathsf{G}^{0,1,0,0}$, and $\mathsf{G}^{0,1,2,0}$. Its corresponding justification logic is JT4.2. These were discussed in Sections 2.7.4, 4.3.6, and 4.5.5. As a special case of Theorem 8.15 we know that modal logic S4.2 and justification logic JT4.2 correspond. It is not known, as of this writing, whether this has a constructive proof. Nonetheless, we now give a concrete instance of an S4.2 theorem, and a corresponding realization. For this example we essentially translate an S4.2 axiomatic proof in its entirety. More will be said about this way of doing things in the next section.

It is convenient to assume we have not only an axiomatically appropriate constant specification, but also one that is *schematic*. This means that the same constant justifies all instances of an axiom schema. The proof of Theorem 2.14 actually shows that a justification term t exists that is variable free and justifies X, and if the constant specification is schematic, it will also justify the result of replacing justification variables in X with more complex justification terms. We assume this in what follows. We use $v_1, v_2 \ldots$ as justification variables.

Modal formula (8.3) is a theorem of S4.2; it is rewritten without $\Diamond$ in (8.4).

$$[\Diamond\Box A \wedge \Diamond\Box B] \rightarrow \Diamond\Box(A \wedge B) \tag{8.3}$$

$$[\neg\Box\neg\Box A \wedge \neg\Box\neg\Box B] \rightarrow \neg\Box\neg\Box(A \wedge B) \tag{8.4}$$

It is (8.4) that we will realize, but we generally maintain the use of $\Diamond$ operators, because it makes reading easier.

Here is a sketch of a proof for (8.3). First, $[\Diamond\Box X \wedge \Box\Diamond Y] \rightarrow \Diamond\Diamond(X \wedge Y)$ is a theorem of K, so as a special case we have, $[\Diamond\Box\Box A \wedge \Box\Diamond\Box B] \rightarrow \Diamond\Diamond(\Box A \wedge \Box B)$. Because $(\Box A \wedge \Box B) \leftrightarrow \Box(A \wedge B)$, we then have $[\Diamond\Box\Box A \wedge \Box\Diamond\Box B] \rightarrow \Diamond\Diamond\Box(A \wedge B)$ in K. In K4 we have $\Diamond\Diamond X \rightarrow \Diamond X$, and in particular, $\Diamond\Diamond\Box(A \wedge B) \rightarrow \Diamond\Box(A \wedge B)$. Consequently we have $[\Diamond\Box\Box A \wedge \Box\Diamond\Box B] \rightarrow \Diamond\Box(A \wedge B)$ in K4. Likewise we have $\Box A \rightarrow \Box\Box A$, and so $\Diamond\Box A \rightarrow \Diamond\Box\Box A$. Then $[\Diamond\Box A \wedge \Box\Diamond\Box B] \rightarrow \Diamond\Box(A \wedge B)$ is a theorem of K4. In S4.2 we have $\Diamond\Box X \rightarrow \Box\Diamond X$, so in particular, $\Diamond\Box\Box B \rightarrow \Box\Diamond\Box B$, and so $[\Diamond\Box A \wedge \Diamond\Box\Box B] \rightarrow \Diamond\Box(A \wedge B)$ is a theorem of S4.2. Finally, using $\Box B \rightarrow \Box\Box B$, we have $[\Diamond\Box A \wedge \Diamond\Box B] \rightarrow \Diamond\Box(A \wedge B)$. (We didn't use $\Box X \rightarrow X$. This is not important, but it is interesting.) Our proof in JT4.2 of a formula that realizes this will incorporate analogs of each part of the proof just sketched.

We begin with $[\Diamond\Box\Box A \wedge \Box\Diamond\Box B] \rightarrow \Diamond\Diamond\Box(A \wedge B)$, which is a theorem of K. We produce a sequence of theorems of J using an axiomatically appropriate, schematic constant specification. In this, j_1, j_2, and j_3 are justification terms supplied by Theorem 2.14.

$A \rightarrow (B \rightarrow (A \wedge B))$	tautology
$j_1{:}\{A \rightarrow (B \rightarrow (A \wedge B))\}$	Theorem 2.14
$\neg[j_1 \cdot v_1 \cdot v_3]{:}(A \wedge B) \rightarrow (v_1{:}A \rightarrow \neg v_3{:}B)$	Application Axioms and classical logic
$j_2{:}\{\neg[j_1 \cdot v_1 \cdot v_3]{:}(A \wedge B) \rightarrow (v_1{:}A \rightarrow \neg v_3{:}B)\}$	Theorem 2.14
$v_5{:}\neg[j_1 \cdot v_1 \cdot v_3]{:}(A \wedge B) \rightarrow \{\neg[j_2 \cdot v_5 \cdot v_2]{:}\neg v_3{:}B \rightarrow \neg[v_2 \cdot v_1]{:}A\}$	Application Axioms and classical logic
$j_3{:}\{v_5{:}\neg[j_1 \cdot v_1 \cdot v_3]{:}(A \wedge B) \rightarrow \{\neg[j_2 \cdot v_5 \cdot v_2]{:}\neg v_3{:}B \rightarrow \neg[v_2 \cdot v_1]{:}A\}\}$	Theorem 2.14

$\{\neg[j_3 \cdot v_6 \cdot v_4]{:}\neg v_2{:}v_1{:}A \wedge v_4{:}\neg[j_2 \cdot v_5 \cdot v_2]{:}\neg v_3{:}B\} \rightarrow$
$\neg v_6{:}v_5{:}\neg[j_1 \cdot v_1 \cdot v_3]{:}(A \wedge B)$ Application Axioms and classical logic

We now have that for the two formulas,

$$[\Diamond\Box\Box A \wedge \Box\Diamond\Box B] \rightarrow \Diamond\Diamond\Box(A \wedge B) \quad (8.5)$$

$$\{\neg[j_3 \cdot v_6 \cdot v_4]{:}\neg v_2{:}v_1{:}A \wedge v_4{:}\neg[j_2 \cdot v_5 \cdot v_2]{:}\neg v_3{:}B\} \rightarrow \neg v_6{:}v_5{:}\neg[j_1 \cdot v_1 \cdot v_3]{:}(A \wedge B) \quad (8.6)$$

justification formula (8.6) is a provable realization of modal formula (8.5).

Next we realize the K4 theorem $\Diamond\Diamond\Box(A \wedge B) \rightarrow \Diamond\Box(A \wedge B)$. We have the LP axiom $v_7{:}\neg v_8{:}(A \wedge B) \rightarrow {!v_7}{:}v_7{:}\neg v_8{:}(A \wedge B)$, so we immediately have the following realizer theorem.

$$\neg{!v_7}{:}v_7{:}\neg v_8{:}(A \wedge B) \rightarrow \neg v_7{:}\neg v_8{:}(A \wedge B) \quad (8.7)$$

To match the antecedent of (8.7) with the consequent of (8.6) we set $v_6 = {!v_7}$, $v_7 = v_5$, and $v_8 = j_1 \cdot v_1 \cdot v_3$. (This is actually a unification problem.) When this is done, and antecedent and consequent match, they can be "cut out", yielding (8.9), which realizes (8.8).

$$[\Diamond\Box\Box A \wedge \Box\Diamond\Box B] \rightarrow \Diamond\Box(A \wedge B) \quad (8.8)$$

$$\{\neg[j_3{\cdot}{!v_5} \cdot v_4]{:}\neg v_2{:}v_1{:}A \wedge v_4{:}\neg[j_2 \cdot v_5 \cdot v_2]{:}\neg v_3{:}B\} \rightarrow \neg v_5{:}\neg[j_1 \cdot v_1 \cdot v_3]{:}(A \wedge B) \quad (8.9)$$

The next step is to realize $\Diamond\Box A \rightarrow \Diamond\Box\Box A$. We have the LP axiom $v_9{:}A \rightarrow {!v_9}{:}v_9{:}A$. This gives the theorem $\neg{!v_9}{:}v_9{:}A \rightarrow \neg v_9{:}A$. Using Theorem 2.14, there is a justification term, call it j_4, provably justifying this. Then using distributivity, we get $\neg[j_4 \cdot v_{10}]{:}\neg v_9{:}A \rightarrow \neg v_{10}{:}\neg{!v_9}{:}v_9{:}A$. We match the consequent of this with the A part of the antecedent of (8.9) by setting $v_{10} = j_3{\cdot}{!v_5} \cdot v_4$, $v_2 = {!v_9}$, and $v_1 = v_9$. Then, again cutting out the common antecedent/consequent, we have (8.11), which realizes (8.10).

$$\{\Diamond\Box A \wedge \Box\Diamond\Box B] \rightarrow \Diamond\Box(A \wedge B) \quad (8.10)$$

$$\{\neg[j_4 \cdot j_3{\cdot}{!v_5} \cdot v_4]{:}\neg v_9{:}A \wedge v_4{:}\neg[j_2 \cdot v_5{\cdot}{!v_9}]{:}\neg v_3{:}B\} \rightarrow \neg v_5{:}\neg[j_1 \cdot v_9 \cdot v_3]{:}(A \wedge B) \quad (8.11)$$

Next we realize $\Diamond\Box\Box B \rightarrow \Box\Diamond\Box B$. This, at last, makes use of J4.2 axiom scheme (2.7.4). By it we have $\neg f(v_{11}, v_{13}){:}\neg v_{11}{:}v_{12}{:}B \rightarrow g(v_{11}, v_{13})\neg v_{13}{:}\neg v_{12}{:}B$.

We match the consequent of this with the B part of the antecedent of (8.11) by setting $v_4 = g(v_{11}, v_{13})$, $v_{13} = j_2 \cdot v_5 \cdot !v_9$, and $v_3 = v_{12}$. Then cutting out common parts, we get (8.13), which realizes (8.12).

$$[\Diamond\Box A \wedge \Diamond\Box\Box B] \rightarrow \Diamond\Box(A \wedge B) \tag{8.12}$$

$$\begin{aligned}\{\neg[j_4 \cdot j_3 \cdot !v_5 \cdot g(v_{11}, j_2 \cdot v_5 \cdot !v_9)]{:}\neg v_9{:}A \wedge \\ \neg f(v_{11}, j_2 \cdot v_5 \cdot !v_9){:}\neg v_{11}{:}v_3{:}B\} \rightarrow \neg v_5{:}\neg[j_1 \cdot v_9 \cdot v_3]{:}(A \wedge B)\end{aligned} \tag{8.13}$$

Finally we realize $\Diamond\Box B \rightarrow \Diamond\Box\Box B$. Using Theorem 2.14, let us say j_5 justifies the LP theorem $\neg !v_3{:}v_3{:}B \rightarrow \neg v_3{:}B$. Then using the Application Axiom, we can get $\neg[j_5 \cdot v_{14}]{:}\neg v_3{:}B \rightarrow \neg v_{14}{:}\neg[!v_3 \cdot v_3]{:}B$. We match consequent of this with the B part of the antecedent of (8.13) by setting $v_{14} = f(v_{11}, j_2 \cdot v_5 \cdot !v_9)$ and $v_{11} = !v_3$. Then cutting out the common portion, we finally arrive at the following, in which (8.15) realizes (8.14) in J4.2.

$$[\Diamond\Box A \wedge \Diamond\Box B] \rightarrow \Diamond\Box(A \wedge B) \tag{8.14}$$

$$\begin{aligned}\{\neg[j_4 \cdot j_3 \cdot !v_5 \cdot g(!v_3, j_2 \cdot v_5 \cdot !v_9)]{:}\neg v_9{:}A \wedge \\ \neg[j_5 \cdot f(!v_3, j_2 \cdot v_5 \cdot !v_9)]{:}\neg v_3{:}B\} \rightarrow \neg v_5{:}\neg[j_1 \cdot v_9 \cdot v_3]{:}(A \wedge B)\end{aligned} \tag{8.15}$$

8.8 Why Cut-Free Is Needed

Nonconstructive proofs of realization apparently have a broader scope than constructive versions. All known constructive realization proofs build on some cut-free proof system: sequents, tableaus, hypersequents, nested sequents, and so on. Why not axiom systems? After all, many more modal logics have axiomatic characterizations than are known to have one that is cut-free. One might think it would be relatively easy to extract a realization from an axiomatic proof, as we did in the previous section. Just work line by line. But, nobody knows how to do this in general. The problem is with modus ponens. In this section we discuss just what that problem is.

We work with the simplest normal modal logic K, and the corresponding justification logic J, but there is nothing special about these choices. Suppose X and Y are modal formulas, we have K proofs of X and of $X \rightarrow Y$, and we have constructed normal realizations for both, provable in J. Say these are X^{r_1} and $(X \rightarrow Y)^{r_2}$. In K modus ponens allows us to infer Y from X and $X \rightarrow Y$; our problem is to construct a provable normal realization for Y from X^{r_1} and $(X \rightarrow Y)^{r_2}$.

$(X \to Y)^{r_2}$ will have the form $X^{r_2} \to Y^{r_2}$, but there is no general reason why X^{r_1} and X^{r_2} should be the same, so a simple application of modus ponens in J is not available to us. However, subformula occurrences in X will have opposite polarities in X and in $X \to Y$. Then a subformula $\Box Z$ of X must occur negatively in one of X and $X \to Y$, and so its counterpart in one of X^{r_1} and $X^{r_2} \to Y^{r_2}$ must begin with a variable. The problem of making X^{r_1} and X^{r_2} match up thus gives rise to a set of equations of the form $v_i = t_i$, where v_i is a justification variable occurring in a negative position in one of X^{r_1} and X^{r_2} and t_i is the justification term at the corresponding positive position in the other. All we need to do is solve this system of equations, a unification problem, to get a substitution σ. Then we apply σ to both X^{r_1} and $X^{r_2} \to Y^{r_2}$ getting $X^{r_1}\sigma$ and $X^{r_2}\sigma \to Y^{r_2}\sigma$, and because $X^{r_1}\sigma = X^{r_2}\sigma$ we can apply modus ponens in J. Because variables in negative positions of normal realizers must be distinct, it can be arranged that $Y^{r_2}\sigma$ will not have its negative position variables affected by σ, so it will meet the conditions for being a normal realizer, and it will be provable in J.

The problem is, how do we know the system of equations is solvable? That is, how do we know unification is possible. Here is a very simple example that illustrates our problem. Suppose that in K we have proved $\Box P \to \Box P$ and $(\Box P \to \Box P) \to Z$. And let us suppose we have constructed provable normal realizations for both. It is certainly possible that the realizer for $(\Box P \to \Box P) \to Z$ has the form $(v_1{:}P \to v_1{:}P) \to Z^r$. In fact, this has a provable antecedent, so modus ponens is available to us, at least in principle. But in J, because $P \to P$ is provable, by Internalization there will be some justification term t so that $t{:}(P \to P)$ is provable. But then $v_2{:}P \to [t \cdot v_2]{:}P$ is provable, and it too is a normal realization of $\Box P \to \Box P$. If our hypothetical axiomatic realization algorithm happens to have chosen this realizer, we will need to match up $v_2{:}P \to [t \cdot v_2]{:}P$ and $v_1{:}P \to v_1{:}P$, which means solving the unification problem

$$\begin{aligned} v_2 &= v_1 \\ v_1 &= t \cdot v_2, \end{aligned}$$

which leads us to $v_2 = t \cdot v_2$, in violation of the "occurs check" condition for unification. This is not a solvable unification problem.

If we make the "wrong" choice for a realizer, it is possible that we block modus ponens applications. How do we know this can be avoided, that a "right" choice is always possible? In Wang (2011b, 2014) *noncircular* axiomatic proofs are introduced. Roughly, these are the ones for which a right choice is guaranteed. But no general methodology is known for constructively showing noncircular proofs for modal logics always exist or for deciding which logics always have such proofs. In fact, this is something that is guaranteed by the existence of a cut-free proof system for a modal logic, and we are back to where we started.

9
Arithmetical Completeness and BHK Semantics

In this section we study arithmetical provability semantics, which, from the beginning, was the intended Gödel semantics for S4 and for the Logic of Proofs. Because the latter is the only justification logic discussed in this chapter, we will be using terminology associated with the Logic of Proofs. In particular, justification terms will generally be called *proof terms* or *proof polynomials*.

9.1 Arithmetical Semantics of the Logic of Proofs

We will work in Peano Arithmetic PA. By Π_1, Σ_1, and Δ_1 we mean the corresponding classes of arithmetical predicates. We will use $x, y, z \ldots$ to denote individual variables in arithmetic and leave it to the reader to distinguish them from proof variables.

We assume here that PA contains terms for all primitive recursive functions (Smoryński, 1985; Takeuti, 1975). These are called primitive recursive terms. Formulas $f(\vec{x}) = 0$ where $f(\vec{x})$ is a primitive recursive term are standard primitive recursive formulas. A standard Σ_1-formula is a formula $\exists x \varphi(x, \vec{y})$ where $\varphi(x, \vec{y})$ is a standard primitive recursive formula. An arithmetical formula φ is provably Σ_1 if φ is provably equivalent in PA to a standard Σ_1-formula; φ is provably Δ_1 iff both φ and $\neg\varphi$ are provably Σ_1. For each natural number n there is a corresponding numeral, $\overline{n}$. To simplify notation we will omit the overbar, writing n for both. Also we will often omit the Gödel number corner brackets when safe and write φ instead of $\ulcorner\varphi\urcorner$.

Definition 9.1 A *proof predicate* is a provably Δ_1-formula $\mathsf{Prf}(x, y)$ such that for every arithmetical sentence φ,

$$\mathsf{PA} \vdash \varphi \quad \textit{iff} \quad \textit{for some } n \in \omega,\ \mathsf{Prf}(n, \ulcorner\varphi\urcorner)\ \textit{holds}.$$

$\mathsf{Prf}(x, y)$ is *normal* if it satisfies the following two conditions:

(1) (finiteness of proofs) For any k, the set $T(k) = \{l \mid \mathsf{Prf}(k, l)\}$ is finite, and the function from k to the code of $T(k)$ is computable.
(2) (conjoinability of proofs) For any k and l, there is some n such that

$$T(k) \cup T(l) \subseteq T(n).$$

The conjoinability property yields that normal proof predicates are multi-conclusion ones.

Example 9.2 The natural arithmetical proof predicate $\mathsf{Proof}(x, y)$

"*x is the code of a derivation containing a formula with the code y*"

is the standard example of a normal proof predicate.

Note, that every normal proof predicate can be straightforwardly transformed into a single-conclusion one by changing from

"p proves $F_1, \ldots, F_n$" to "(p, i) proves F_i, $i = 1, \ldots, n$."

Moreover, every single-conclusion proof predicate may be regarded as normal multiconclusion, e.g., by reading

"p proves $F_1 \wedge \ldots \wedge F_n$" as "$p$ proves each of F_i, $i = 1, \ldots, n$."

Proposition 9.3 *For every normal proof predicate* Prf *there are computable functions* $\mathbf{m}(x, y)$, $\mathbf{a}(x, y)$, *and* $\mathbf{c}(x)$ *such that for all arithmetical formulas* φ, ψ *and all natural numbers* k, n *the following formulas are valid:*

- $\mathsf{Prf}(k, \varphi \to \psi) \wedge \mathsf{Prf}(n, \varphi) \to \mathsf{Prf}(\mathbf{m}(k, n), \psi)$
- $\mathsf{Prf}(k, \varphi) \to \mathsf{Prf}(\mathbf{a}(k, n), \varphi)$ *and* $\mathsf{Prf}(n, \varphi) \to \mathsf{Prf}(\mathbf{a}(k, n), \varphi)$
- $\mathsf{Prf}(k, \varphi) \to \mathsf{Prf}(\mathbf{c}(k), \mathsf{Prf}(k, \varphi))$.

Proof The following function can be taken as $\mathbf{m}$: $\mathbf{m}(k, n) = \mu z.$"$\mathsf{Prf}(z, \psi)$ *holds for all* ψ *such that there are* $\ulcorner \varphi \to \psi \urcorner \in T(k)$ *and* $\ulcorner \varphi \urcorner \in T(n)$."

For $\mathbf{a}$ one can take $\mathbf{a}(k, n) = \mu z. T(k) \cup T(n) \subseteq T(z)$. Such a z exists by conjoinability.

Finally, $\mathbf{c}$ may be given by: $\mathbf{c}(k) = \mu z.$"$\mathsf{Prf}(z, \mathsf{Prf}(k, \varphi))$ for all $\ulcorner \varphi \urcorner \in T(k)$." Such a z always exists. Indeed, $\mathsf{Prf}(k, \varphi)$ is a true Δ_1-sentence for every $\ulcorner \varphi \urcorner \in T(k)$, therefore each such $\mathsf{Prf}(k, \varphi)$ is provable in PA. Use conjoinability to find a uniform proof of all of them. □

Definition 9.4 An *arithmetical interpretation* $*$ of the LP-language has the following parameters:

- a normal proof predicate Prf with the functions $\mathbf{m}(x, y)$, $\mathbf{a}(x, y)$ and $\mathbf{c}(x)$ as in Proposition 9.3,

- an evaluation of propositional letters by sentences of arithmetic,
- an evaluation of proof variables and constants by natural numbers.

We assume that $*$ commutes with Boolean connectives, and

$$(t \cdot s)^* = \mathbf{m}(t^*, s^*), \quad (t + s)^* = \mathbf{a}(t^*, s^*),$$

$$(!t)^* = \mathbf{c}(t^*), \quad (t{:}F)^* = \mathsf{Prf}(t^*, \ulcorner F^* \urcorner).$$

Under an interpretation $*$ a proof polynomial t becomes the natural number t^*, an LP-formula F becomes the arithmetical sentence F^*. A formula $(t{:}F)^*$ is always provably Δ_1. For a set X of LP-formulas, by X^* we mean $\{F^* \mid F \in X\}$.

Given a constant specification CS, an arithmetical interpretation $*$ is a CS-*interpretation* if all formulas from CS^* are true (equivalently, are provable in PA). An LP-formula F is *valid* (with respect to the arithmetical semantics), if F^* is true under all interpretations $*$. F is *provably valid* if $\mathsf{PA} \vdash F^*$ for any interpretation $*$. F is *valid under constant specification* CS, if F^* is true under all CS-interpretations $*$. F is *provably valid under constant specification* CS if $\mathsf{PA} \vdash F^*$ for any CS-interpretation $*$. It is obvious that "provably valid" yields "valid."

Proposition 9.5 (Arithmetical soundness of LP_0) *If* $\mathsf{LP}_0 \vdash F$ *then* F *is provably valid (hence, valid).*

Proof A straightforward induction on the derivation in LP_0. Let us check the reflexivity axiom $t{:}F \to F$. Under an interpretation $*$,

$$(t{:}F \to F)^* \;=\; \mathsf{Prf}(t^*, F^*) \to F^*.$$

Consider two possibilities. Either $\mathsf{Prf}(t^*, F^*)$ is true, in which case t^* is indeed a proof of F^*, thus, $\mathsf{PA} \vdash F^*$ and $\mathsf{PA} \vdash \mathsf{Prf}(t^*, F^*) \to F^*$, i.e., $\mathsf{PA} \vdash (t{:}F \to F)^*$. Otherwise $\mathsf{Prf}(t^*, F^*)$ is false, in which case it is refutable in PA, as it is a Δ_1-formula. Hence, $\mathsf{PA} \vdash \neg\mathsf{Prf}(t^*, F^*)$ and again $\mathsf{PA} \vdash (t{:}F \to F)^*$. □

Corollary 9.6 (Arithmetical soundness of LP(CS)) *If* $\mathsf{LP(CS)} \vdash F$, *then* F *is provably valid under the constant specification* CS.

The provability semantics for LP, given earlier, may be characterized as a call-by-value semantics because the evaluation F^* of a given LP-formula F depends upon the value of participating functions. A different call-by-name provability semantics for LP was introduced in Artemov (1995) and then used in Krupski (1997) and Sidon (1997). In the latter semantics, F^* depends upon the particular programs for the functions participating in $*$.

9.2 A Constructive Canonical Model for the Logic of Proofs

Following Artemov (2001), we proceed with establishing arithmetical completeness of the logic of proofs by building a decidable version of the canonical model for LP and then embedding this model into PA.

Though the logic of proofs has a basic model, cf. Chapter 3, we need more information about the structure of this model for its arithmetization. For historical reasons, we present the original construction of a *decidable canonical model* for the logic of proofs from Artemov (1995, 2001) because this was the first nonarithmetical model of the logic of proofs.

Sequents were discussed in Chapter 5, though primarily with respect to modal logics. We now need a corresponding version for the justification logic LP itself. By a sequent we mean a pair $\Gamma \Rightarrow \Delta$ where Γ and Δ are finite sets of LP-formulas. By Γ, F we mean $\Gamma \cup \{F\}$. Without loss of generality we assume a Boolean basis $\rightarrow, \bot$ and treat the remaining Boolean connectives as defined.

Axioms of LP_0^G are sequents of the form $A \Rightarrow A$ and $\bot \Rightarrow$. Along with the usual Gentzen sequent rules of classical propositional logic, cf. Chapter 5, including the cut and weakening rules, the system LP_0^G contains the rules

$$\frac{A, \Gamma \Rightarrow \Delta}{t{:}A, \Gamma \Rightarrow \Delta}\,(:\Rightarrow) \qquad\qquad \frac{\Gamma \Rightarrow \Delta, t{:}A}{\Gamma \Rightarrow \Delta, !t{:}t{:}A}\,(\Rightarrow !)$$

$$\frac{\Gamma \Rightarrow \Delta, t{:}A}{\Gamma \Rightarrow \Delta, [t+s]{:}A}\,(+\Rightarrow) \qquad\qquad \frac{\Gamma \Rightarrow \Delta, t{:}A}{\Gamma \Rightarrow \Delta, [s+t]{:}A}\,(\Rightarrow +)$$

$$\frac{\Gamma \Rightarrow \Delta, s{:}(A \rightarrow B) \qquad \Gamma \Rightarrow \Delta, t{:}A}{\Gamma \Rightarrow \Delta, [s \cdot t]{:}B}\,(\Rightarrow \cdot)$$

LP_0^{G-} is the corresponding systems without the rule Cut.

By a straightforward induction in both directions it is easy to establish the following.

Proposition 9.7 $\mathsf{LP}_0^G \vdash \Gamma \Rightarrow \Delta$ *iff* $\mathsf{LP}_0 \vdash \bigwedge \Gamma \rightarrow \bigvee \Delta$.

Corollary 9.8 $\mathsf{LP}(\mathsf{CS}) \vdash F$ *iff* $\mathsf{LP}_0 \vdash \bigwedge \mathsf{CS} \rightarrow F$.

Definition 9.9 The sequent $\Gamma \Rightarrow \Delta$ is *saturated* if

(1) $A \rightarrow B \in \Gamma$ implies $B \in \Gamma$ or $A \in \Delta$,
(2) $A \rightarrow B \in \Delta$ implies $A \in \Gamma$ and $B \in \Delta$,
(3) $t{:}A \in \Gamma$ implies $A \in \Gamma$,

(4) $!t{:}t{:}A \in \Delta$ implies $t{:}A \in \Delta$,
(5) $[s+t]{:}A \in \Delta$ implies $s{:}A \in \Delta$ and $t{:}A \in \Delta$,
(6) $[s \cdot t]{:}B \in \Delta$ implies for each $X \to B$ occurring as a subformula in Γ, Δ either $s{:}(X \to B) \in \Delta$ or $t{:}X \in \Delta$.

Lemma 9.10 (Saturation lemma) *Suppose* $\mathsf{LP}_0^{G-} \nvdash \Gamma \Rightarrow \Delta$. *Then there exists a saturated sequent* $\Gamma' \Rightarrow \Delta'$ *such that*

(1) $\Gamma \subseteq \Gamma', \Delta \subseteq \Delta'$,
(2) $\mathsf{LP}_0^{G-} \nvdash \Gamma' \Rightarrow \Delta'$.

Proof A saturated sequent is obtained by the following Saturation Algorithm $\mathcal{A}$. Given $\Gamma \Rightarrow \Delta$, for each undischarged formula S, that is, formulas that are available, from $\Gamma \cup \Delta$ nondeterministically perform one of the following steps.[1] At time 0 all formulas from $\Gamma \cup \Delta$ are available (undischarged). After a step is performed, discharge S (make it unavailable). If none of the clauses (i)–(vii) is applicable, terminate.

(i) if $S = (A \to B) \in \Gamma$, then put A into Δ or B into Γ,
(ii) if $S = (A \to B) \in \Delta$, then put A into Γ and B into Δ,
(iii) if $S = t{:}A \in \Gamma$, then put A into Γ,
(iv) if $S = !t{:}t{:}A \in \Delta$, then put $t{:}A$ into Δ,
(v) if $S = [s+t]{:}A \in \Delta$, then put both $s{:}A$ and $t{:}A$ into Δ,
(vi) if $S = [s \cdot t]{:}B \in \Delta$, then for each $X_1, \dots, X_n$ such that $X_i \to B$ is a subformula in $\Gamma \cup \Delta$ put either $s{:}(X_i \to B)$ or $t{:}X_i$ into Δ,
(vii) If none of (i)–(vi) is applicable, backtrack.

The Saturation Algorithm $\mathcal{A}$ always terminates. Indeed, the computation tree in finitely branching and each step decomposes either a subformula of $\Gamma \Rightarrow \Delta$ or a formula of the type $t{:}F$, where both t and F occur in $\Gamma \Rightarrow \Delta$. There are only finitely many of those formulas, which guarantees termination.

By the description of $\mathcal{A}$, each leaf node of the computation tree $\mathcal{T}$ of $\mathcal{A}$ is a saturated sequent: we claim that at least one of them, $\Gamma' \Rightarrow \Delta'$, is not derivable in LP_0^{G-}. Indeed, otherwise, by the following standard induction on the depth of a node in $\mathcal{T}$ one can prove that every sequent in $\mathcal{T}$ is derivable in LP_0^{G-}, which contradicts the assumption that $\mathsf{LP}_0^{G-} \nvdash \Gamma \Rightarrow \Delta$.

The nodes corresponding to the steps (i)–(v) are trivial.

Let us consider a node that corresponds to (vi). Such a node is labeled by a sequent $\Pi \Rightarrow \Theta, [s \cdot t]{:}B$, and its children are 2^n sequents of the form

$$\Pi \Rightarrow \Theta, [s \cdot t]{:}B, Y_1^\sigma, \dots, Y_n^\sigma$$

[1] Basically, this means that $\mathcal{A}$ tries all possible paths.

where $\sigma = (\sigma_1, \ldots, \sigma_n)$ is an n-tuple of 0's and 1's, and

$$Y_i^\sigma = \begin{cases} s{:}(X_i \to B) & \text{if } \sigma_i = 0 \\ t{:}X_i & \text{if } \sigma_i = 1. \end{cases}$$

Here $X_1, \ldots, X_n$ is the list of all formulas such that $X_i \to B$ is a subformula in $\Gamma \Rightarrow \Delta$. By the induction hypothesis all the child sequents are derivable in LP_0^{G-}. In particular, among them there are 2^{n-1} pairs of sequents of the form $\Pi \Rightarrow \Theta', s{:}(X_1 \to B)$ and $\Pi \Rightarrow \Theta', t{:}X_1$. To every such pair apply the rule $(\Rightarrow \cdot)$ to obtain $\Pi \Rightarrow \Theta'$ (we assume that $[s \cdot t]{:}B \in \Theta'$). The resulting 2^{n-1} sequents are of the form $\Pi \Rightarrow \Theta, [s \cdot t]{:}B, Y_2^\sigma, \ldots, Y_n^\sigma$. After we repeat this procedure $n - 1$ more times we end up with the sequent $\Pi \Rightarrow \Theta, [s \cdot t]{:}B$, which is thus derivable in LP_0^{G-}.

This $\Gamma' \Rightarrow \Delta'$ satisfies the conditions of Lemma 9.10. Note that $\Gamma' \cap \Delta' = \emptyset$, and Γ' is closed under the rules $t{:}X/X$ and $X \to Y, X/Y$. □

Saturated sequents get us halfway to the desired constructive version of the canonical model for LP_0. What we need is also a closure of the "true set" under operations and LP-principles (which makes such "true set" infinite and forces us to abandon the convenient framework of sequents).

Lemma 9.11 (Completion Lemma) *For each saturated sequent $\Gamma \Rightarrow \Delta$ not derivable in LP_0^{G-} there is a set of LP-formulas $\widetilde{\Gamma}$ (a completion of $\Gamma \Rightarrow \Delta$) such that*

(1) *$\widetilde{\Gamma}$ is provably decidable, for each t the set $I(t) = \{X \mid t{:}X \in \widetilde{\Gamma}\}$ is finite and a function from a code of t to a code of $I(t)$ is provably computable,*
(2) *$\Gamma \subseteq \widetilde{\Gamma}$, $\Delta \cap \widetilde{\Gamma} = \emptyset$,*
(3) *if $t{:}A \in \widetilde{\Gamma}$, then $A \in \widetilde{\Gamma}$,*
(4) *if $s{:}(X \to Y) \in \widetilde{\Gamma}$ and $t{:}X \in \widetilde{\Gamma}$, then $[s \cdot t]{:}Y \in \widetilde{\Gamma}$,*
(5) *if $t{:}X \in \widetilde{\Gamma}$, then $!t{:}t{:}X \in \widetilde{\Gamma}$,*
(6) *if $t{:}X \in \widetilde{\Gamma}$, then $[t + s]{:}X \in \widetilde{\Gamma}$ and $[s + t]{:}X \in \widetilde{\Gamma}$.*

Proof Let $\Gamma \Rightarrow \Delta$ be a saturated sequent not derivable in LP_0^G. Within this proof we restrict ourselves to proof polynomials built from the finite set of constants and variables occurring in $\Gamma \Rightarrow \Delta$. Therefore, there is only a finite number of proof polynomials of any given length.

We describe a completion algorithm $\mathcal{COM}$ that produces an increasing sequence of finite sets of LP-formulas $\Gamma_0, \Gamma_1, \ldots$. Let $\Gamma_0 = \Gamma$.

For each $i \geq 0$ let $\mathcal{COM}$ perform the following:

- if $i = 3k$, then $\mathcal{COM}$ sets

$$\Gamma_{i+1} = \Gamma_i \cup \{[s \cdot t]{:}Y \mid s{:}(X \to Y), t{:}X \in \Gamma_i\},$$

- if $i = 3k + 1$, then $\mathcal{COM}$ sets

$$\Gamma_{i+1} = \Gamma_i \cup \{!t{:}t{:}X \mid t{:}X \in \Gamma_i\},$$

- if $i = 3k + 2$, then $\mathcal{COM}$ sets

$$\Gamma_{i+1} = \Gamma_i \cup \{[s + t]{:}X, [t + s]{:}X \mid t{:}X \in \Gamma_i, |s| < i\}.$$

Let

$$\widetilde{\Gamma} = \bigcup_i \Gamma_i.$$

It is easy to see that at step $i > 0$, new formulas in Γ_i are of the form $t{:}X$ with the length of t greater than $i/3$ (and $X = s{:}Y$ for some s, Y, or else $X \in \Gamma$). This observation secures the decidability of Γ. Indeed, given a formula F of length n wait until step $i = 3n$ of $\mathcal{COM}$; $F \in \Gamma_n$ if and only if $F \in \widetilde{\Gamma}$. A similar argument establishes the finiteness of $I(t)$ from which one can construct the desired provable computable arithmetical term for $I(t)$. We thus have *(1)*.

In order to establish *(2)* and *(3)* we prove by induction on i that for all $i = 0, 1, 2, \ldots$,

A. $\Gamma_i \cap \Delta = \emptyset$,

B. $t{:}X \in \Gamma_i$ implies $X \in \Gamma_i$,

C. $X \to Y, X \in \Gamma_i$ implies $Y \in \Gamma_i$.

The base case $i = 0$ holds because of the saturation properties of $\Gamma_0 = \Gamma$. For the induction step assume as induction hypothesis that the properties A, B, and C hold for i and consider Γ_{i+1}.

A. Suppose there is $F \in \Gamma_{i+1} \cap \Delta$ but $F \notin \Gamma_i$. There are three possibilities.

If $i = 3k$, then F is $[s \cdot t]{:}Y$ such that $s{:}(X \to Y), t{:}X \in \Gamma_i$ for some X. From the description of $\mathcal{COM}$ it follows that $(X \to Y) \in \Gamma$. By the saturation properties of $\Gamma \Rightarrow \Delta$, because $[s \cdot t]{:}Y \in \Delta$ and $X \to Y$ occurs in $\Gamma \cup \Delta$ either $s{:}(X \to Y) \in \Delta$ or $t{:}X \in \Delta$. In either case $\Gamma_i \cap \Delta \neq \emptyset$, which is impossible by the induction hypothesis.

If $i = 3k + 1$ then F is $!t{:}t{:}X$ such that $t{:}X \in \Gamma_i$. By assumption, $!t{:}t{:}X \in \Delta$ and, by the saturation properties of $\Gamma \Rightarrow \Delta$, $t{:}X \in \Delta$. Again $\Gamma_i \cap \Delta \neq \emptyset$, which is impossible by the induction hypothesis.

If $i = 3k + 2$ then F is $[t + s]{:}X$ such that either $t{:}X \in \Gamma_i$ or $s{:}X \in \Gamma_i$. By the saturation properties of $\Gamma \Rightarrow \Delta$, from $[t + s]{:}X \in \Delta$ conclude that both $t{:}X \in \Delta$ and $s{:}X \in \Delta$. Once again, $\Gamma_i \cap \Delta \neq \emptyset$, which is impossible by the induction hypothesis.

Thus $\Gamma_{i+1} \cap \Delta = \emptyset$.

B. Suppose $p{:}B \in \Gamma_{i+1}$ and $p{:}B \notin \Gamma_i$. We conclude that in this case $B \in \Gamma_{i+1}$. Again there are three possibilities.

If $i = 3k$ then $p{:}B$ is $[s \cdot t]{:}Y$ such that $s{:}(X \to Y)$, $t{:}X \in \Gamma_i$ for some X. By the induction hypothesis for Γ_i, $(X \to Y), X \in \Gamma_i$ and thus $Y \in \Gamma_i$. By the inclusion $\Gamma_i \subseteq \Gamma_{i+1}$, $Y \in \Gamma_{i+1}$.

If $i = 3k + 1$ then $p{:}B$ is $!t{:}t{:}X$ such that $t{:}X \in \Gamma_i$. Then $t{:}X \in \Gamma_{i+1}$.

If $i = 3k + 2$ then $p{:}B$ is $[t + s]{:}B$ such that either $t{:}B \in \Gamma_i$ or $s{:}B \in \Gamma_i$. By the induction hypothesis, in either case $B \in \Gamma_i$, therefore $B \in \Gamma_{i+1}$.

C. Suppose $X \to Y, X \in \Gamma_{i+1}$. From the description of $\mathcal{COM}$ it follows that $(X \to Y) \in \Gamma$. By the saturation properties of $\Gamma \Rightarrow \Delta$, either $Y \in \Gamma$ or $X \in \Delta$. In the former case we are done. If $X \in \Delta$, then $\Gamma_{i+1} \cap \Delta \neq \emptyset$, which is impossible by item A of the induction step.

Items (4), (5), and (6) of Lemma 9.11 are guaranteed by the definition of $\mathcal{COM}$. Indeed, if some *if* condition is fulfilled, then it occurs at some step i and $\mathcal{COM}$ necessarily puts the *then* formula into Γ_{i+3} at the latest. □

9.3 Arithmetical Completeness of the Logic of Proofs

According to Corollary 9.6, the logic of proofs is sound with respect to the intended arithmetical interpretation. In particular, if an interpretation $*$ makes a constant specification CS true, then all theorems of $\mathsf{LP(CS)}$ are also true (provable in PA) under interpretation $*$.

This section is devoted to establishing the *arithmetical completeness* of the logic of proofs with finite constant specifications. A fundamental corollary of the completeness theorem states that any finite constant specification can be made true/provable in PA under an appropriate arithmetical interpretation. This applies also to constant specifications containing formulas of the type $c{:}A(c)$. Such self-referential specifications cannot be true under arithmetical interpretations based on the standard provability predicate "from the textbook" because the Gödel number of a proof c^* should be greater that the Gödel number of a formula A^* containing this very c^*, which is inconsistent with the monotonicity of Gödel numbering. So, a provability interpretation of $c{:}A(c)$ requires special proof predicates obtained by an arithmetical fixed-point construction.

Theorem 9.12 *If $\mathsf{LP}_0 \nvdash F$, then F^* is false for some arithmetical interpretation $*$.*

Proof Because $\mathsf{LP}_0 \nvdash F$, $\mathsf{LP}_0^{G-} \nvdash \neg F \Rightarrow$. Perform the saturation procedure on $(\neg F \Rightarrow)$, Lemma 9.10, to get a saturated sequent $\Gamma \Rightarrow \Delta$ in which $\neg F \in \Gamma$ such that $\mathsf{LP}_0^{G-} \nvdash \Gamma \Rightarrow \Delta$. Then do a completion to get a set of formulas $\widetilde{\Gamma}$ satisfying Lemma 9.11.

We define the desired interpretation $*$ on propositional letters S_i, proof variables x_j, and proof constants a_j. We assume that the Gödel numbering of the joint language of LP and PA is injective, i.e.,

$$\ulcorner E_1 \urcorner = \ulcorner E_2 \urcorner \;\Leftrightarrow\; E_1 \equiv E_2$$

for any expressions E_1, E_2, and that 0 is not a Gödel number of any expression.

For a propositional letter S, proof variable x and proof constant a let

$$S^* = \begin{cases} \ulcorner S \urcorner = \ulcorner S \urcorner & \textit{if } S \in \widetilde{\Gamma} \\ \ulcorner S \urcorner = 0 & \textit{if } S \notin \widetilde{\Gamma} \end{cases}$$

$$x^* = \ulcorner x \urcorner$$

$$a^* = \ulcorner a \urcorner.$$

The remaining parts of $*$ are constructed by an arithmetical fixed point equation that follows.

For any arithmetical formula $\mathsf{Prf}(x, y)$ define an auxiliary translation $\dagger$ (depending on $\mathsf{Prf}(x, y)$) of proof polynomials to numerals and LP-formulas to PA-formulas such that $S^\dagger = S^*$ for any propositional letter S, and $t^\dagger = \ulcorner t \urcorner$ for any proof polynomial t.

It is clear that if $\mathsf{Prf}(x, y)$ contains quantifiers, then $\dagger$ is injective, i.e., $F^\dagger \equiv G^\dagger$ yields $F \equiv G$. Indeed, from $F^\dagger \equiv G^\dagger$ it follows that the principal connectives in F and G coincide. We consider one case: $(F_1 \to F_2)^\dagger \equiv (s{:}G)^\dagger$ is impossible. Because $(s{:}G)^\dagger \equiv \mathsf{Prf}(k, n)$ for the corresponding k and n, this formula contains quantifiers. Therefore the formula $(F_1 \to F_2)^\dagger$ (which coincides with $F_1^\dagger \to F_2^\dagger$) also contains quantifiers and thus contains a subformula of the form $\mathsf{Prf}(k', n')$. However, $(s{:}G)^\dagger \equiv F_1^\dagger \to F_2^\dagger$ is impossible because the complexity (number of logical connectives and quantifiers) in both parts of this equivalence are different. Indeed, the complexity of $(s{:}G)^\dagger$ is the complexity of $\mathsf{Prf}(k, n)$ whereas the complexity of $F_1^\dagger \to F_2^\dagger$ is the complexity of $\mathsf{Prf}(k', n')$ (the same as of $\mathsf{Prf}(k, n)$) plus at least one.

Now the injectivity of $\dagger$ can be shown by an easy induction on the construction of an LP-formula. Moreover, one can construct primitive recursive functions f and g such that

$$f(\ulcorner B \urcorner, \ulcorner \mathsf{Prf} \urcorner) = \ulcorner B^\dagger \urcorner$$
$$g(\ulcorner B^\dagger \urcorner, \ulcorner \mathsf{Prf} \urcorner) = \ulcorner B \urcorner.$$

Let $(\mathsf{Proof}, \otimes, \oplus, \Uparrow)$ be the standard multiconclusion proof predicate with $\otimes$ standing for “application,” $\oplus$ for “sum” and $\Uparrow$ for “proof checker” operations associated with Proof. In particular, for any arithmetical formulas φ, ψ and any natural numbers k, n the following formulas are true:

$$\mathsf{Proof}(k, \varphi \to \psi) \to \mathsf{Proof}(n, \varphi) \to \mathsf{Proof}(k \otimes n, \psi),$$

$$\mathsf{Proof}(k, \varphi) \to \mathsf{Proof}(k \oplus n, \varphi), \quad \mathsf{Proof}(n, \varphi) \to \mathsf{Proof}(k \oplus n, \varphi),$$

$$\mathsf{Proof}(k, \varphi) \to \mathsf{Proof}(\uparrow k, Proof(k, \varphi)).$$

Without loss of generality we assume that $\mathsf{Proof}(\ulcorner t \urcorner, k)$ is false for any proof polynomial t and any $k \in \omega$.

Let $\varphi(\vec{y}, z)$ be a provably Σ_1 arithmetical formula. Again without loss of generality we assume that $\varphi(\vec{y}, z)$ is provably equivalent to $\exists x \psi(x, \vec{y}, z)$ for some provably Δ_1-formula $\psi(x, \vec{y}, z)$. By $\mu z.\varphi(\vec{y}, z)$ we mean a function $z = f(\vec{y})$ that, given $\vec{y}$,

1. calculates the first pair of natural numbers (k, l) such that $\psi(k, \vec{y}, l)$ holds,
2. puts $z = l$.

It is clear that $\mu z.\varphi(\vec{y}, z)$ is computable (though not necessarily total).

By a fixed point argument we construct a formula $\mathsf{Prf}(x, y)$ such that PA proves the following *fixed point equation (FPE)*:

$$\begin{aligned} &\mathsf{Prf}(x, y) \leftrightarrow \\ &\qquad \mathsf{Proof}(x, y) \vee (\text{“}x = \ulcorner t \urcorner \textit{ for some } t \textit{ and } y = \ulcorner B^\dagger \urcorner \textit{ for some } B \in I(t)\text{”}). \end{aligned}$$

The preceding arithmetical formula "..." describes a primitive recursive procedure: given x and y recover t and B such that $x = \ulcorner t \urcorner$ and $y = \ulcorner B^\dagger \urcorner$, then verify $B \in I(t)$. From *FPE* it is immediate that Prf is a provably Δ_1-formula because $\mathsf{Proof}(x, y)$ is provably Δ_1. It also follows from *FPE* that $\mathsf{PA} \vdash \psi$ yields $\mathsf{Prf}(k, \psi)$ for some $k \in \omega$.

We define the arithmetical formulas $M(x, y, z)$, $A(x, y, z)$, $C(x, z)$ as follows. Here s, t denote proof polynomials.

$$\begin{aligned} M(x, y, z) \leftrightarrow\ & (\text{“}x = \ulcorner s \urcorner \textit{ and } y = \ulcorner t \urcorner \textit{ for some } s \textit{ and } t\text{”} \wedge z = \ulcorner s \cdot t \urcorner) \\ & \vee \\ & (\text{“}x = \ulcorner s \urcorner \textit{ for some } s \textit{ and } y \neq \ulcorner t \urcorner \textit{ for any } t\text{”} \wedge \\ & \exists v [v = \mu w.(\textstyle\bigwedge\{\mathsf{Proof}(w, B^\dagger) \mid B \in I(s)\}) \wedge z = v \otimes y]) \\ & \vee \\ & (\text{“}x \neq \ulcorner s \urcorner \textit{ for any } s \textit{ and } y = \ulcorner t \urcorner \textit{ for some } t\text{”} \wedge \\ & \exists v [v = \mu w.(\textstyle\bigwedge\{\mathsf{Proof}(w, B^\dagger) \mid B \in I(t)\}) \wedge z = x \otimes v]) \\ & \vee \\ & (\text{“}x \neq \ulcorner s \urcorner \textit{ and } y \neq \ulcorner t \urcorner \textit{ for any } s \textit{ and } t\text{”} \wedge z = x \otimes y). \end{aligned}$$

$$A(x,y,z) \leftrightarrow (\text{“}x = \ulcorner s \urcorner \textit{ and } y = \ulcorner t \urcorner \textit{ for some } s \textit{ and } t\text{”} \wedge z = \ulcorner s + t \urcorner)$$
$$\vee$$
$$(\text{“}x = \ulcorner s \urcorner \textit{ for some } s \textit{ and } y \neq \ulcorner t \urcorner \textit{ for any } t\text{”} \wedge$$
$$\exists v[v = \mu w.(\bigwedge\{\mathsf{Proof}(w, B^\dagger) \mid B \in I(s)\}) \wedge z = v \oplus y])$$
$$\vee$$
$$(\text{“}x \neq \ulcorner s \urcorner \textit{ for any } s \textit{ and } y = \ulcorner t \urcorner \textit{ for some } t\text{”} \wedge$$
$$\exists v[v = \mu w.(\bigwedge\{\mathsf{Proof}(w, B^\dagger) \mid B \in I(t)\}) \wedge z = x \oplus v])$$
$$\vee$$
$$(\text{“}x \neq \ulcorner s \urcorner \textit{ and } y \neq \ulcorner t \urcorner \textit{ for any } s \textit{ and } t\text{”} \wedge z = x \oplus y)$$

$$C(x,z) \leftrightarrow (\text{“}x = \ulcorner t \urcorner \textit{ for some } t\text{”} \wedge z = \ulcorner !t \urcorner) \vee (\text{“}x \neq \ulcorner t \urcorner \textit{ for any } t\text{”} \wedge$$
$$\exists v[v = \mu w.(\bigwedge\{\mathsf{Proof}(w, \mathsf{Proof}(x,\psi) \rightarrow \mathsf{Prf}(x,\psi)) \mid \psi \in T(x)\}) \wedge z = v \otimes \uparrow x]).$$

Here each of "..." denotes a natural arithmetical formula representing in PA the corresponding condition. Note that in the definitions of $M(x,y,z)$, $A(x,y,z)$ and $C(x,z)$ earlier, all the functions of sort $\mu w.\varphi$ are computable because all the corresponding φ formulas are Δ_1. Therefore, $M(x,y,z)$, $A(x,y,z)$ and $C(x,z)$ are provably Δ_1. Let

$$\mathbf{m}(x,y) \coloneqq \mu z.M(x,y,z), \quad \mathbf{a}(x,y) \coloneqq \mu z.A(x,y,z), \quad \mathbf{c}(x) \coloneqq \mu z.C(x,z).$$

As follows from the preceding, the functions $\mathbf{m}(x,y)$, $\mathbf{a}(x,y)$, and $\mathbf{c}(x)$ are computable. Moreover, Lemma 9.17 yields that these functions are total on Prf-proofs.

We continue defining the interpretation $*$. Let Prf for $*$ be the one from *FPE*, the functions $\mathbf{m}(x,y)$, $\mathbf{a}(x,y)$, and $\mathbf{c}(x)$ are as earlier, and $*$ is based on Prf, $\mathbf{m}(x,y)$, $\mathbf{a}(x,y)$, and $\mathbf{c}(x)$. The following lemma is an immediate corollary of definitions.

Lemma 9.13

(a) $t^* \equiv t^\dagger$ *for any proof polynomial t,*
(b) $B^* \equiv B^\dagger$ *for any* LP*–formula B.*

It is also immediate that the mapping $*$ is injective on terms and formulas of LP. In particular, for all expressions E_1 and E_2,

$$E_1^* \equiv E_2^* \quad \Leftrightarrow \quad E_1 \equiv E_2.$$

Corollary 9.14 X^* *is provably* Δ_1*, for any* LP*–formula X.*

Indeed, if X is atomic, then X^* is provably Δ_1 by the definition of $*$. If X is $t{:}Y$, then $(t{:}Y)^*$ is $\mathsf{Prf}(t^*, Y^*)$, which is provably Δ_1 as well. Because the set of provably Δ_1-formulas is closed under Boolean connectives, X^* is provably Δ_1 for each X.

Lemma 9.15 *If $X \in \widetilde{\Gamma}$, then* $\mathsf{PA} \vdash X^*$. *If $X \in \Delta$, then* $\mathsf{PA} \vdash \neg X^*$.

Proof By induction on the length of X. Let X be atomic. By the definition of $*$, X^* is true iff $X \in \widetilde{\Gamma}$. Because $\widetilde{\Gamma} \cap \Delta = \emptyset$, $X \in \Delta$ yields $X \notin \widetilde{\Gamma}$, which yields that X^* is false.

Let $X = t{:}Y \in \widetilde{\Gamma}$. Then $\mathsf{PA} \vdash$ "$Y \in I(t)$". By *FPE*, $\mathsf{PA} \vdash \mathsf{Prf}(t, Y^\dagger)$. By Lemma 9.13, $\mathsf{PA} \vdash \mathsf{Prf}(t^*, Y^*)$. Therefore $\mathsf{PA} \vdash (t{:}Y)^*$.

If $t{:}Y \in \Delta$, then $t{:}Y \notin \widetilde{\Gamma}$ and "$Y \in I(t)$" is false. The formula $\mathsf{Proof}(t^*, Y^*)$ is also false because t^* is $\ulcorner t \urcorner$ (by Lemma 9.13) and $\mathsf{Proof}(t, k)$ is false for any k by assumption. By *FPE*, $(t{:}Y)^*$ is false. Because $(t{:}Y)^*$ is provably Δ_1 (Corollary 9.14), $\mathsf{PA} \vdash \neg(t{:}Y)^*$.

The induction steps corresponding to Boolean connectives are standard and based on the saturation properties of $\Gamma \Rightarrow \Delta$. For example, let X be $Y \rightarrow Z$ and $X \in \widetilde{\Gamma}$. Then $(Y \rightarrow Z) \in \Gamma$, because the completion process from Γ to $\widetilde{\Gamma}$ does not introduce any new implications. By saturation, $Y \in \Delta$ or $Z \in \Gamma$, hence $Z \in \widetilde{\Gamma}$. By induction hypothesis, Y^* is false or Z^* is true, and in either case, $(Y \rightarrow Z)^*$ is true.

If $(Y \rightarrow Z) \in \Delta$, then, by saturation, $Y \in \Gamma$ and $Z \in \Delta$. By induction hypothesis, Y^* is true, Z^* is false, hence $(Y \rightarrow Z)^*$ is false. □

Lemma 9.16 $\mathsf{PA} \vdash \varphi \;\Leftrightarrow\; \mathsf{Prf}(n, \varphi)$ *for some $n \in \omega$.*

Proof Because "$\Rightarrow$" follows immediately from *FPE*, it remains to establish "$\Leftarrow$." Let $\mathsf{Prf}(n, \varphi)$ hold for some $n \in \omega$. By *FPE*, either $\mathsf{Proof}(n, \varphi)$ holds or $\varphi \equiv B^\dagger$ for some B such that $t{:}B \in \widetilde{\Gamma}$. In the latter case by Lemma 9.11(3), $B \in \widetilde{\Gamma}$. By Lemma 9.15, $\mathsf{PA} \vdash B^*$, hence $\mathsf{PA} \vdash \varphi$. □

Lemma 9.17 *For all arithmetical formulas φ, ψ and natural numbers k, n,*

(a) $\mathsf{Prf}(k, \varphi \rightarrow \psi) \wedge \mathsf{Prf}(n, \varphi) \rightarrow \mathsf{Prf}(\mathbf{m}(k, n), \psi)$,
(b) $\mathsf{Prf}(k, \varphi) \rightarrow \mathsf{Prf}(\mathbf{a}(k, n), \varphi)$, $\quad \mathsf{Prf}(n, \varphi) \rightarrow \mathsf{Prf}(\mathbf{a}(k, n), \varphi)$,
(c) $\mathsf{Prf}(k, \varphi) \rightarrow \mathsf{Prf}(\mathbf{c}(k), \mathsf{Prf}(k, \varphi))$.

Proof
(a) Assume $\mathsf{Prf}(k, \varphi \rightarrow \psi)$ and $\mathsf{Prf}(n, \varphi)$. There are four possibilities.

(i) Neither of k, n is a Gödel number of a proof polynomial. By *FPE*, both $\mathsf{Proof}(k, \varphi \rightarrow \psi)$ and $\mathsf{Proof}(n, \varphi)$ hold, so $\mathsf{Proof}(k \otimes n, \psi)$ also does. Because in this case $k \otimes n = \mathbf{m}(k, n)$, we have $\mathsf{Proof}(\mathbf{m}(k, n), \psi)$, hence also $\mathsf{Prf}(\mathbf{m}(k, n), \psi)$.

(ii) Both k and n are equal to Gödel numbers of some proof polynomials, say $k = \ulcorner s \urcorner$ and $n = \ulcorner t \urcorner$. By *FPE*, φ is F^* and ψ is G^* for some LP-formulas F, G such that $F \to G \in I(s)$ and $F \in I(t)$. By the closure property of $\widetilde{\Gamma}$ (Lemma 9.11(4)), $G \in I(s \cdot t)$. By *FPE*, $\mathsf{Prf}(s \cdot t, G^*)$. By Lemma 9.13 and by definitions,

$$\mathbf{m}(k, n) = \mathbf{m}(\ulcorner s \urcorner, \ulcorner t \urcorner) = \mathbf{m}(s^*, t^*) = (s \cdot t)^* = \ulcorner s \cdot t \urcorner.$$

Thus, $\mathbf{m}(k, n) = \ulcorner s \cdot t \urcorner$ and $\mathsf{Prf}(\mathbf{m}(k, n), \psi)$ is true.

(iii) k is not equal to the Gödel number of a proof polynomial, $n = \ulcorner t \urcorner$ for some proof polynomial t. By *FPE*, $\mathsf{Proof}(k, \varphi \to \psi)$ and $\varphi \equiv F^*$ for some LP-formula F such that $F \in I(t)$. Compute the number

$$j = \mu w.(\bigwedge\{\mathsf{Proof}(w, B^\dagger) \mid B \in I(t)\})$$

by the following method. Take $I(t) = \{B_1, \ldots B_l\}$. By definition, $B_i \in \widetilde{\Gamma}$, $i = 1, \ldots, l$. By Lemma 9.15, $\mathsf{PA} \vdash B_i^*$ hence $\mathsf{PA} \vdash B_i^\dagger$ for all $i = 1, \ldots, l$. By the conjoinability property of Proof there exists w such that $\mathsf{Proof}(w, B_i^\dagger)$ for all $i = 1, \ldots, l$. Let j be the least such w.

In particular, $\mathsf{Proof}(j, F^\dagger)$, i.e., $\mathsf{Proof}(j, \varphi)$. By the definition of $\otimes$, $\mathsf{Proof}(k \otimes j, \psi)$. By the definition of M, $\mathbf{m}(k, n) = k \otimes j$, therefore $\mathsf{Proof}(\mathbf{m}(k, n), \psi)$ holds, hence $\mathsf{Prf}(\mathbf{m}(k, n), \psi)$.

(iv) "s is a Gödel number of a proof polynomial, but t is not a Gödel number of any proof polynomial" is similar to iii).

(b) can be checked in the same way as (a).

(c) Given $\mathsf{Prf}(k, \varphi)$ there are two possibilities.

(i) $k = \ulcorner t \urcorner$ for some proof polynomial t. By *FPE*, $\varphi \equiv F^\dagger$ for some $F \in I(t)$. By the closure property from Lemma 9.11(5), $!t{:}t{:}F \in \widetilde{\Gamma}$. By Lemma 9.15, $(!t{:}t{:}F)^*$ holds. By definitions,

$$(!t{:}t{:}F)^* \equiv \mathsf{Prf}(\mathbf{c}(t^*), \mathsf{Prf}(t^*, F^*)).$$

Because $t^* = \ulcorner t \urcorner$ and $F^* \equiv F^\dagger$, then $t^* = k$, $F^* \equiv \varphi$ and we conclude

$$\mathsf{Prf}(\mathbf{c}(k), \mathsf{Prf}(k, \varphi)).$$

(ii) $k \neq \ulcorner t \urcorner$ for any proof polynomial t. By *FPE*, $\mathsf{Proof}(k, \varphi)$ holds. By definition of the proof checking operation $\uparrow$ for Proof,

$$\mathsf{Proof}(\uparrow k, \mathsf{Proof}(k, \varphi)).$$

By the definition of $C(x, z)$, in this case $\mathsf{PA} \vdash \mathbf{c}(k) = l \otimes \uparrow k$ where

$$l = \mu w.(\bigwedge\{\mathsf{Proof}(w, \mathsf{Proof}(k, \psi) \to \mathsf{Prf}(k, \psi)) \mid \mathsf{Proof}(k, \psi)\}).$$

Note that because $\mathsf{PA} \vdash \mathsf{Proof}(k, \Theta) \to \mathsf{Prf}(k, \Theta)$ for all Θ's and $\mathsf{Proof}(k, \varphi)$ holds, the set

$$\{\mathsf{Proof}(w, \mathsf{Proof}(k, \psi) \to \mathsf{Prf}(k, \psi)) \mid \mathsf{Proof}(k, \psi)\}$$

is not empty. So l is well-defined and

$$\mathsf{Proof}(l, \mathsf{Proof}(k, \varphi) \to \mathsf{Prf}(k, \varphi)),$$

therefore

$$\mathsf{Proof}(l \otimes \uparrow k, \mathsf{Prf}(k, \varphi)).$$

By *FPE*,

$$\mathsf{Prf}(l \otimes \uparrow k, \mathsf{Prf}(k, \varphi)),$$

therefore

$$\mathsf{Prf}(\mathbf{c}(k), \mathsf{Prf}(k, \varphi)).$$

□

Lemma 9.18 Prf *is a normal proof predicate.*

Proof By *FPE*, $\mathsf{Prf}(x.y)$ is provably Δ_1. It follows from *FPE*, and Lemma 9.16 that for any arithmetical sentence φ

$$\mathsf{PA} \vdash \varphi \quad \textit{if and only if} \quad \mathsf{Prf}(n, \varphi) \textit{ holds for some } n.$$

Finiteness of proofs. For each k, the set

$$T(k) = \{l \mid \mathsf{Prf}(k, l)\}$$

is finite. Indeed, if k is a Gödel number of a proof polynomial, we can use the finiteness of $I(t)$, otherwise we use the normality of Proof. An algorithm for the function from k to the code of $T(k)$ for Prf can be easily constructed from those for Proof, and from the decision algorithm for $I(t)$, Lemma 9.11(1).

Conjoinability of proofs for Prf is realized by the function $\mathbf{a}(x, y)$ because by Lemma 9.17(b)

$$T(k) \cup T(n) \subseteq T(\mathbf{a}(k, n)).$$

□

Let us finish the proof of Theorem 9.12. Given an LP-formula F not provable in LP_0 we have constructed a saturated sequent $\Gamma \Rightarrow \Delta$ not provable in LP_0^G such that $\neg F \in \Gamma$ and a "constructive canonical model" $\widetilde{\Gamma}$ satisfying Lemma 9.11; in particular, $F \in \Gamma \subseteq \widetilde{\Gamma}$. Given such $\widetilde{\Gamma}$ we have built a normal proof predicate Prf (Lemma 9.18) and an arithmetical interpretation $*$ based on Prf such that all formulas from $\widetilde{\Gamma}$ are true, Lemma 9.15. Therefore $(\neg F)^*$ is true and F^* is false. □

Corollary 9.19 (Arithmetical completeness of LP(CS)) *For any finite constant specification* CS, LP(CS) $\vdash F$ *iff* F *is (provably) valid under* CS.

Proof Arithmetical soundness of LP(CS) was established in Corollary 9.6. Suppose LP(CS) $\nvdash F$. Then $\mathsf{LP}_0 \nvdash \bigwedge(\mathsf{CS}) \to F$. By Theorem 9.12, under some arithmetical interpretation $*$, $\bigwedge(\mathsf{CS}) \to F$ is false, i.e., $(\mathsf{CS})^*$ is true and F^* is false. □

Corollary 9.20 (Arithmetical completeness of LP) LP $\vdash F$ *iff* F *is (provably) valid under some finite constant specification.*

Proof Indeed, if LP $\vdash F$ then, by compactness, LP(CS) $\vdash F$ for some finite constant specification CS. By Corollary 9.6, F is (provably) valid under CS. Conversely, if F is (provably) valid under some constant specification CS, then, by Corollary 9.19, LP(CS) $\vdash F$, hence LP $\vdash F$. □

Corollary 9.21 (Cut elimination in LP_0^G) $\mathsf{LP}_0^G \vdash \Theta \Rightarrow \Xi$ *implies* $\mathsf{LP}_0^{G-} \vdash \Theta \Rightarrow \Xi$.

Proof By contrapositive: suppose $\mathsf{LP}_0^{G-} \nvdash \Theta \Rightarrow \Xi$. By Theorem 9.12, for some arithmetical interpretation $*$, $(\bigwedge \Theta \to \bigvee \Xi)^*$ is false, hence not provable in PA. By Proposition 9.5, LP $\nvdash \bigwedge \Theta \to \bigvee \Xi$ hence, by Proposition 9.7,

$$\mathsf{LP}_0^G \nvdash \Theta \Rightarrow \Xi.$$

□

The obvious conceptual meaning of arithmetical completeness theorems is that invariant (i.e., independent of specifics of the numeration) properties of proof predicates are exactly the ones that follow from the postulates of the logic of proofs. Consider two examples. Formula $\neg x{:}x{:}\bot$ is provable in LP_0 (by reflexivity, twice):

$$x{:}x{:}\bot \ \to \ x{:}\bot \ \to \ \bot.$$

Formula $\neg x{:}x{:}\top$ where $\top$ is the propositional constant *true* (or, alternatively, $\top \equiv \bot \to \bot$) is not derivable in LP_0. Indeed, it is easy to check that the sequent $x{:}x{:}\top \Rightarrow$ is not derivable in LP_0^G without cut.

Model theoretically, $\neg x{:}x{:}\top$ fails in the basic model $\mathcal{M}$ of LP_0 in which for each term t, $t^* = \Omega$ with

$$\Omega = \{t_n{:}t_{n-1}{:}\ldots t_0{:}\top \mid n \in \omega,\ t_i \in Tm,\ i = 0, \ldots, n\}.$$

Informally, in $\mathcal{M}$, every term is a proof that every term is a proof of … every term is a proof of $\top$. Axioms for operations $\cdot, +, !$ obviously hold. Because all formulas from Ω are true in $\mathcal{M}$, the reflection principle $t{:}F \to F$ also holds in $\mathcal{M}$. Therefore, $\mathcal{M}$ is a model of LP_0 in which $x{:}x{:}\top$ holds. Therefore, $\mathsf{LP}_0 \nvdash \neg x{:}x{:}\top$.

However, no arithmetical interpretation $\star$ based on the usual Gödel numbering can make $x{:}x{:}\top$ arithmetically true. Indeed, the standard Gödel numbering is monotonic: a code of a proper subexpression of an expression is strictly less than a code of the whole expression, a code of a numeral $\overline{n}$ is strictly greater than n. Under these constraints, $x^\star$ should be strictly greater than the code of $\mathsf{Prf}(x^\star, \top^\star)$, which is impossible.

The proof of arithmetical completeness, Theorem 9.12, provides an example of a normal proof predicate Prf under which $x{:}x{:}\top$ is arithmetically true. Such a proof predicate is built by a fixed-point construction and is not based on the usual monotonic Gödel numbering.

There is another fundamental corollary of arithmetical completeness: it provides a comprehensive class of arithmetical interpretations necessary for in-depth justification of derivations from given constant specifications. The arithmetical completeness for LP(CS) guarantees that F^* is true under each interpretation $*$, which makes all formulas from the constant specification CS true. But what if $(\mathsf{CS})^*$ is false under all arithmetical interpretations $*$? The first suspect could be

$$\{c{:}(c{:}\bot \to \bot)\},$$

which is a legitimate constant specification. The same monotonicity argument as earlier shows that this constant specification cannot be made true by any proof predicate based on the standard Gödel numbering.

However, Theorem 9.12 provides a desired arithmetical interpretation, which makes this specification true. Moreover, by this theorem, any constant specification CS is true under some normal arithmetical interpretation.

Theorem 9.22 (Nonemptiness of provability semantics for LP) *For any finite constant specification* CS *there exists an arithmetical interpretation* $*$ *such that* $(\mathsf{CS})^*$ *is true (provable in* PA*).*

Proof Indeed, first note that LP(CS) is consistent for any constant specification CS. Indeed, an easy inspection of the rules of LP_0^{G-} shows that the sequent $\mathsf{CS} \Rightarrow$ is not derivable in LP_0^{G-}. This can be shown by induction on the number n of formulas in CS. If $n = 1$, then we speak of the sequent $c{:}A \Rightarrow$ with a constant c and an LP-axiom A. This sequent could be introduced in LP_0^{G-} be weakening from the empty sequent $\Rightarrow$ of by $({:}\Rightarrow)$ from $A \Rightarrow$. Because $\mathsf{LP}_0^{G-} \vdash\Rightarrow A$, in the latter case, by cut, we conclude that $\mathsf{LP}_0^G \vdash\Rightarrow$, hence, by cut-elimination, $\mathsf{LP}_0^G \vdash\Rightarrow$, which is impossible.

Induction step: $\mathsf{CS} = \{c_1{:}A_1, \ldots, c_n{:}A_n\}$. Again, the sequent $c_1{:}A_1, \ldots, c_n{:}A_n \Rightarrow$ can be introduced in LP_0^{G-} either by weakening (then we are done by the induction hypothesis) or by $({:}\Rightarrow)$. In the latter case, without loss of generality, we can assume that $c_1{:}A_1, \ldots, c_n{:}A_n \Rightarrow$ is obtained from $A_1, c_2{:}A_2 \ldots, c_n{:}A_n \Rightarrow$. Because $\mathsf{LP}_0^G \vdash\ \Rightarrow A_1$, by cut, $\mathsf{LP}_0^G \vdash c_2{:}A_2 \ldots, c_n{:}A_n \Rightarrow$ hence $\mathsf{LP}_0^{G-} \vdash c_2{:}A_2 \ldots, c_n{:}A_n \Rightarrow$, which is impossible by the induction hypothesis.

So $\mathsf{LP}_0 \nvdash \neg \bigwedge \mathsf{CS}$. By Theorem 9.12, there exists an interpretation $*$ such that $(\neg \bigwedge \mathsf{CS})^*$ is false, i.e., all formulas in $\mathsf{CS})^*$ are true. □

The significance of self-referential constant specifications will be also demonstrated in Section 9.5 where it is shown that neither S4 nor IPC can be realized without using self-referential constant specifications, which contain formulas of type $c{:}A(c)$. None of such formulas can be realized by proof predicates based on monotonic Gödel numbering.

9.4 BHK Semantics

We recall that the intended meaning of intuitionistic logic is given by the informal *Brouwer–Heyting–Kolmogorov (BHK) semantics* of constructive proofs, cf. Section 1.2:

- a proof of $A \wedge B$ consists of a proof of proposition A and a proof of proposition B,
- a proof of $A \vee B$ is given by presenting either a proof of A or a proof of B,
- a proof of $A \rightarrow B$ is a construction transforming proofs of A into proofs of B,
- falsehood $\bot$ is a proposition which has no proof; $\neg A$ is shorthand for $A \rightarrow \bot$.

Informally there are objects of two sorts in BHK: proofs and (computable) functions. Note that, conceptually, proofs/provability in a broad mathematical context can express computability, but the converse is not immediate. For example, in true arithmetic TA, which is the set of all arithmetical formulas true in the standard model of arithmetic, all computable functions are represented by corresponding terms. However, there is no conventional notion of a proof/provability represented in TA. So, it is *a priori* unlikely that computational programs alone could represent BHK adequately, whereas proofs alone might do.

If we want to capture some notion of a BHK proof then a natural question might be "proofs where"? Derivations in an intuitionistic system itself make such semantics immediately circular, which undermines foundational ambitions. Derivations in the usual classical system, if applied naively, do not satisfy BHK conditions for $\vee$.

Historically, we observe two distinct classes of BHK-style semantics:

- *Computational BHK*, originating from Kleene's discovery (1945) of a computational content of intuitionistic logic. Kleene, according to his students' recollections, thought about formalizing proof-based BHK but found it to be too hard. However, Kleene found a computational version of BHK, which was an outstanding discovery: a computational content of constructive reasoning. A good example of a computational BHK semantics is given by Martin–Löf type theory. Though it uses a BHK proof terminology, Martin-Löf "proofs" or "constructions" are not identified with formal proofs, but rather have a natural computational interpretation.
- *Provability BHK*, originating from Gödel's works on modal logic of provability S4 (Gödel, 1933), on systems with explicit proofs (Gödel, 1938), and completed within the framework of the Logic of Proofs LP in the mid 1990s. Since the modal logic S4 embeds to the logic of proofs (Realization theorem) and LP enjoys the arithmetical proof interpretation, S4 receives an exact arithmetical provability semantics consistent with BHK and Gödel's sketches.

In addition Kreisel (1962) tried to develop a provability BHK from scratch but the system turned out to be inconsistent, cf. Dean and Kurokawa (2016) for a complete survey of this thread.

Kleene realizability, Kleene (1945), as a formalization of "constructively true" is reminiscent of BHK semantics; here the role of BHK proofs is played by computational programs (indices of recursive functions). In particular, a realizer of an implication $A \to B$ is a program p, which when applied to any realizer x of A returns a realizer of B. Symbolically:

$$p{:}(A \to B) \to (x{:}A \to [p{\cdot}x]{:}B)$$

All BHK clauses are satisfied by the computational semantics except for disjunction. In computational semantics for disjunction there is a requirement of a bit selector that points at the proper disjunct:

3$'$. p proves $A \vee B$ iff $p = (p_0, p_1)$ with $p_0 \in \{0, 1\}$, and p_1 proves A if $p_0 = 0$ and p_1 proves B if $p_0 = 1$.

This adjustment illustrates a difference between proofs and computational programs in the BHK setting: the proof predicate

$$p \textit{ is a proof of } F \tag{9.1}$$

is *decidable*, whereas the realizability assertion

$$p \textit{ realizes } F \tag{9.2}$$

is *not decidable.* The "selector" is needed for (9.2) but is redundant for (9.1) because given proof p, one can compute the right disjunct.

Let us discuss how the provability BHK works in the context of the logic of proofs, LP. For each modal formula F, let F^r denote a realization of F in the logic of proofs by some r (Theorem 6.5). According to Corollary 6.6, $\mathsf{S4} \vdash F$ if and only if $\mathsf{LP} \vdash F^r$ for some normal realization r.

Corollary 9.23 (Arithmetical completeness of S4) $\mathsf{S4} \vdash F \quad \Leftrightarrow \quad F^r$ *is (provably) valid for some normal realization r.*

Recall that by Gödel's translation $tr(F)$ of an intuitionistic formula F we understand the result of prefixing every subformula of F with the modality $\Box$. According to Gödel–McKinsey–Tarski, cf. Section 1.2,

$$\mathsf{IPC} \vdash F \quad \Leftrightarrow \quad \mathsf{S4} \vdash tr(F),$$

which defines IPC inside S4.

Definition 9.24 A propositional intuitionistic formula F is *proof realizable* if $(tr(F))^r$ is arithmetically valid under some normal realization r.

Theorem 9.25 (Provability completeness of IPC) *For any propositional intuitionistic formula F,* $\mathsf{IPC} \vdash F \;\Leftrightarrow\; F$ *is proof realizable.*

Proof A straightforward combination of the aforementined Gödel–McKinsey–Tarski equivalence and Corollary 9.23. □

Theorem 9.25 provides an exact specification of IPC by means of a classical notion of proof that is consistent with *BHK* semantics. In addition to Gödel's translation $tr(\cdot)$ one could consider the *McKinsey–Tarski translation* that prefixes only atoms and implications in F. A result similar to Theorem 9.25 holds for proof realizability based on such a McKinsey–Tarski translation too.

IPC is sound and compete with respect to the class of proof systems in arithmetic in the following precise sense:

1. Any derivation of F in IPC produces a realization consisting of a proof term assignment supported by a constant specification CS. This realization

of F is a proof tautology, i.e., holds for any arithmetical interpretation that validates CS.

2. If F is not derivable in IPC, then such a realization is impossible.

We shall now explain that the logic of proofs, in combination with Gödel's translation:

(a) is BHK-compliant in the original formulation of the latter, with BHK proofs interpreted as proof objects in Peano Arithmetic;

(b) naturally straightens known omissions of BHK in implication/negation and "for all" clauses.

Assuming a certain amount of good will from the reader, we will check that LP complies with BHK clauses.

Implication *A proof of $A \to B$ is a construction that, given a proof of A, returns a proof of B.*

An intuitionistic implication $A \to B$ is realized in LP as $t{:}(\widetilde{A} \to \widetilde{B})$ where $\widetilde{A}$ and $\widetilde{B}$ are LP-versions of A and B respectively. By "application" axiom B2,

$$t{:}(\widetilde{A} \to \widetilde{B}) \wedge u{:}\widetilde{A} \to [t{\cdot}u]{:}\widetilde{B},$$

and t is indeed a construction that, given a proof u of the antecedent of the implication, returns a proof $t{\cdot}u$ of the succedent.

Conjunction *A proof of $A \wedge B$ consists of a proof of A and a proof of B.*

An intuitionistic conjunction $A \wedge B$ is realized in LP as $t{:}(\widetilde{A} \wedge \widetilde{B})$ where $\widetilde{A}$ and $\widetilde{B}$ are, as before, LP-versions of A and B. This t contains sufficient information to recover both a proof of $\widetilde{A}$ and a proof of $\widetilde{B}$. Indeed, given such t and commonly known proofs a and b such that

$$a{:}((\widetilde{A} \wedge \widetilde{B}) \to \widetilde{A}) \quad \text{and} \quad b{:}((\widetilde{A} \wedge \widetilde{B}) \to \widetilde{B}),$$

one can find a proof of $\widetilde{A}$, $a{\cdot}t$, and a proof of $\widetilde{B}$, $b{\cdot}t$. Likewise, having a proof of $\widetilde{A}$ and a proof of $\widetilde{B}$, one can construct a proof of $\widetilde{A} \wedge \widetilde{B}$ within LP.

Disjunction *A proof of $A \vee B$ is given by presenting either a proof of A or a proof of B.*

We argue in LP. Suppose $u{:}A$ or $u{:}B$. We have to construct a proof term $t(u)$ such that $t(u){:}(A \vee B)$. Consider the internalized disjunction principles

$$a{:}(A \to (A \vee B)) \text{ and } b{:}(B \to (A \vee B)),$$

both obviously provable in LP, with commonly available proof a and b. Using application axiom B2, we conclude that either $[a\cdot u]{:}(A \vee B)$ or $[b\cdot u]{:}(A \vee B)$. In either case,

$$[a\cdot u + b\cdot u]{:}(A \vee B),$$

and we can set $t(u)$ to $[a\cdot u + b\cdot u]$.

Negation. *$\neg A$ is shorthand for $A \rightarrow \bot$ with a proposition $\bot$, which has no proof*

This formulation leaves a loophole (which created a lasting foundational problem in computational BHK semantics in general and in Kleene realizability in particular). If $\neg A$ constructively true (has a BHK proof) then *any* proof object t is a constructive proof of $\neg A$. Indeed, the BHK requirement for t is "*for any proof u of A, $t\cdot u$ should provide a proof of $\bot$*" holds vacuously because there is no such u at all. Any such t is a dummy and does not help to establish the constructive truth of $\neg A$, hence is not a correct BHK proof of $\neg A$. So, the right notion of a constructive proof of $\neg A$ is missing in this analysis.

The provability BHK semantics in the logic of proofs provides a more refined account of this phenomenon and closes this loophole. Gödel's translation of a propositional formula $\neg A$ (assume that A is atomic, to keep notations simpler) is $\Box(\neg\Box A)$. Its realization in LP is

$$p(x){:}[\neg x{:}A]$$

where $p(x)$ is a proof term depending on x. Therefore, in addition to *any* t from the naïve BHK reading of $\neg A$, our BHK proof of $\neg A$ requires a meaningful proof term $p(x)$ such that

for each x, $p(x)$ is a proof that x cannot be a proof of A.

This $p(x)$ is the real constructive witness of $\neg A$, which was completely missing in the naïve analysis and in Kleene realizability.

Implication, revised. The trouble with the constructive proof for negation is, of course, a special case of a more general problem with the right notion of the constructive proof for implication. Again, consider atomic A and B. The original BHK required a constructive proof of $A \rightarrow B$ to be a construction (computable function) $f(x)$ such that for any proof x of A, $f(x)$ is a proof of B. Symbolically, $x{:}A \rightarrow f(x){:}B$.

As we have seen, this led to problems with the constructive semantics of negation and needed a refinement. Provability BHK offers a natural fix:

> *a constructive proof of $A \rightarrow B$ is a pair of constructions (f, g) such that for each x, $g(x)$ is a classical proof that $x{:}A$ implies $f(x){:}B$.*

Symbolically this can be written as

$$g(x){:}[x{:}A \to f(x){:}B].$$

Again, we see that provability BHK semantics via the logic of proofs "automatically" fixes loopholes in the set of original BHK clauses. Historically, the discussion of these loopholes can be traced back to Kreisel (1962) and his so-called second clause, cf. Dean and Kurokawa (2016).

In summary: the Logic of Proofs offers an adequate explicit version of S4 and can be regarded as a reasonable formalization of provability BHK for IPC. BHK proofs admit arithmetical interpretation as combinations of classical proofs in the base theory, e.g., in Peano Airthmetic PA.

9.5 Self-Referentiality of Justifications

LP admits *self-referential types* of the sort $t{:}F(t)$ stating that t is a proof of a sentence F, which explicitly contains t. This self-referentiality is supported by the provability semantics that includes an arithmetical fixed-point argument. But is self-referentiality actually needed for realization of S4? For BHK semantics?

An inspection of the realization theorem, Theorem 6.5 shows that self-referential proof terms can appear as realizers of modalities by this particular realization algorithm. This, however, does not answer the question of whether self-referentiality can be avoided here.

The BHK clause for implication, though informal, also appears to be self-referential because a proof s of an implication $A \to B$ should be a construction that accepts *any* proof of A and returns a proof of B, and among the proofs of A could possibly be s itself.

These considerations: partial for S4 and informal for BHK, point at the possibility of a negative answer. It is time for us to be precise.

Definition 9.26 (Direct Self-Referentiality) A constant specification formula $c{:}A$ is *directly self-referential* if A contains an occurrence of c.

Consider the so-called Moore sentence: *It rains but I don't know it.* If p stands for *it rains* and $\Box$ denotes "knowledge" then a modal formalization of the Moore sentence is

$$M = p \wedge \neg\Box p.$$

M is easily satisfiable in S4, hence consistent, e.g., when p is true but not known. However, it is impossible to know Moore's sentence.

Here is a derivation of $\neg\Box M$ in S4:

(1) $(p \wedge \neg\Box p) \rightarrow p$, *propositional axiom*
(2) $\Box((p \wedge \neg\Box p) \rightarrow p)$, *Necessitation*
(3) $\Box(p \wedge \neg\Box p) \rightarrow \Box p$, *from 2, by Distribution*
(4) $\Box(p \wedge \neg\Box p) \rightarrow (p \wedge \neg\Box p)$, *Reflexivity*
(5) $\neg\Box(p \wedge \neg\Box p)$, *from 3 and 4, in Boolean logic.*

Its natural realization in LP is directly self-referential:

(1) $(p \wedge \neg[c\cdot x]{:}p) \rightarrow p$, *logical axiom*
(2) $c{:}((p \wedge \neg[c\cdot x]{:}p) \rightarrow p)$, *directly self-referential* CS
(3) $x{:}(p \wedge \neg[c\cdot x]{:}p) \rightarrow [c\cdot x]{:}p$, *from 2, by Application*
(4) $x{:}(p \wedge \neg[c\cdot x]{:}p) \rightarrow (p \wedge \neg[c\cdot x]{:}p)$, *Reflexivity*
(5) $\neg x{:}(p \wedge \neg[c\cdot x]{:}p)$, *from 3 and 4, in Boolean logic.*

As we see, step 2 of this proof introduced a directly self-referential constant specification $c{:}\,((p \wedge \neg[c\cdot x]{:}p) \rightarrow p)$.

Kuznets's Theorem *Any realization of* $\neg\Box M$ *in* LP *requires self-referential constant specifications*, Brezhnev and Kuznets (2006).

A general theorem from Kuznets (2009) states that self-referentiality is unavoidable in realizations of K4, D4, and T, but can be avoided in realizations of K and D.

The question of the self-referentiality of BHK-semantics for IPC has been answered by Junhua Yu in Yu (2014). Extending Kuznets' method, he established

Yu's Theorem: *Each* LP *realization of the intuitionistic law of double negation* $\neg\neg(\neg\neg p \rightarrow p)$ *requires directly self-referential constant specifications.*

More generally, Yu has proved that any double negation of a classical tautology (by Glivenko's Theorem all of them are theorems of IPC) needs directly self-referential constant specifications for its realization in LP. Another example of unavoidable self-referentiality was found by Yu in the purely implicational fragment of IPC. This suggests that the BHK semantics of intuitionistic logic (even just of intuitionistic implication) is intrinsically self-referential.

These results by Kuznets and Yu indicate that provability BHK semantics for S4 and IPC is essentially self-referential and needs a fixed-point construction to connect it to formal proofs in PA or similar systems. This might explain, in part, why any attempt to build provability BHK semantics in a direct inductive manner without self-referentiality was doomed to fail.

10

Quantifiers in Justification Logic

Propositional justification logic went through a historical process of development taking over twenty years from its beginnings to the stage represented in this book. Quantified justification logic is much more recent and has been in development for less than half that time. Interestingly, it seems to be passing through stages analogous to the propositional version. The initial quantified justification logic is a first-order version of LP and is presented in this chapter. Its intended goal is to connect quantified intuitionistic logic to an arithmetic semantics, and this is also presented here. Its first (and so far only) realization proof is constructive, using a sequent calculus for first-order S4. So far, other quantified justification logics have been confined to those connected with well-known sublogics of S4. A possible world semantics has been created. And that's where things are now. Propositionally, the range of modal logics with justification counterparts was gradually extended. That has yet to happen in the quantificational setting, but it seems likely that it will.

Quantificational modal logic brings complexities that do not exist propositionally, or purely classically. With possible world semantics, each possible world can have its own domain of quantification, or they can all share the same one. Other possibilities exist. So for each propositional modal logic there is generally more than one quantified version. So far all development for quantified LP has been connected with the so-called monotonic version of quantified S4. This is just beginning to broaden, and a semantics for a constant domain version now exists (Fitting and Salvatore, 2018). Clearly there is much here that awaits exploration.

The initial paper on the first-order Logic of Proofs was Artemov and Yavorskaya (2001), cf. also a short discussion in Artemov (2001), in which it was shown that, unlike LP, the first-order Logic of Proofs cannot be completely axiomatized. In Artemov and Yavorskaya (Sidon) (2011), the First-Order Logic of Proofs, FOLP, was introduced as an explicit counterpart of first-order S4, the realization theorem was established, and several natural versions of arithmetical semantics for FOLP were outlined. Because first-order intuitionistic logic HPC can be faithfully embedded in FOS4 via Gödel's translation, this also provided HPC with a BHK-style semantics in FOLP.

In this chapter we present the current state of the subject. The only modal logic considered is S4 with a first-order semantics that is monotonic. There are no general results of the kind we saw propositionally. We hope the work presented here serves to motivate a larger community of people to continue our investigations in first-order justification logic.

10.1 Free Variables in Proofs

Let $A(x)$ be a formula with a free variable x. Then, in FOS4, $\Box A(x)$ also has x free. Using the provability reading of $\Box$, we have that $\Box A(x)$ is read as

given a natural parameter $x = n$, formula $A(n)$ is provable.

On the other hand, FOS4 cannot directly express the notion

formula $A(x)$ with a free variable x is provable.

In FOLP, there are tools for representing both of the aforementioned readings of individual variables: as global parameters or local variables not accessible from outside the proof/provability operator. Suppose p is interpreted as a specific proof (in PA) and A as a specific formula (of PA) with a free variable x. We can read $p{:}A$ in two ways:

- p proves A for a given value of the parameter x and thus the formula $p{:}A$ and its truth value depend on x. For example, p is $\{0 = 0\}$, and A is $x = x$: then $p{:}A$ holds only when x is substituted by 0.
- p is a proof of a formula with a free variable x and so the truth value of $p{:}A$ does not depend on x. For example, p is $\{x = x\}$ and A is, as before, $x = x$.

In the language FOLP, the proof predicate is represented by formulas of the form

$$t{:}_X A$$

where X is a finite set of individual variables that are interpreted as global parameters, free in $t{:}_X A$. All free variables of A, which are not in X are considered local and are not free in $t{:}_X A$. For example, if $A(x, y)$ is an atomic formula, then in $p{:}_{\{x\}}A(x, y)$, variable x is global/free, and variable y is local/not free.

Likewise, in $p{:}_{\{x,y\}}A(x,y)$ both variables are free, and in $p{:}_{\emptyset}A(x,y)$, neither x nor y is free.

Proofs are represented by proof terms that do not contain individual variables.

Definition 10.1 (FOLP Language) Let $\mathcal{L}$ denote the first-order language with individual variables that contains a countable set of predicate symbols of any arity, without functional symbols or equality. *The language of* FOLP is the extension of $\mathcal{L}$ with special means to represent proofs and proof assertions. Namely, the language FOLP contains individual variables $x_0, x_1, x_2, \ldots$, the usual Boolean connectives, quantifiers over individual variables, predicate symbols Q_i^n of any arity n ($i, n = 0, 1, 2, \ldots$) and

(1) proof variables p_k, $k = 0, 1, 2\ldots$ and proof constants $c_0, c_1, c_2, \ldots$;
(2) functional symbols for operations on proofs:
  (a) those of LP: binary $+$, $\cdot$ and unary $!$,
  (b) unary gen_x for each individual variable x;
(3) an operational symbol $(\cdot){:}_X(\cdot)$ for each finite set X of individual variables.

Definition 10.2 (FOLP Proof Term) *Proof terms* are constructed as follows:

(1) each proof constant and proof variable is a proof term;
(2) if t, s are proof terms, then $t \cdot s$, $!t$, $t + s$, and $\mathsf{gen}_x(t)$ are proof terms.

Notation 10.3 By X, Y, etc., we denote finite sets of individual variables. If y is an individual variable, then we will write Xy for $X \cup \{y\}$. As an additional convention, the notation Xy assumes that $y \notin X$. Note also that in gen_x, the variable x is merely a syntactic label of this operation and is not considered to be an occurrence of a variable. Terms in FOLP do not contain individual variables.

Definition 10.4 (FOLP Formulas) *Formulas* are defined in the standard way with an additional clause for the proof operator. Namely,

(1) If Q_i^n is a predicate symbol of arity n and $x_1, \ldots, x_n$ are individual variables, then $Q_i^n(x_1, \ldots, x_n)$ is an atomic formula; all occurrences of individual variables are free.
(2) If A, B are formulas, then $\neg A$, $A\alpha B$ (α being a binary Boolean connective) are formulas; Boolean connectives preserve free and bound occurrences of variables.
(3) If A is a formula and x is an individual variable, then $\forall x A$ is a formula; $\forall x$ binds all occurrences of x in A and preserves free and bound occurrences of all other variables.

(4) If t is a proof term, A is a formula, and X is a finite set of individual variables, then $t{:}_X A$ is a formula. The free individual variable occurrences in $t{:}_X A$ are the free individual variable occurrences in A, provided the variables also occur in X, together with all variable occurrences in X itself.

The set of free variables of a formula G is denoted by $FVar(G)$. According to Definition 10.4,

$$FVar(t{:}_X A) = X.$$

We use the abbreviation $t{:}A$ for $t{:}_\emptyset A$.

Definition 10.5 (Substitution) There are two types of substitutions in the language of FOLP: individual variables and proof variables.

Substitution for individual variables: We need the familiar notion of an individual variable y being free for x in a formula. This is defined as usual, but with one more case that reads as follows. An individual variable y is free for x in $t{:}_X A$ if, first, y is free for x in A and, second, if y occurs free in A then $y \in X$.

If x, y are individual variables and A is a formula, then by $A(x/y)$ we denote the result of substituting y for all free occurrences of x in A. We always assume that substitution is correct, that is, y is free for x in A.

Substitution for proof terms: Let p be a proof variable and t be a proof term. By $A(p/t)$ we denote the result of substitution of t for all occurrences of p in A.

Definition 10.6 (FOLP Axiom System) The first-order logic of proofs FOLP_0 is axiomatized as follows, where A, B are formulas, s, t are terms, X is a set of individual variables, and y is an individual variable.

Axiom Schemes:

(A1) classical axioms of first-order logic
(A2) $t{:}_{Xy} A \to t{:}_X A, \quad y \notin FVar(A)$
(A3) $t{:}_X A \to t{:}_{Xy} A$

(B1) $t{:}_X A \to A$
(B2) $s{:}_X(A \to B) \to (t{:}_X A \to (s \cdot t){:}_X B)$
(B3) $t{:}_X A \to (t+s){:}_X A, \quad s{:}_X A \to (t+s){:}_X A$
(B4) $t{:}_X A \to\ !t{:}_X t{:}_X A$
(B5) $t{:}_X A \to \mathsf{gen}_x(t){:}_X \forall x A,\ x \notin X$

Inference Rules:

(R1) $\vdash A, A \to B \ \Rightarrow\ \vdash B$ modus ponens

(R2) $\vdash A \;\Rightarrow\; \vdash \forall xA$ *generalization.*

As in the propositional case, we define a *Constant Specification* to be a set of formulas

$$c_1{:}A_1, c_2{:}A_2, \ldots$$

where all A_i are FOLP_0-axioms and all c_i are proof constants.

We reserve the name FOLP for the special case when the constant specification is *total*, i.e., contains $c{:}A$ for all c and A. Axiomatically, FOLP is obtained from FOLP_0 by adding to the former the rule

(R3) $\vdash c{:}A$, A is an axiom, c is a proof constant *axiom necessitation.*

We define derivations in FOLP and derivations from the hypothesis in the standard way. Let us recall that in a derivation from the set of hypotheses Γ, the generalization rule may not be applied to variables that are free in Γ.

Example 10.7 Let us derive (in FOLP) an explicit counterpart of the converse Barcan Formula

$$\Box\forall xA \to \forall x\Box A.$$

1. $\forall xA \to A$ - logical axiom;
2. $c{:}(\forall xA \to A)$ - axiom necessitation;
3. $c{:}_{\{x\}}(\forall xA \to A)$ - from 2, by axiom (A3);
4. $c{:}_{\{x\}}(\forall xA \to A) \to (u{:}_{\{x\}}\forall xA \to (c\cdot u){:}_{\{x\}}A)$ - axiom (B2);
5. $u{:}_{\{x\}}\forall xA \to (c\cdot u){:}_{\{x\}}A$ - from 3, 4, by modus ponens;
6. $u{:}\forall xA \to u{:}_{\{x\}}\forall xA$ - by axiom (A3);
7. $u{:}\forall xA \to (c\cdot u){:}_{\{x\}}A$ - from 5, 6;
8. $\forall x[u{:}\forall xA \to (c\cdot u){:}_{\{x\}}A]$ - from 7, by generalization;
9. $u{:}\forall xA \to \forall x(c\cdot u){:}_{\{x\}}A$ - from 8, because the antecedent of 8 does not contain x free.

The following two lemmas can be proved in the standard way.

Lemma 10.8 (Substitution) *If* FOLP $\vdash F$, *p is a proof variable and t is a proof term, then* FOLP $\vdash F(p/t)$.

Lemma 10.9 (Deduction) *If* $\Gamma, A \vdash F$ *in* FOLP, *then* $\Gamma \vdash A \to F$.

Theorem 10.10, which follows, establishes the internalization property for FOLP (though not in its most general form, but nonetheless sufficient for purposes of this book).

Theorem 10.10 (Internalization) *Let $p_0, \ldots, p_k$ be proof variables, $X_0, \ldots, X_k$ be sets of individual variables, and $X = X_0 \cup X_1 \cup \ldots \cup X_k$. Suppose that in* FOLP

$$p_0{:}_{X_0}A_0, \ldots, p_k{:}_{X_k}A_k \vdash F.$$

Then there exists a proof term $t(p_0, p_1, \ldots, p_k)$ such that

$$p_0{:}_{X_0}A_0, \ldots, p_k{:}_{X_k}A_k \vdash t{:}_X F.$$

Proof Induction on derivation of F from $p_0{:}_{X_0}A_0, \ldots, p_k{:}_{X_k}A_k$.

Case 1. F is an axiom of FOLP. By the axiom necessitation rule, $c{:}F$ is derivable for a proof constant c. By (A3), $c{:}_X F$ as well. Take $t = c$.

Case 2. F is $p_i{:}_{X_i}A_i$ for some i. By (B4), $!p_i{:}_{X_i}p_i{:}_{X_i}A_i$ is derivable, and then by (A3), $!p_i{:}_X p_i{:}_{X_i}A_i$ is also derivable. Take $t = !p_i$.

Case 3. F follows by modus ponens. Then, by the Induction Hypothesis, from $p_0{:}_{X_0}A_0, \ldots, p_k{:}_{X_k}A_k$ it is derivable that $s_1{:}_X(G \rightarrow F)$ and $s_2{:}_X G$ for some G, s_1, and s_2. By (B2), $(s_1 \cdot s_2){:}_X F$ is derivable. Take $t = s_1 \cdot s_2$.

Case 4. F follows by *generalization*, i.e., $F = \forall x G$ for some x not occurring free in the set of hypotheses. In particular, $x \notin X$. By the Induction Hypothesis, $s{:}_X G$ is derivable for some s. By (B5), $s{:}_X G \rightarrow \mathsf{gen}_x(s){:}_X \forall x G$, hence $\mathsf{gen}_x(s){:}_X \forall x G$ is derivable. Take $t = \mathsf{gen}_x(s)$.

Case 5. F follows by *axiom necessitation*, i.e., $F = c{:}A$ for some axiom A and constant c. By (B4), $!c{:}c{:}A$ is derivable, and then by (A3),so is $!c{:}_X c{:}A$. Take $t = !c$. □

In particular, given $\vdash F$, there is a proof term t containing no proof variables such that $\vdash t{:}F$. Such t can be chosen $+$-free.

10.2 Realization of FOS4 in FOLP

Definition 10.11 (Realization) Let A be a first-order modal formula. By a *realization* of a formula A we mean a formula A^r of the language of FOLP that is obtained from A by replacing all occurrences of subformulas of the form $\Box B$ by $t{:}_X B$ for some proof terms t where $X = FVar(B)$. To avoid unnecessary formalism, we suggest thinking of a realization as a result of an iterated procedure, which always replaces an innermost $\Box B$ by $t{:}_X B$. A realization is *normal* if all negative occurrences of $\Box$ are assigned proof variables.

Remark 10.12 If A^r is a realization of A, then for every subformula B of A,

$$FVar(B^r) = FVar(B).$$

As in the propositional case, we define a forgetful projection $(\cdot)^\circ$ of FOLP to modal logic. The straightforward definition of $(t{:}_X F)^\circ$ as $\Box F$ does not work in first-order logic. The problem is due to the fact that $\Box$ does not bind individual variables while the proof operator can bind them. For example, in order to define the appropriate modal reading of $t{:}_{\{x\}}F(x,y)$, we should respect the provability reading of this formula: It states that t is a derivation of a formula F with a "local" variable y bound in this formula for a given value of a "global" variable x. The naive projection of $t{:}_{\{x\}}F(x,y)$ as $\Box F(x,y)$ changes the meaning of this formula: both variables of this formula are now global and open for substitution, which gives a distorted account of the original formula. A more adequate definition of the forgetful projection should not change the status of individual variables.

A way to meet this condition is to require that forgetful projection binds local variables by the universal quantifier. Provability of a formula F with a free variable y is equivalent to the provability of $\forall yF$. So a forgetful projection of $t{:}_{\{x\}}F(x,y)$ that respects binding could be $\Box\forall yF(x,y)$. This example leads us to the following definition.

Definition 10.13 (Forgetful Functor) We define the forgetful projection $(\cdot)^\circ$ of FOLP formulas to the first-order modal language by induction on a formula complexity. For atomic formula F we have $F^\circ = F$. Forgetful projection commutes with Boolean connectives and quantifiers. For proof assertions we require the following,

$$(t{:}_X F)^\circ = \Box\forall y_0 \ldots \forall y_k F^\circ, \text{ where } \{y_0,\ldots,y_k\} = FVar(F)\setminus X.$$

Lemma 10.14 *If* FOLP $\vdash F$*, then* F° *is derivable in* FOS4.

Proof By straightforward induction on derivations in FOLP. □

Theorem 10.15 (Realization Theorem) *If* FOS4 $\vdash A$*, then there is a normal realization* A^r *such that* FOLP $\vdash A^r$.

Proof The proof is similar to that in Chapter 6 with additional care given to individual variables. We consider a Gentzen-style calculus for FOS4 and prove that for every sequent $\Gamma \Rightarrow \Delta$ that is provable in FOS4, there exists a realization r of all formulas from Γ and Δ such that FOLP $\vdash (\bigwedge\Gamma^r \rightarrow \bigvee\Delta^r)$. For this purpose, we take a cut-free derivation of $\Gamma \Rightarrow \Delta$ and construct realization for the whole derivation.

According to Troelstra and Schwichtenberg (1996), section 9.1.3, in addition to structural rules, the sequential calculus for FOS4, GFOS4, has the following axioms:

$$\bot \Rightarrow \qquad \text{and} \qquad P(\vec{x}) \Rightarrow P(\vec{x})$$

where P is a predicate letter, and logical rules:

$$\frac{\Gamma, A \Rightarrow \Delta, B}{\Gamma \Rightarrow \Delta, A \to B}\ (R\to), \qquad \frac{\Gamma \Rightarrow \Delta, A, \quad \Gamma, B \Rightarrow \Delta}{\Gamma, A \to B \Rightarrow \Delta}\ (L\to),$$

$$\frac{\Gamma \Rightarrow \Delta, A}{\Gamma \Rightarrow \Delta, \forall x\, A}\ (R\forall), \qquad \frac{\Gamma, A(y/x) \Rightarrow \Delta}{\Gamma, \forall x\, A \Rightarrow \Delta}\ (L\forall),$$

$$\frac{\Box\Gamma \Rightarrow A}{\Box\Gamma \Rightarrow \Box A}\ (R\Box), \qquad \frac{\Gamma, A \Rightarrow \Delta}{\Gamma, \Box A \Rightarrow \Delta}\ (L\Box).$$

In $R\forall$, we suppose that $x \notin FVar(\Gamma, \Delta)$.

The following connection between FOS4 and its Gentzen-style version GFOS4 takes place:

$$\mathsf{GFOS4} \vdash \Gamma \Rightarrow \Delta \ \text{ iff } \ \mathsf{FOS4} \vdash \bigwedge \Gamma \to \bigvee \Delta.$$

Cut-elimination holds in GFOS4 (Troelstra and Schwichtenberg, 1996): If GFOS4 $\vdash \Gamma \Rightarrow \Delta$, then $\Gamma \Rightarrow \Delta$ can be derived in GFOS4 without using the cut-rule.

Lemma 10.16 *If* GFOS4 $\vdash \Gamma \Rightarrow \Delta$*, then there exists a normal realization r such that* FOLP $\vdash (\bigwedge \Gamma^r \to \bigvee \Delta^r)$.

Proof Suppose that $\mathcal{D}$ is a cut-free derivation in GFOS4. We will construct a realization for each sequent $\Gamma \Rightarrow \Delta$ in $\mathcal{D}$ in such a way that the formula $\bigwedge \Gamma^r \to \bigvee \Delta^r$ is provable in FOLP.

Following the proof schema for the Realization Theorem from Chapter 6, Section 6.2, we split all occurrences of $\Box$'s in derivation $\mathcal{D}$ into families of related ones, cf. the proof of Theorem 6.5. Namely, two occurrences of $\Box$ are related if they occur in related subformulas of premises and conclusions of rules; we extend this relationship by reflexivity and transitivity. All the rules of GFOS4 respect polarities, hence all $\Box$'s in every family have the same polarity. So we can speak about *positive and negative families of* $\Box$*'s*. If f is a positive family, then all $\Box$'s from f are introduced either by weakening on the right or by the rule $(R\Box)$. If at least one $\Box$ in f is introduced by $(R\Box)$, then we call f an *essential* family, otherwise f is called *inessential family*.

Step 1. Initialization. To every negative or inessential positive family f we assign a fresh proof variable p_f. Replace all $\Box A$, where $\Box$ is from f, by $p_f{:}_X A$ with $X = FVar(A)$.

Suppose that f is an essential positive family. We enumerate the rules $(R\Box)$, which introduce $\Box$'s from the family f. Let $n(f)$ be the total number of such rules for f. For the $(R\Box)$–rule number k in the family f, where $k = 1, \ldots, n(f)$, we take a fresh proof variable u_k called *a provisional variable*. Finally, replace all $\Box A$ from the family f by

$$[u_1 + \ldots + u_{n(f)}]{:}_X A$$

with $X = FVar(A)$.

After initialization is completed, all nodes in the resulting tree $\mathcal{D}'$ are assigned formulas of the logic FOLP.

Step 2. Realization. Now we travel along the tree $\mathcal{D}'$ from leaves to root and replace all provisional variables by FOLP–terms. We retain the notation u_j for both provisional variables and terms substituted for them. The resulting tree is denoted by $\mathcal{D}^r$. By induction on the depth of a node in $\mathcal{D}'$, we prove that after the process passes the node $\Gamma \Rightarrow \Delta$ in $\mathcal{D}'$ and replaces it by $\Gamma^r \Rightarrow \Delta^r$,

(1) sequent $\Gamma^r \Rightarrow \Delta^r$ is derivable in FOLP[1];
(2) for every subformula B occurring in Γ, Δ, we have $FVar(B^r) = FVar(B)$.

We do not change the realization when the process passes sequents that are not conclusions of an $(R\Box)$–rule. All rules except $(R\Box)$ are admissible in FOLP, therefore the conclusions of those rules are derivable in FOLP as long as the premises are derivable.

The only case in which we alter realization is rule $(R\Box)$. Suppose that $\Gamma \Rightarrow \Delta$ is obtained by rule $(R\Box)$:

$$\frac{\Box A_1, \ldots, \Box A_k \Rightarrow A}{\Box A_1, \ldots, \Box A_k \Rightarrow \Box A}.$$

The $\Box$ symbol introduced by this rule belongs to an essential positive family f. Let this rule have the number i among rules $(R\Box)$, which introduce $\Box$'s from this family f, and $n = n(f)$.

Currently in $\mathcal{D}^r$, the node corresponding to the premise of this rule is assigned a sequent $q_1{:}_{X_1} B_1, \ldots, q_k{:}_{X_k} B_k \Rightarrow B$, which, by the Induction Hypothesis, is provable in FOLP. The node corresponding to the conclusion is assigned a sequent

$$q_1{:}_{X_1} B_1, \ldots, q_k{:}_{X_k} B_k \Rightarrow [u_1 + \ldots + u_i + \ldots + u_n]{:}_X B$$

where all q_j are proof variables, all u_j are either provisional variables or terms, u_i is a provisional variable, and $X = FVar(B)$.

By the Internalization Lemma, 10.10, there exists a term t such that FOLP

[1] Which means that FOLP $\vdash \wedge\Gamma^r \rightarrow \vee\Delta^r$, or equivalently, $\Gamma^r \vdash \vee\Delta^r$.

derives

$$q_1{:}_{X_1}B_1, \ldots, q_k{:}_{X_k}B_k \Rightarrow t{:}_Y B$$

where $Y = X_1 \cup X_2 \cup \ldots \cup X_n$. Using axiom (A2), we remove from Y all variables that are not in $FVar(B)$ and obtain $Y' = Y \cap FVar(B)$. Then, by (A3), add to Y' all free variables of B that were not yet there and obtain X. The resulting sequent

$$q_1{:}_{X_1}B_1, \ldots, q_k{:}_{X_k}B_k \Rightarrow t{:}_X B$$

is provable in FOLP. Therefore, by (B3),

$$q_1{:}_{X_1}B_1, \ldots, q_k{:}_{X_k}B_k \Rightarrow [u_1 + \ldots + u_{i-1} + t + u_{i+1} + \ldots + u_n]{:}_X B$$

is also provable in FOLP. Replace provisional variable u_i by t everywhere in $\mathcal{D}^r$. By the Substitution Lemma, this substitution respects provability in FOLP. □

Given the Lemma, the proof of the Theorem is now immediate. Because FOS4 $\vdash A$, there is a cut-free proof of sequent $\Rightarrow A$ in GFOS4. By Lemma 10.16, there exists its normal realization $\Rightarrow A^r$ provable in FOLP, i.e., FOLP $\vdash A^r$. □

Similar to Theorem 6.5, this proof fits for all *axiomatically appropriate and schematic* constant specifications CS, cf. Section 6.3.

Corollary 10.17 FOS4 *is the forgetful projection of* FOLP.

Corollary 10.18 *F is derivable in* HPC *if and only if its Gödel translation is realizable in* FOLP.

10.2.1 Implications for First-Order BHK Semantics

Gödel's translation of HPC into FOS4, followed by realization of FOS4 in FOLP, provide a formal analysis of first-order BHK semantics which respects Kreisel's critique (Kreisel, 1962; cf. also Dean and Kurokawa, 2016). Along with fixing clauses for implication and negation, cf. Section 9.4, FOLP offers a fix to the case of universal quantifier. In the original BHK, this case was presented as

- a proof of $\forall x A(x)$ is a function converting c into a proof of $A(c)$.

The deficiency of this definition is vividly demonstrated by the following example which is due to Helmut Schwichtenberg (Schwichtenberg, n.d.).

> Here is a simple "constructive proof" S of the Fermat's Last Theorem (FLT): Let u range over quadruples of integers (x,y,z,n), and $F(u)$ is the standard Fermat's condition that if $x,y,z > 0$ and $n > 2$, then $x^n + y^n \neq z^n$, which is clearly algorithmically verifiable for each specific u. Algorithm S takes any specific quadruple $u = d$, substitutes it to $F(u)$, and presents a straightforward PA-derivation of $F(d)$. Apparently, S satisfies the aforementioned BHK $\forall$-clause, but could not by any stretch of imagination be called a proof of FLT.

It is clear that in this way any true Π_1-sentence has a "constructive" BHK proof, which is unacceptable.

FOLP realization offers a natural fix. To simplify the notation, assume that $A(x)$ is atomic. An intuitionistic statement $\forall x A(x)$ is represented in FOS4 by

$$\Box\forall x\Box A(x).$$

Its realization in FOLP

$$u{:}\forall x[v{:}_{\{x\}}A(x)].$$

states that there is a uniform proof u that for each c, the substitution of c for x produces a proof $v(c)$ of $A(c)$.

This reading is the one that naturally comes to mind after inspecting the Schwichtenberg Paradox: Algorithm S fails this new BHK test because it does not provide a required proof that for all u, $S(u)$ is indeed a proof of $F(u)$.

Here is the corrected FOLP-compliant reading of the BHK clause for $\forall$, as suggested by Gödel's embedding and its subsequent realization in FOLP:

- a constructive proof of $\forall x A(x)$ is a pair (f,d) where f is a function and d is a classical proof such that for each c, $f(c)$ is a classical proof of $A(c)$.

10.3 Possible World Semantics for FOLP

Starting in Section 10.4 an arithmetic semantics is presented for FOLP—an essential step for an arithmetic interpretation of quantified intuitionistic logic. Of course this is closely related to the material found in Chapter 9. But first, in this section, a possible world semantics is given, combining features of first-order Kripke models for S4 with features of the possible world semantics for propositional justification logic given in Chapter 4.

10.3.1 The Ideas Informally

Fitting models for propositional justification logics are built on top of propositional Kripke models. In a similar way first-order justification models are based on the possible world semantics appropriate for quantified modal logics. Because we want a justification analog of first-order S4, we build on possible world frames having a transitive, reflexive accessibility relation. The most common first-order S4 models also have domains of quantification associated with each possible world, meeting a monotonicity condition. Monotonicity says that if possible world Δ is accessible from possible world Γ, then the quantification domain associated with Γ is a subset of that associated with Δ. Why monotonicity? As was the case historically with LP, the ultimate goal is to provide a semantics for intuitionistic logic, though now quantifiers are involved. But why should intuitionistic logic have something to do with monotonicity?

Classical mathematics is Platonic—its structures are simply there, changeless and timeless. Many mathematicians think they are discovered and not created. Constructive mathematics is decidedly otherwise. Brouwer talked about the creative subject, who actually constructs mathematical objects. Free choice in the process is allowed. In fact, these ideas are not really foreign to the classical mathematician. Mathematical structures that are not known to us are as if they aren't there at all. At one time there were no such things as complex numbers within the known mathematical universe. Historically, they gradually moved into our view. Whether they were there all along is not epistemically important. What is important is that, in the realm of what we know, complex numbers once were not, and then were. It is reasonable to assume that after creation as in intuitionism, or discovery as in Platonism, a mathematical structure continues to exist, or to be known. We do not forget. Whether we consider things constructively or epistemically, monotonicity and not constant domains is appropriate.

Propositional connectives will be truth functional at each world. Quantification at each possible world is over the domain associated with that world. The central issue is how proof terms, or justification terms, behave. As an example we consider the formula $t{:}_{\{x,y\}}Q(x, y, z, w)$ at possible world Γ. In this formula occurrences of x and y are free, but not those of z or w. We could use the machinery of valuations to assign values to free variables, but it is simpler and more perspicuous to allow members of domains to appear directly in formulas. Say a and b are in the quantification domain associated with Γ; we will talk about $t{:}_{\{a,b\}}Q(a, b, z, w)$ instead of talking about $t{:}_{\{x,y\}}Q(x, y, z, w)$ under the valuation that maps x to a and y to b. With this notational convention understood, what should it mean for $t{:}_{\{a,b\}}Q(a, b, z, w)$ to be true at Γ? Much as was the case with propositional justification logics, there will be two conditions, one syntactic, one semantic.

We use the *evidence function* machinery we have already seen in the propositional setting, Section 4.2. Recall that for an evidence function $\mathcal{E}$, for each proof term t and each formula A, $\mathcal{E}(t, A)$ is the set of possible worlds at which t can serve as meaningful evidence for A. Of course $\mathcal{E}$ must meet certain closure conditions, and this will be taken care of in Definition 10.21.

In propositional Fitting models we take $t{:}A$ to be true at a possible world if A is true at all accessible worlds (the Kripkean condition) and also t serves as meaningful evidence for A at that world. In the first-order setting things are more complicated because of the distinction between the two roles that variables can play in proofs. Recall, in $t{:}_{\{x,y\}}Q(x, y, z, w)$ the variables in $\{x, y\}$ are supposed to be those that can be substituted for, but the variables z and w are the ones to which universal generalization can be applied. Then for the variables in $\{x, y\}$ we will only talk about truth at a possible world Γ for *instances*, such as $t{:}_{\{a,b\}}Q(a, b, z, w)$, where a and b are in the domain of Γ—in effect, we confine things to what proof term t says about the results of assignment to those variables that are, in fact, subject to substitution. There remains the idea that z and w are universally quantifiable. We incorporate this by saying $Q(a, b, c, d)$ is true at every possible world Δ accessible from Γ, for every c, d in the quantificational domain of Δ. Roughly, the z and w play universal roles in $t{:}_{\{a,b\}}Q(a, b, z, w)$ because, no matter what future work we might carry out (the move from Γ to an accessible Δ), and no matter what mathematical objects we might encounter (any c and d available at Δ), we will have $Q(a, b, c, d)$.

In brief, we will take $t{:}_{\{a,b\}}Q(a, b, z, w)$ to be true at possible world Γ provided t is meaningful evidence for $Q(a, b, z, w)$ at Γ, that is, $\Gamma \in \mathcal{E}(t, Q(a, b, z, w))$, and for every Δ accessible from Γ, and for every c, d in the quantificational domain of Δ, $Q(a, b, c, d)$ is true at Δ.

10.3.2 FOLP Fitting Models

We begin with a small word about terminology. In Definition 10.6 FOLP_0 was characterized axiomatically, then FOLP was introduced as FOLP_0 combined with the *total* constant specification. Here we will be a little more nuanced in our requirements for constant specifications, and so we will write FOLP(CS) for FOLP_0 with constant specification CS. Then FOLP_0 itself is the same as FOLP($\emptyset$), and FOLP is FOLP($\mathcal{T}$), using the empty constant specification and the total constant specification respectively.

The following formal definitions incorporate, quite directly, what we discussed informally in the previous section.

Definition 10.19 (Skeleton) $\langle \mathcal{G}, \mathcal{R}, \mathcal{D}\rangle$ is an FOLP *skeleton* provided:

(1) $\langle \mathcal{G}, \mathcal{R}\rangle$ is an S4 frame: $\mathcal{G}$ is a nonempty set of possible worlds, and $\mathcal{R}$ is a reflexive and transitive accessibility relation on $\mathcal{G}$,

(2) $\mathcal{D}$ is a monotonic domain function mapping each member of $\mathcal{G}$ to a nonempty set, where monotonicity means that if $\Gamma\mathcal{R}\Delta$ then $\mathcal{D}(\Gamma) \subseteq \mathcal{D}(\Delta)$.
(3) We set $\mathcal{D}^* = \cup\{\mathcal{D}(\Gamma) \mid \Gamma \in \mathcal{G}\}$, and call $\mathcal{D}^*$ the domain of the skeleton.

Before building models on skeletons, we have some remarks about languages. It is common in model theory to use formulas in which members of the domain of the model appear as if they were individual constant symbols of the language itself. This is mathematically justifiable because a formula is a sequence of symbols, and members of the domain are as much entitled to be symbols as anything else. It also eliminates the need for valuation functions. We have already been doing this in our informal discussion earlier.

Definition 10.20 (*D*-formulas) Let D be a nonempty set. A D-formula is the result of replacing some (possibly all) free occurrences of individual variables in an FOLP formula with members of D. In a D formula members of D act like individual constant symbols, not free variables, and we refer to them as *domain constants*, or *D constants* if we need to be specific. We call a D-formula *closed* if it contains no free occurrences of individual variables, though domain constants are allowed.

Suppose a, b are domain constants, x, y are individual variables, and Q is a two-place relation symbol. In the formula $t{:}_{\{x\}}Q(x, y)$ the occurrences of x are free, and the occurrence of y is bound. Then $t{:}_{\{a\}}Q(a, y)$ is a D formula, though y is not free because it does not occur in the subscript set, and a is not free simply because it is not an individual variable. Finally $t{:}_{\{a\}}Q(a, b)$ is not well-formed, informally because the occurrence of b would also count as not free, but domain constants cannot be bound. More formally, the expression $t{:}_{\{a\}}Q(a, b)$ cannot arise by substituting domain constants for free individual variables in an FOLP formula.

When working with D-formulas we systematically use $\vec{x}, \vec{y}, \ldots$ as sequences of individual variables, and $x, y, \ldots$ as single individual variables. Likewise we use $\vec{a}, \vec{b}, \ldots$ as sequences of D constants, and $a, b, \ldots$ as single D constants. Whenever possible we are informal about substitution notation. If we write $A(\vec{x})$, and later we write $A(\vec{a})$, we mean that free occurrences of the individual variables in $\vec{x}$ (if any) have been replaced with corresponding occurrences of domain constants in $\vec{a}$. In more complicated circumstances we will use notation like $\{\vec{x}/\vec{a}, y/b\}$ to indicate the substitution that replaces free occurrences of variables in $\vec{x}$ with D constants in $\vec{a}$, and replaces free occurrences of y with occurrences of b. We assume D constants can always be substituted for free occurrences of individual variables—in effect a D constant is always free for a variable in a formula.

Models are built on skeletons, Definition 10.19, and we refer to the domain of a *model*, meaning the domain of its underlying skeleton. We work with formulas *of the model*, $\mathcal{D}^*$-formulas where $\mathcal{D}^*$ is the domain of the model, or with $\mathcal{D}(\Gamma)$-formulas, where $\mathcal{D}(\Gamma)$ is the domain associated with possible world Γ.

Definition 10.21 (FOLP Fitting Models) Let $\langle \mathcal{G}, \mathcal{R}, \mathcal{D}\rangle$ be an FOLP skeleton. A model *based on* $\langle \mathcal{G}, \mathcal{R}, \mathcal{D}\rangle$ is a structure $\mathcal{M} = \langle \mathcal{G}, \mathcal{R}, \mathcal{D}, \mathcal{I}, \mathcal{E}\rangle$ where:

(1) $\mathcal{I}$ is an *interpretation function*—for each n-place relation symbol Q and each $\Gamma \in \mathcal{G}$, $\mathcal{I}(Q, \Gamma)$ is an n-place relation on $\mathcal{D}(\Gamma)$.

(2) $\mathcal{E}$ is an *evidence function*—for each proof term t and each $\mathcal{D}^*$-formula A, $\mathcal{E}(t, A)$ is some set of possible worlds. $\mathcal{E}$ must meet the condition that, if $\Gamma \in \mathcal{E}(t, A)$ then all domain constants occurring in A are from $\mathcal{D}(\Gamma)$ (or as we say here, A *lives in* Γ). Evidence functions satisfy the following closure conditions.

$\cdot$ **Condition** $\mathcal{E}(s, A \to B) \cap \mathcal{E}(t, A) \subseteq \mathcal{E}(s \cdot t, B)$.

$+$ **Condition** $\mathcal{E}(s, A) \cup \mathcal{E}(t, A) \subseteq \mathcal{E}(s + t, A)$.

! **Condition** $\mathcal{E}(t, A) \subseteq \mathcal{E}(!t, t{:}_X A)$, where X is the set of domain constants in A.

$\mathcal{R}$ **Closure Condition** $\Gamma \in \mathcal{E}(t, A)$ and $\Gamma\mathcal{R}\Delta$ imply $\Delta \in \mathcal{E}(t, A)$.

Instantiation Condition $\Gamma \in \mathcal{E}(t, A(x))$ and $a \in \mathcal{D}(\Gamma)$ imply $\Gamma \in \mathcal{E}(t, A(a))$.

gen_x **Condition** $\mathcal{E}(t, A) \subseteq \mathcal{E}(\mathsf{gen}_x(t), \forall x A)$.

Condition (1) listed earlier is essentially semantic, while condition (2) is syntactic. The idea behind the evidence function is as it was propositionally: if $\Gamma \in \mathcal{E}(t, A)$, then informally Γ is a possible world in which t serves as relevant evidence for the formula A. Of the closure conditions imposed on evidence functions, the first four come from LP, and the last two are new to FOLP.

Note that the monotonicity condition on domains figures into the $\mathcal{R}$ Closure Condition earlier. If $\Gamma \in \mathcal{E}(t, A)$, A must live in Γ. If also $\Gamma\mathcal{R}\Delta$ then using monotonicity, $\mathcal{D}(\Gamma) \subseteq \mathcal{D}(\Delta)$, so A lives in Δ too, and thus part of the requirement for $\Delta \in \mathcal{E}(t, A)$ is automatic. The Instantiation Condition is connected with the idea that a proof of $\varphi(x)$ is a template from which we can generate proofs of $\varphi(a), \varphi(b), \ldots$, with all proofs having the same structure.

We remind the reader that in Definition 10.4 the notation $t{:}A$ was introduced as an abbreviation for $t{:}_\emptyset A$. We make use of it in the following.

Definition 10.22 (Constant Specifications) Let $\mathcal{M} = \langle \mathcal{G}, \mathcal{R}, \mathcal{D}, \mathcal{I}, \mathcal{E}\rangle$ be a Fitting model. The evidence function $\mathcal{E}$ *meets the constant specification* CS provided, if $c{:}A \in \mathsf{CS}$ then $\Gamma \in \mathcal{E}(c, A)$, for each $\Gamma \in \mathcal{G}$ where A lives in Γ. The model meets a constant specification if its evidence function does. If $\mathcal{M}$ meets constant specification CS, we say $\mathcal{M}$ is an FOLP(CS) model.

Now we define truth at possible worlds of models. Truth is first defined *for formulas having no free individual variables*, though they can contain domain constants. These are the closed $\mathcal{D}^*$ formulas. Subsequently the definition is extended.

Definition 10.23 (Truth At Worlds) Let $\mathcal{M} = \langle \mathcal{G}, \mathcal{R}, \mathcal{D}, \mathcal{I}, \mathcal{E} \rangle$ be an FOLP(CS) model, and let A be a closed $\mathcal{D}^*$ formula that lives in $\Gamma \in \mathcal{G}$. We write $\mathcal{M}, \Gamma \Vdash A$ to symbolize that A is true at world Γ. This meets the following conditions:

(1) For an n-place predicate symbol Q, $\mathcal{M}, \Gamma \Vdash Q(\vec{d}) \Longleftrightarrow \langle \vec{d} \rangle \in \mathcal{I}(\Gamma, Q)$;
(2) $\mathcal{M}, \Gamma \nVdash \bot$;
(3) $\mathcal{M}, \Gamma \Vdash A \rightarrow B \Longleftrightarrow \mathcal{M}, \Gamma \nVdash A$ or $\mathcal{M}, \Gamma \Vdash B$, and similarly for other propositional connectives;
(4) $\mathcal{M}, \Gamma \Vdash \forall x A(x) \Longleftrightarrow \mathcal{M}, \Gamma \Vdash A(a)$ for every $a \in \mathcal{D}(\Gamma)$;
(5) Assume $t{:}_X A(\vec{x})$ is closed and $\vec{x}$ are all the free variables of A. (Note that because the formula is closed, members of X must be domain constants.) $\mathcal{M}, \Gamma \Vdash t{:}_X A(\vec{x}) \Longleftrightarrow$
   (a) $\mathcal{M}, \Delta \Vdash A(\vec{d})$ for every $\Delta \in \mathcal{G}$ such that $\Gamma \mathcal{R} \Delta$ and for every $\vec{d}$ in $\mathcal{D}(\Delta)$ and
   (b) $\Gamma \in \mathcal{E}(t, A(\vec{x}))$.

Truth in models is defined for formulas allowing domain constants. But basically, we are interested in formulas of FOLP itself, and these do not contain domain constants. The following covers this.

Definition 10.24 (Validity) Let $\mathcal{M} = \langle \mathcal{G}, \mathcal{R}, \mathcal{D}, \mathcal{I}, \mathcal{E} \rangle$ be an FOLP Fitting model. A closed FOLP formula A (without domain constants, because it is an FOLP formula) is *valid in* $\mathcal{M}$ if $\mathcal{M}, \Gamma \Vdash A$ for every $\Gamma \in \mathcal{G}$, and A is *valid* using constant specification CS if it is valid in every model FOLP(CS). An FOLP formula that is not closed is valid, or valid in a model, if its universal closure is.

The definition of validity for FOLP formulas with free individual variables needs a few small comments. Consider $A(x)$ as an example, where only x has a free occurrence. To show validity of $A(x)$ we must show validity of $\forall x A(x)$. To show $\forall x A(x)$ is true at a possible world of an FOLP model we must show the truth of $A(a)$ for each a in the domain of that possible world. Thus showing validity of an FOLP formula with free individual variables amounts to showing the truth, at each possible world Γ, of all instances of the formula that live in Γ.

10.3.3 Nonvalidity Examples

As we've noted several times, models have both a semantic and a syntactic component. It often happens that nonvalidity can be shown by appropriate use of only one of these. In the following examples we concentrate on the semantic, possible world, side and trivialize the evidence function. Of course once soundness is shown, in Section 10.28, examples in this section become nonprovability examples as well.

Definition 10.25 Let $\mathcal{M} = \langle \mathcal{G}, \mathcal{R}, \mathcal{D}, \mathcal{I}, \mathcal{E} \rangle$ be an FOLP Fitting model. We say the evidence function $\mathcal{E}$ is *universal* provided $\Gamma \in \mathcal{E}(t, A)$ whenever A lives in Γ, for every justification term t.

Note that a universal evidence function meets every constant specification. When using a universal evidence function, a justification term t serves as evidence for A wherever $t{:}A$ makes sense. It is easy to check that a universal evidence function meets all the evidence closure conditions of Definition 10.21.

We sometimes give diagrams representing models. In these we systematically omit the display of arrows representing reflexivity, but they may play a significant role, and their implicit presence should be remembered.

Example 10.26 Axiom (A2) asserts $t{:}_{Xy}A \to t{:}_X A$, provided y does not occur free in A. The proviso is necessary. We will show the nonvalidity of $t{:}_{\{x,y\}}Q(x,y) \to t{:}_{\{x\}}Q(x,y)$, where $Q(x,y)$ is atomic and the only individual variables are the ones displayed.

Recall the comments at the end of Section 10.3.2. To show nonvalidity of an FOLP formula containing free variables, we must find an FOLP model, a possible world Γ of it, and a closed instance of the formula, involving domain constants, that lives in Γ but is not true at Γ.

Let $\langle \mathcal{G}, \mathcal{R}, \mathcal{D} \rangle$ be a skeleton given by: $\mathcal{G} = \{\Gamma, \Delta\}$; $\mathcal{R}$ is reflexive on $\mathcal{G}$ and also $\Gamma \mathcal{R} \Delta$; $\mathcal{D}(\Gamma) = \{a, b\}$ and $\mathcal{D}(\Delta) = \{a, b, c\}$. We build a model $\mathcal{M} = \langle \mathcal{G}, \mathcal{R}, \mathcal{D}, \mathcal{I}, \mathcal{E} \rangle$ on this. First we set $\mathcal{I}(\Gamma, Q) = \mathcal{I}(\Delta, Q) = \{\langle a, b \rangle\}$. Second we set $\mathcal{E}$ to be the universal evidence function. Here is the model schematically. Because the evidence function is universal, it is not shown. Also, recall we do not indicate reflexivity of the accessibility relation.

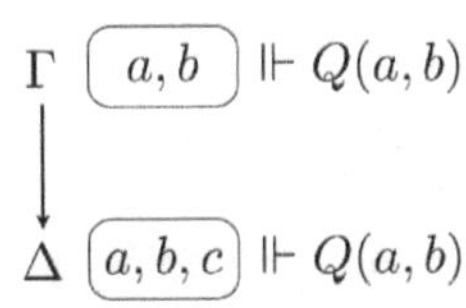

Consider the instance of $t{:}_{\{x,y\}}Q(x,y) \to t{:}_{\{x\}}Q(x,y)$ resulting from the substitution $\{x/a, y/b\}$, which is a formula that lives in Γ. We show this instance is not true at Γ by showing the following.

$$\mathcal{M}, \Gamma \Vdash t{:}_{\{a,b\}}Q(a, b) \text{ but } \mathcal{M}, \Gamma \nVdash t{:}_{\{a\}}Q(a, y)$$

We have $\mathcal{M}, \Gamma \Vdash t{:}_{\{a,b\}}Q(a, b)$ because $\Gamma \in \mathcal{E}(t, Q(a, b))$, and also $\mathcal{M}, \Gamma \Vdash Q(a, b)$ and $\mathcal{M}, \Delta \Vdash Q(a, b)$. We have $\mathcal{M}, \Gamma \nVdash t{:}_{\{a\}}Q(a, y)$ because, although $\Gamma \in \mathcal{E}(t, Q(a, y))$, we do not have that $\mathcal{M}, \Delta \Vdash Q(a, c)$, violating (5)(a) of Definition 10.23. We thus have $\mathcal{M}, \Gamma \nVdash t{:}_{\{a,b\}}Q(a, b) \to t{:}_{\{a\}}Q(a, y)$.

We showed nonvalidity by constructing a two-world model in which the evidence function is universal, and hence trivial. All the work is done by the modal structure. The reader might try constructing a one-world counter model in which all the work is done by the evidence function.

Example 10.27 In the next section we show validity of axiom scheme (B5), $t{:}_X A \to \mathsf{gen}_x(t){:}_X \forall x A$, where $x \notin X$. Here we construct a model to show validity does not hold if the proviso that $x \notin X$ is not met. Specifically, we show nonvalidity of the special case $t{:}_{\{x\}}Q(x) \to \mathsf{gen}_x(t){:}_{\{x\}}\forall x Q(x)$ (where $Q(x)$ is atomic).

Let $\mathcal{M} = \langle \mathcal{G}, \mathcal{R}, \mathcal{D}, \mathcal{I}, \mathcal{E} \rangle$ where: $\mathcal{G} = \{\Gamma\}$, with $\Gamma \mathcal{R} \Gamma$. Also let $\mathcal{D}(\Gamma) = \{a, b\}$, and $\mathcal{I}(Q, \Gamma) = \{a\}$. Finally, let $\mathcal{E}$ be the universal evidence function.

$$\Gamma \ \boxed{a, b} \ \Vdash Q(a)$$

We have $\mathcal{M}, \Gamma \Vdash t{:}_{\{a\}}Q(a)$ because $\Gamma \in \mathcal{E}(t, Q(a))$ (because $\mathcal{E}$ is universal) and we have $\mathcal{M}, \Gamma \Vdash Q(a)$ (and Γ is the only possible world accessible from Γ). If we had $\mathcal{M}, \Gamma \Vdash \mathsf{gen}_x(t){:}_{\{a\}}\forall x Q(x)$ we should also have $\mathcal{M}, \Gamma \Vdash \forall x Q(x)$ (reflexivity again), but $\mathcal{M}, \Gamma \nVdash Q(b)$ and $b \in \mathcal{D}(\Gamma)$. Thus $\mathcal{M}, \Gamma \nVdash t{:}_{\{a\}}Q(a) \to \mathsf{gen}_x(t){:}_{\{a\}}\forall x Q(x)$, from which nonvalidity of $t{:}_{\{x\}}Q(x) \to \mathsf{gen}_x(t){:}_{\{x\}}\forall x Q(x)$ follows.

10.3.4 Soundness

Axiomatic soundness follows the standard pattern. We omit explicit discussion of constant specifications, which is straightforward. Each of the FOLP_0 axioms is valid in all Fitting FOLP models, the rules preserve validity, and hence each theorem of FOLP_0 is valid. We show this for four representative axioms and omit discussion of the rest, and to keep notation simple we work with special cases that are sufficiently typical. In what follows, assume $\mathcal{M} = \langle \mathcal{G}, \mathcal{R}, \mathcal{D}, \mathcal{E}, \mathcal{I} \rangle$ is a Fitting FOLP(CS) model.

(A2) $t{:}_{Xy}A \rightarrow t{:}_X A$, where y does not occur free in A. We consider the special case with $X = \{x\}$ and $A = A(x,z)$, where $y \neq x$ and $y \neq z$. Thus we show validity of $t{:}_{\{x,y\}}A(x,z) \rightarrow t{:}_{\{x\}}A(x,z)$. Let $\Gamma \in \mathcal{G}$ and $a, b \in \mathcal{D}(\Gamma)$; we will show that $\mathcal{M}, \Gamma \Vdash t{:}_{\{a,b\}}A(a,z) \rightarrow t{:}_{\{a\}}A(a,z)$. The reasoning is quite simple.

The Evidence Function condition to be met for $\mathcal{M}, \Gamma \Vdash t{:}_{\{a,b\}}A(a,z)$ is that $\Gamma \in \mathcal{E}(t, A(a,z))$, and this is also the condition for $\mathcal{M}, \Gamma \Vdash t{:}_{\{a\}}A(a,z)$. The modal condition for $\mathcal{M}, \Gamma \Vdash t{:}_{\{a,b\}}A(a,z)$ is that $\mathcal{M}, \Delta \Vdash A(a,d)$ for every $\Delta \in \mathcal{G}$ with $\Gamma\mathcal{R}\Delta$ and every $d \in \mathcal{D}(\Delta)$, and this is the same modal condition for $\mathcal{M}, \Gamma \Vdash t{:}_{\{a\}}A(a,z)$.

(A3) $t{:}_X A \rightarrow t{:}_{Xy}A$. (Recall, Xy presupposes that $y \notin X$.) Again we discuss a special case, this time where $X = \{x\}$ and $A = A(x,y,z)$, so we consider $t{:}_{\{x\}}A(x,y,z) \rightarrow t{:}_{\{x,y\}}A(x,y,z)$. Let $\Gamma \in \mathcal{G}$ and instantiate with $\{x/a, y/b\}$ where $a, b \in \mathcal{D}(\Gamma)$; we show $\mathcal{M}, \Gamma \Vdash t{:}_{\{a\}}A(a,y,z) \rightarrow t{:}_{\{a,b\}}A(a,b,z)$.

Assume $\mathcal{M}, \Gamma \Vdash t{:}_{\{a\}}A(a,y,z)$. By part (5) of Definition 10.23 this has two consequences, which we consider separately.

First, $\Gamma \in \mathcal{E}(t, A(a,y,z))$. It follows from the Instantiation Condition of Definition 10.21 that $\Gamma \in \mathcal{E}(t, A(a,b,z))$.

Second, for every $\Delta \in \mathcal{G}$ with $\Gamma\mathcal{R}\Delta$, $\mathcal{M}, \Delta \Vdash A(a,y,z))$ for all substitutions of members of $\mathcal{D}(\Delta)$ for y and z. Because $b \in \mathcal{D}(\Gamma)$, by monotonicity $b \in \mathcal{D}(\Delta)$. Then $\mathcal{M}, \Delta \Vdash A(a,b,z)$ for all substitutions of members of $\mathcal{D}(\Delta)$ for z.

By part (5) of Definition 10.23 again, it follows that $\mathcal{M}, \Gamma \Vdash t{:}_{\{a,b\}}A(a,b,z)$.

(B4) $t{:}_X A \rightarrow\ !t{:}_X t{:}_X A$. Our representative special case is $X = \{x\}$ and $A = A(x,y)$, so our formula is $t{:}_{\{x\}}A(x,y) \rightarrow\ !t{:}_{\{x\}}t{:}_{\{x\}}A(x,y)$. Let $\Gamma \in \mathcal{G}$ and instantiate with $\{x/a\}$ where $a \in \mathcal{D}(\Gamma)$. We show $\mathcal{M}, \Gamma \Vdash t{:}_{\{a\}}A(a,y) \rightarrow\ !t{:}_{\{a\}}t{:}_{\{a\}}A(a,y)$. Suppose $\mathcal{M}, \Gamma \Vdash t{:}_{\{a\}}A(a,y)$.

First, $\Gamma \in \mathcal{E}(t, A(a,y))$ so by the **!** Condition of Definition 10.21, we have $\Gamma \in \mathcal{E}(!t, t{:}_{\{a\}}A(a,y))$.

Suppose $\Gamma\mathcal{R}\Delta$ and $\Delta\mathcal{R}\Omega$. Because $\mathcal{R}$ is transitive, $\Gamma\mathcal{R}\Omega$ and because $\mathcal{M}, \Gamma \Vdash t{:}_{\{a\}}A(a,y)$ then $\mathcal{M}, \Omega \Vdash A(a,y)$ for every instance of y from $\mathcal{D}(\Omega)$. Also because $\Gamma \in \mathcal{E}(t, A(a,y))$ then $\Delta \in \mathcal{E}(t, A(a,y))$ by the $\mathcal{R}$ Closure Condition of Definition 10.21. Because Ω is arbitrary, $\mathcal{M}, \Delta \Vdash t{:}_{\{a\}}A(a,y)$. And because Δ is arbitrary, $\mathcal{M}, \Gamma \Vdash\ !t{:}_{\{a\}}t{:}_{\{a\}}A(a,y)$.

(B5) $t{:}_X A \rightarrow \mathsf{gen}_x(t){:}_X \forall x A$, where $x \notin X$. Again, a representative case: $X = \{y\}$ and $A = A(x,y,z)$, so the formula is $t{:}_{\{y\}}A(x,y,z) \rightarrow \mathsf{gen}_x(t){:}_{\{y\}} \forall x A(x,y,z)$, where $x \neq y$. Let $\Gamma \in \mathcal{G}$ and consider the instantiation $\{y/b\}$ where $b \in \mathcal{D}(\Gamma)$. We show that $\mathcal{M}, \Gamma \Vdash t{:}_{\{b\}}A(x,b,z) \rightarrow \mathsf{gen}_x(t){:}_{\{b\}} \forall x A(x,b,z)$. Assume $\mathcal{M}, \Gamma \Vdash t{:}_{\{b\}}A(x,b,z)$.

We have that $\Gamma \in \mathcal{E}(t, A(x,b,z))$. By the gen_x Condition of Definition 10.21, it follows that $\Gamma \in \mathcal{E}(\mathsf{gen}_x(t), (\forall x)A(x,b,z))$.

Next, for every $\Delta \in \mathcal{G}$ with $\Gamma\mathcal{R}\Delta$, $\mathcal{M}, \Delta \Vdash A(a,b,c)$ for all a, c in $\mathcal{D}(\Delta)$.

By (4) of Definition 10.23, $\mathcal{M}, \Delta \Vdash \forall x A(x,b,c)$ for all c in $\mathcal{D}(\Delta)$. Now by (5) of Definition 10.23, we have $\mathcal{M}, \Gamma \Vdash \mathsf{gen}_x(t){:}_{\{b\}} \forall x A(x,b,z)$.

Note that Example 10.27 shows the restriction $x \notin X$ is essential for validity of instances of (B5).

The other axioms are valid, and the rules preserve validity—verification is left to the reader. It follows that the axiom system for FOLP_0 is sound with respect to the Fitting semantics. Further, constant specifications are quite straightforward and are left to the reader.

Theorem 10.28 (Soundness) *Let* CS *be a constant specification. If the* FOLP *formula A is provable using constant specification* CS *then A is valid in every Fitting* FOLP(CS) *model.*

10.3.5 Language Extensions

Let us call the basic FOLP language $\mathcal{L}$ from now on. Of course completeness is for formulas of $\mathcal{L}$ itself, but because our completeness proof is Henkin style we will need to extend $\mathcal{L}$ by adding what are often called *witness variables*. Let $\mathbf{V}$ be a countable set of symbols not used in $\mathcal{L}$, fixed for the rest of this chapter. We will use members of $\mathbf{V}$ as our witness variables. In fact, we will need multiple extensions of $\mathcal{L}$ because we will have multiple possible worlds in the models we construct, and we do not have constant domain models. For this we will simply use various subsets of $\mathbf{V}$.

Definition 10.29 Let $W \subseteq \mathbf{V}$, where $W = \emptyset$ is allowed. $\mathcal{L}(W)$ is the language defined like $\mathcal{L}$ except that it also allows individual variables from W, as well as justification operators gen_w for $w \in W$.

Note that $\mathcal{L}$ is the same as $\mathcal{L}(\emptyset)$. Also note that $\mathcal{L}(W)$ differs from $\mathcal{L}$ only in its set of individual variables—the sets of justification variables and justification constants are unchanged. The axiomatization of FOLP is by schemes, and schemes make sense for formulas in a language $\mathcal{L}(W)$, for any W. When working with a language $\mathcal{L}(W)$ we assume instances of axiom schemes can involve individual variables from W, without further comment.

Constant specifications require some care. In the propositional setting soundness holds with respect to any constant specification, and we saw in Section 10.3.4 that this carries over to FOLP. Propositionally, for completeness, we had to require the condition of axiomatic appropriateness for constant specifications, Definition 2.7. Not surprisingly this is also the case when quantifiers are involved. The following is a version appropriate to the present setting.

Definition 10.30 (Axiomatically Appropriate) A constant specification CS is *axiomatically appropriate* if, for every axiom A, there is a proof constant c such that $c{:}A \in \mathsf{CS}$.

Our central problem with justification constants is how to extend constant specifications from the base language $\mathcal{L}$ to the various languages $\mathcal{L}(W)$ that we will need in our model construction. We could, of course, simply work with universal constant specifications, but that is actually more stringent than we need. Instead we will require that constant specifications for $\mathcal{L}$ meet a condition that amounts to saying it is the *pattern* of individual variable usage that matters, and not what we *call* particular variables. This is a reasonable requirement that behaves well with respect to language extensions. The universal constant specification meets this condition, but so do others.

Definition 10.31 (Variable Renaming) Let A be a formula in the language $\mathcal{L}(\mathbf{V})$ and let $\vec{x} = \langle x_1, \ldots, x_n \rangle$ and $\vec{y} = \langle y_1, \ldots, y_n \rangle$ be variables of $\mathcal{L}(\mathbf{V})$. By $A(\vec{x} \mapsto \vec{y})$ we mean the result of replacing all occurrences of x_i (free or bound) by occurrences of y_i, and also occurrences of gen_{x_i} with gen_{y_i}, for $i = 1, \ldots, n$. It is always assumed that $x_1, \ldots, x_n$ are all distinct, and $y_1, \ldots, y_n$ also are all distinct.

Formulas A and B are *variable variants* if, for some $\vec{x}$ and $\vec{y}$, $B = A(\vec{x} \mapsto \vec{y})$ and $A = B(\vec{y} \mapsto \vec{x})$.

Variable renaming is not simply substitution because bound variables and gen instances are also modified. Informally, formulas are *variable variants* if each can be turned into the other by a renaming of free and bound individual variables. For example, $A = \forall x P(x, y)$ and $B = \forall w P(w, z)$ are variable variants because $B = A(\langle x, y \rangle \mapsto \langle w, z \rangle)$ and $A = B(\langle w, z \rangle \mapsto \langle x, y \rangle)$. But $C = \forall x P(x, y)$ and $D = \forall y P(y, y)$ are not variable variants because, although $D = C(\langle x \rangle \mapsto \langle y \rangle)$, it is not the case that $C = D(\langle y \rangle \mapsto \langle x \rangle)$.

It is easy, but nonetheless important, to see that the relation of being variable variants is an equivalence relation.

Definition 10.32 (Constant Specification Extension) Let CS be a constant specification for axioms in the language $\mathcal{L}$, and let $W \subseteq \mathbf{V}$. We extend CS to a constant specification $\mathsf{CS}(W)$, appropriate for $\mathcal{L}(W)$, as follows. For each axiom A in the language $\mathcal{L}(W)$, $c{:}A \in \mathsf{CS}(W)$ provided $c{:}B \in \mathsf{CS}$ for some B in the language $\mathcal{L}$, where A and B are variable variants.

Constant specification extensions behave nicely provided we impose the following condition.

Definition 10.33 (Variant Closed) A constant specification $\mathsf{CS}(W)$ for a language $\mathcal{L}(W)$ is *variant closed* provided that whenever A and B are formulas of $\mathcal{L}(W)$ that are variable variants, $c{:}A \in \mathsf{CS}(W)$ if and only if $c{:}B \in \mathsf{CS}(W)$.

Note that W can be $\emptyset$, so the preceding definition also specifies what it means for CS to be variant closed for $\mathcal{L}$. Here is what we meant when we noted that variant closure entailed nice behavior.

Proposition 10.34 *Let* CS *be an axiomatically appropriate, variant closed constant specification for the language* $\mathcal{L}$*, and let* $W \subseteq \mathbf{V}$.

(1) $\mathsf{CS}(W)$ *conservatively extends* CS. *That is,* $\mathsf{CS} \subseteq \mathsf{CS}(W)$*, and if* $c{:}A \in \mathsf{CS}(W)$ *where* A *contains only individual variables of* $\mathcal{L}$*, then* $c{:}A \in \mathsf{CS}$.
(2) $\mathsf{CS}(W)$ *is variant closed.*
(3) $\mathsf{CS}(W)$ *is axiomatically appropriate with respect to* $\mathcal{L}(W)$.
(4) *The Internalization Theorem, 10.10, extends to* $\mathcal{L}(W)$ *with respect to* $\mathsf{CS}(W)$.

Proof

(1) $\mathsf{CS}(W)$ extends CS using the reflexivity of the variable variant relation. The extension is conservative by the following. Suppose $c{:}A \in \mathsf{CS}(W)$ where A is a formula of $\mathcal{L}$. Then for some B in the language $\mathcal{L}$, A and B are variable variants, and $c{:}B \in \mathsf{CS}$. Because CS is variant closed, $c{:}A \in \mathsf{CS}$.
(2) Suppose A_1 and A_2 are axioms in the language $\mathcal{L}(W)$ that are variable variants, and suppose $c{:}A_1 \in \mathsf{CS}(W)$. For some B in the language $\mathcal{L}$, A_1 and B are variable variants and $c{:}B \in \mathsf{CS}$. Using transitivity of the variable variant relation, A_2 and B are variable variants, hence $c{:}A_2 \in \mathsf{CS}(W)$.
(3) Suppose A is an axiom of FOLP in the language $\mathcal{L}(W)$. Let $\vec{w}$ be the individual variables from W that occur in A, and let $\vec{x}$ be distinct individual variables of the base language $\mathcal{L}$ that do not occur in A. Let $B = A(\vec{w} \mapsto \vec{x})$. It follows that $A = B(\vec{x} \mapsto \vec{w})$. Because FOLP is axiomatized using schemes, B is an axiom, but in the language $\mathcal{L}$. Because CS is axiomatically appropriate, for some justification constant c, $c{:}B \in \mathsf{CS}$. But then $c{:}A \in \mathsf{CS}(W)$. It follows that $\mathsf{CS}(W)$ is axiomatically appropriate.
(4) Left to the reader.

□

10.3.6 The Canonical Fitting Model

In this section we construct a kind of *canonical* model, and then use it to prove completeness in the following section. Completeness is for formulas in the language $\mathcal{L}$, using a constant specification CS that is axiomatically complete and variant closed. The construction is Henkin style. Possible worlds are, essentially, maximally consistent sets of formulas whose domains contain witnesses for the true existential formulas. Providing for the existence of witnesses requires that we extend the language $\mathcal{L}$, and in Section 10.3.5 we introduced a new set $\mathbf{V}$ of witness variables for this purpose. Quantification domains can vary from world to world, and so we work with $\mathcal{L}(W)$ for various subsets W of $\mathbf{V}$. But now, variables from $\mathbf{V}$ play a dual role. Semantically they serve as domain members and, as such, are not subject to quantification. But possible worlds are sets of formulas drawn from the language $\mathcal{L}(\mathbf{V})$, these sets must be consistent, no derivation from them should lead to falsehood, but members of $\mathbf{V}$ are individual variables in a formal language and so can appear quantified in such derivations. We begin by introducing some notation to distinguish these two roles.

For each $W \subseteq \mathbf{V}$ we defined a language $\mathcal{L}(W)$, Definition 10.29, and we discussed extending a constant specification CS meeting appropriate conditions from $\mathcal{L}$ to $\mathcal{L}(W)$, Definition 10.32. Consistency is a syntactic, proof-theoretic notion, and it is defined relative to $\mathcal{L}(W)$ which allows quantification of witness variables.

Definition 10.35 ($\mathcal{L}(W)$ Consistency) Let CS be an axiomatically complete, variant closed constant specification for $\mathcal{L}$. A set S of $\mathcal{L}(W)$ formulas is $\mathcal{L}(W)$ *inconsistent using* CS if there is a finite subset $\{A_1, \ldots, A_n\}$ of S such that $(A_1 \wedge \ldots \wedge A_n) \to \bot$ is provable, where the proof allows any formula from $\mathcal{L}(W)$, and members of CS(W). S is $\mathcal{L}(W)$ *consistent using* CS if it is not $\mathcal{L}(W)$ inconsistent using CS.

The preceding definition involves the syntactic role of $\mathcal{L}(W)$. Semantically, witness variables will make up quantification domains in the model we construct, and are not subject to quantification when playing this role. For this we introduce a restricted version of $\mathcal{L}(W)$, which we denote by $\mathcal{L}^M(W)$, where the superscript is intended to suggest this is relevant to the *model* we will construct.

Definition 10.36 (The Language $\mathcal{L}^M(W)$) $\mathcal{L}^M(W)$ is the subset of $\mathcal{L}(W)$ consisting of those formulas in which members of W *can not occur bound* or as subscripts on gen. We say A is $\mathcal{L}^M(W)$ *closed* if A is a formula of $\mathcal{L}^M(W)$ in which all variables from $\mathcal{L}$ occur bound.

If A is $\mathcal{L}^M(W)$ closed, any variables from W must occur free, and any variables from $\mathcal{L}$ must occur bound. For example, suppose $t{:}_{\{x,y,z\}}A(x, y, w)$ is a closed $\mathcal{L}^M(W)$ formula. Then w must be an individual variable from the base language $\mathcal{L}$, and not a witness variable from W, because it occurs bound, while x, y, and z must be from W because they occur free.

Definition 10.37 (E-Complete) A *witness* for the negated universal formula $\neg(\forall x)A(x)$ of $\mathcal{L}(\mathbf{V})$ is a formula $\neg A(v)$ where v is an individual variable, possibly from $\mathbf{V}$. Let S be a set of $\mathcal{L}^M(W)$ formulas. S is *E-complete* if every negated universal formula in S has a witness in S. (We assume the universal quantifier is primitive and the existential quantifier is defined.)

At last we have all the machinery in place and can proceed to the specification of the canonical model. Because it may aid understanding of the $\mathcal{R}$ and $\mathcal{E}$ parts of the definition, we note the following. If the formula $t{:}_X A$ is $\mathcal{L}^M(W)$ closed, Definition 10.36, then X must include the free variables of A that are in W, and must not include any $\mathcal{L}$ variables.

Definition 10.38 (Canonical Model) Let CS be an axiomatically complete, variant closed constant specification for the language $\mathcal{L}$. Relative to CS, the *canonical model* is $\mathcal{M} = \langle \mathcal{G}, \mathcal{R}, \mathcal{D}, \mathcal{I}, \mathcal{E} \rangle$, which is specified as follows.

Specification of $\mathcal{G}$: Call $\Gamma = \langle S, W \rangle$ an $\mathcal{M}$-world if:

(1) $W \subseteq \mathbf{V}$ where W omits countably many members of $\mathbf{V}$.
(2) S is a set of $\mathcal{L}^M(W)$ closed formulas.
(3) S is $\mathcal{L}(W)$ consistent using $\mathsf{CS}(W)$, maximally so among consistent sets of $\mathcal{L}^M(W)$ closed formulas, and E-complete, with members of W as witnesses.

$\mathcal{G}$ is the collection of all $\mathcal{M}$-worlds. For each $\Gamma \in \mathcal{G}$, if $\Gamma = \langle S, W \rangle$ we write $\mathsf{form}(\Gamma)$ for S and $\mathsf{var}(\Gamma)$ for W.

Specification of $\mathcal{R}$: For a set S of $\mathcal{L}^M(W)$ closed formulas, where $W \subseteq \mathbf{V}$, let $S^\sharp$ be the set of all formulas $(\forall \vec{y})A$ such that $t{:}_X A \in S$ for some t, where $\vec{y}$ are all the free $\mathcal{L}$-variables of A. For $\Gamma, \Delta \in \mathcal{G}$, set $\Gamma \mathcal{R} \Delta$ provided:

(i) $\mathsf{var}(\Gamma) \subseteq \mathsf{var}(\Delta)$
(ii) $(\mathsf{form}(\Gamma))^\sharp \subseteq \mathsf{form}(\Delta)$

Specification of $\mathcal{D}$: For $\Gamma \in \mathcal{G}$, set $\mathcal{D}(\Gamma) = \mathsf{var}(\Gamma)$.

Specification of $\mathcal{I}$: For an n-place relation symbol Q and for $\Gamma \in \mathcal{G}$, let $\mathcal{I}(Q, \Gamma)$ be the set of $\langle v_1, \ldots, v_n \rangle$ where $v_i \in \mathcal{D}(\Gamma) = \mathsf{var}(\Gamma)$, and $Q(v_1, \ldots, v_n) \in \mathsf{form}(\Gamma)$.

Specification of $\mathcal{E}$: For $\Gamma \in \mathcal{G}$, set $\Gamma \in \mathcal{E}(t, A)$ provided $t{:}_X A \in \mathsf{form}(\Gamma)$ where X is the set of free variables in A from $\mathsf{var}(\Gamma)$.

We have finished the definition of a canonical model $\mathcal{M} = \langle \mathcal{G}, \mathcal{R}, \mathcal{D}, \mathcal{I}, \mathcal{E} \rangle$.

It is not hard to see that the domain of the model, $\mathcal{D}^*$, is exactly $\mathbf{V}$ (see Definition 10.19). It still must be checked that the canonical model actually is an FOLP model, and that it does the job we require of it. We take up the first issue now, and the second in the following section.

Theorem 10.39 *Let* CS *be an axiomatically appropriate, variant closed constant specification for the language* $\mathcal{L}$*. The canonical model relative to* CS *(Definition 10.38) is an* FOLP *model meeting* CS *(Definition 10.21).*

Proof First we verify that $\langle \mathcal{G}, \mathcal{R}, \mathcal{D} \rangle$ is an FOLP skeleton, Definition 10.19. Monotonicity is immediate because if $\Gamma\mathcal{R}\Delta$, then by definition of $\mathcal{R}$, $\mathcal{D}(\Gamma) = \mathsf{var}(\Gamma) \subseteq \mathsf{var}(\Delta) = \mathcal{D}(\Delta)$.

Reflexivity of $\mathcal{R}$ requires two things: $\mathsf{var}(\Gamma) \subseteq \mathsf{var}(\Gamma)$, which is trivial, and $(\mathsf{form}(\Gamma))^\sharp \subseteq \mathsf{form}(\Gamma)$, which we show. Suppose $F \in (\mathsf{form}(\Gamma))^\sharp$. Then $F = (\forall \vec{y})A$ where $t{:}_X A \in \mathsf{form}(\Gamma)$ for some t and X, and $\vec{y}$ are the free $\mathcal{L}$-variables of A. Let us say $\vec{y}$ is $\langle y_1, \ldots, y_n \rangle$. No y_i can occur in X. Then by repeated use of axiom (B5) the following is provable

$$t{:}_X A \rightarrow \mathsf{gen}_{y_1}(\mathsf{gen}_{y_2}(\cdots \mathsf{gen}_{y_n}(t))){:}_X (\forall \vec{y})A$$

so by maximal consistency of $\mathsf{form}(\Gamma)$, $\mathsf{gen}_{y_1}(\mathsf{gen}_{y_2}(\cdots \mathsf{gen}_{y_n}(t))){:}_X (\forall \vec{y})A \in \mathsf{form}(\Gamma)$. Then by axiom (B1) and maximal consistency again, $(\forall \vec{y})A \in \mathsf{form}(\Gamma)$, that is, $F \in \mathsf{form}(\Gamma)$.

Transitivity of $\mathcal{R}$ is by a similar argument, but axiom (B4) also comes in. We omit the proof.

Thus $\langle \mathcal{G}, \mathcal{R}, \mathcal{D} \rangle$ is a skeleton.

$\mathcal{I}$ is easily seen to be an interpretation function. Finally we need to verify that the evidence function $\mathcal{E}$ meets the conditions of Definition 10.38 part (2). We check some of the cases.

$\mathcal{R}$ Closure Condition: Suppose $\Gamma, \Delta \in \mathcal{G}$, $\Gamma\mathcal{R}\Delta$, and $\Gamma \in \mathcal{E}(t, A)$; we show $\Delta \in \mathcal{E}(t, A)$. Because $\Gamma \in \mathcal{E}(t, A)$ we have $t{:}_X A \in \mathsf{form}(\Gamma)$ for some X. We show that $t{:}_X A \in \mathsf{form}(\Delta)$. Using Axiom (B4) and maximal consistency, $!t{:}_X t{:}_X A \in \mathsf{form}(\Gamma)$. Then $(\forall \vec{y}) t{:}_X A \in \mathsf{form}(\Delta)$ by definition of $\mathcal{R}$, where $\vec{y}$ are the free $\mathcal{L}$-variables of $t{:}_X A$. But there are no free $\mathcal{L}$-variables in $t{:}_X A$ because it is $\mathcal{L}^M(\mathsf{form}(\Gamma))$ closed. Thus $t{:}_X A \in \mathsf{form}(\Delta)$, and hence $\Delta \in \mathcal{E}(t, A)$.

! Condition: Suppose $\Gamma \in \mathcal{E}(t, A)$; we show $\Gamma \in \mathcal{E}(!t, t{:}_X A)$, where X is the set of domain constants in A. Because $\Gamma \in \mathcal{E}(t, A)$, then $t{:}_X A \in \mathsf{form}(\Gamma)$. Using Axiom (B4) and maximal consistency, $!t{:}_X t{:}_X A \in \mathsf{form}(\Gamma)$, and it follows that $\Gamma \in \mathcal{E}(!t, t{:}_X A)$.

Instantiation Condition: Assume tht $\Gamma \in \mathcal{E}(t, A(x))$, where x is an $\mathcal{L}$-variable, and $a \in \mathcal{D}(\Gamma)$. We must show that $\Gamma \in \mathcal{E}(t, A(a))$. By the first assumption

$t{:}_X A(x) \in \mathsf{form}(\Gamma)$. Because X must be a set of witness variables, $x \notin X$. By Axiom (A3) and universal generalization (R2), $(\forall x)[t{:}_X A(x) \to t{:}_{X\cup\{x\}} A(x)]$ is a theorem of FOLP. Because x does not occur free in $t{:}_X A(x)$, then $t{:}_X A(x) \to t{:}_{X\cup\{a\}} A(a)$ is also a theorem. By maximal consistency, $t{:}_{X\cup\{a\}} A(a) \in \mathsf{form}(\Gamma)$, and hence $\Gamma \in \mathcal{E}(t, A(a))$.

gen_x Condition: Assume $\Gamma \in \mathcal{E}(t, A)$ and x is an $\mathcal{L}$-variable. By definition, $t{:}_X A \in \Gamma$, and Because X is a set of witness variables it cannot contain x. It follows that $\mathsf{gen}_x(t){:}_X \forall x A \in \Gamma$ using maximal consistency of Γ and Axiom (B5), and so $\Gamma \in \mathcal{E}(\mathsf{gen}_x(t), \forall x A)$.

Constant Specification Condition: Let $c{:}A \in \mathsf{CS}$, and let $\Gamma \in \mathcal{G}$. Because A is a formula in the language $\mathcal{L}$, the set of variables it contains that are from $\mathsf{var}(\Gamma)$ is empty. By maximal consistency of $\mathsf{form}(\Gamma)$, $c{:}A \in \mathsf{form}(\Gamma)$. Then by definition, $\Gamma \in \mathcal{E}(c, A)$.

□

10.3.7 Completeness

We are almost ready to show that the canonical model, from Definition 10.38, does what we want. But first we sketch the details of the version of the Henkin–Lindenbaum construction that we will need.

Proposition 10.40 *Let $W_1 \subset W_2 \subseteq \mathbf{V}$, where W_2 is a countable extension of W_1, and W_1 may be empty, and let CS be a variant closed constant specification for $\mathcal{L}$, with $\mathsf{CS}(W_1)$ and $\mathsf{CS}(W_2)$ extensions of it, Definition 10.32. Let S be a set of $\mathcal{L}^M(W_1)$ closed formulas that is $\mathcal{L}(W_1)$ consistent using $\mathsf{CS}(W_1)$. Then S extends to a set S' of $\mathcal{L}^M(W_2)$ closed formulas that is maximally $\mathcal{L}(W_2)$ consistent using $\mathsf{CS}(W_2)$, and E-complete with members of W_2 as witnesses.*

Proof We are given that W_2 is a countable extension of W_1; let $v_1, v_2, \ldots$ be an enumeration of it. S is a set of $\mathcal{L}^M(W_1)$ closed formulas that is $\mathcal{L}(W_1)$ consistent using $\mathsf{CS}(W_1)$. W_2 is a countable extension of W_1, and $\mathsf{CS}(W_2)$ is the corresponding extension of CS.

The set of $\mathcal{L}^M(W_2)$ closed formulas is countable; let $A_1, A_2, \ldots$ be an enumeration of it. We define a sequence, $S_0, S_1, S_2, \ldots$ of sets of $\mathcal{L}^M(W_2)$ closed formulas as follows.

$$S_0 = S$$

$$S_{n+1} = \begin{cases} S_n \text{ if } S_n \cup \{A_n\} \text{ is not } \mathcal{L}(W_2) \text{ consistent using } \mathsf{CS}(W_2) \\ S_n \cup \{A_n\} \text{ if } S_n \cup \{A_n\} \text{ is } \mathcal{L}(W_2) \text{ consistent using } \mathsf{CS}(W_2) \\ \quad \text{and } A_n \text{ is not of the form } \neg\forall x\varphi(x) \\ S_n \cup \{A_n, \neg\varphi(v)\} \text{ if } S_n \cup \{A_n\} \text{ is } \mathcal{L}(W_2) \text{ consistent using } \mathsf{CS}(W_2), \\ \quad A_n \text{ is } \neg\forall x\varphi(x), \text{ and } v \text{ is the first individual variable} \\ \quad \text{in the list } v_1, v_2, \ldots \text{that does not occur in } S_n \text{ or in } A_n \end{cases}$$

Each set S_n is $\mathcal{L}(W_2)$ consistent. The only case that needs checking is the last part of the definition of S_{n+1}. We must verify that if if $S_n \cup \{A_n\}$ is $\mathcal{L}(W_2)$ consistent, so is $S_n \cup \{A_n, \neg\varphi(v)\}$ where v does not occur in S_n or in A_n. We show the contrapositive. Suppose, for some $F_1, \ldots, F_k \in S_n$ the formula $(F_1 \wedge \ldots \wedge F_k \wedge A_n \wedge \neg\varphi(v)) \to \bot$ is $\mathcal{L}(W_2)$ provable using $\mathsf{CS}(W_2)$. Then so is $(F_1 \wedge \ldots \wedge F_k \wedge A_n) \to \varphi(v)$, and hence so is $\forall v[(F_1 \wedge \ldots \wedge F_k \wedge A_n) \to \varphi(v)]$. Because v does not occur in S_n or in A_n, $(F_1 \wedge \ldots \wedge F_k \wedge A_n) \to \forall v\varphi(v)$ is provable. Then so is $(F_1 \wedge \ldots \wedge F_k \wedge A_n) \to \forall x\varphi(x)$, and it follows that $S_n \cup \{\neg\forall x\varphi(x)\}$ is not $\mathcal{L}(W_2)$ consistent using $\mathsf{CS}(W_2)$.

Because each S_n is $\mathcal{L}(W_2)$ consistent using $\mathsf{CS}(W_2)$, it follows in the usual way that $S' = S_0 \cup S_1 \cup S_2 \cup \ldots$ is maximally consistent, and is E complete by its construction. □

With this out of the way, a version of the familiar Truth Lemma can be shown, and then completeness is simple.

Proposition 10.41 (Truth Lemma) *Let $\mathcal{M} = \langle \mathcal{G}, \mathcal{R}, \mathcal{D}, \mathcal{I}, \mathcal{E} \rangle$ be the canonical model meeting an axiomatically appropriate, variant closed constant specification. For each $\Gamma \in \mathcal{G}$, where $\Gamma = \langle \mathsf{form}(\Gamma), \mathsf{var}(\Gamma) \rangle$, and for each $\mathcal{L}^M(\mathsf{var}(\Gamma))$ closed formula A,*

$$\mathcal{M}, \Gamma \Vdash A \iff A \in \mathsf{form}(\Gamma).$$

Proof The proof is by induction on formula degree. Much of this is familiar, so we only give the two most significant cases.

Justification Formulas: Assume $\Gamma \in \mathcal{G}$, $t{:}_X A$ is a closed $\mathcal{L}^M(\mathsf{var}(\Gamma))$ formula, and the result is known for simpler formulas.

First, suppose $t{:}_X A \notin \mathsf{form}(\Gamma)$. The set X must consist entirely of witness variables, and include all those that occur in A. Let X' be the subset of X containing exactly the witness variables that occur in A. Then $t{:}_{X'} A \notin \mathsf{form}(\Gamma)$ because otherwise, by repeated use of Axiom (A3) and the maximal consistency of $\mathsf{form}(\Gamma)$ we would have that $t{:}_X A \in \mathsf{form}(\Gamma)$. Then by definition of $\mathcal{E}$ we have $\Gamma \notin \mathcal{E}(t, A)$ and it follows that $\mathcal{M}, \Gamma \not\Vdash t{:}_X A$.

Second, suppose $t{:}_X A \in \mathsf{form}(\Gamma)$. Let us say the free $\mathcal{L}$ variables of A are $\vec{y}$, and write $A(\vec{y})$ instead of A. As earlier, let X' be the subset of X containing exactly the witness variables that occur in A. Then $t{:}_{X'} A(\vec{y}) \in \mathsf{form}(\Gamma)$ by repeated use of Axiom (A2) and maximal consistency of $\mathsf{form}(\Gamma)$, and hence $\Gamma \in \mathcal{E}(t, A(\vec{y}))$. Further if $\Gamma \mathcal{R} \Delta$, by definition of $\mathcal{R}$ we have $\forall \vec{y} A(\vec{y}) \in \Delta$ where $\vec{y}$ are the free $\mathcal{L}$ variables of $A(\vec{y})$, and hence (maximal consistency again) for every $\vec{d}$ consisting of members of $\mathcal{D}(\Delta) = \mathsf{var}(\Delta)$, we have $A(\vec{d}) \in \mathsf{form}(\Delta)$. By the induction hypothesis, for each such instance, $\mathcal{M}, \Delta \Vdash A(\vec{d})$. We now have the conditions needed to conclude $\mathcal{M}, \Gamma \Vdash t{:}_X A(\vec{y})$.

Quantified Formulas: Assume that $\forall x A(x)$ is a closed $\mathcal{L}^M(\mathsf{var}(\Gamma))$ formula and the result is known for simpler formulas.

Suppose first that $\forall x A(x) \in \mathsf{form}(\Gamma)$. Let a be an arbitrary member of $\mathcal{D}(\Gamma)$. Then a is also a variable of the language $\mathcal{L}(\mathsf{var}(\Gamma))$, and $\forall x A(x) \to A(a)$ is a provable formula. By maximal consistency of $\mathsf{form}(\Gamma)$, $A(a) \in \mathsf{form}(\Gamma)$. By the induction hypothesis, $\mathcal{M}, \Gamma \Vdash A(a)$. Because a was arbitrary, $\mathcal{M}, \Gamma \Vdash \forall x A(x)$.

Finally, suppose that $\forall x A(x) \notin \mathsf{form}(\Gamma)$. By maximal consistency we have $\neg \forall x A(x) \in \mathsf{form}(\Gamma)$. Because $\mathsf{form}(\Gamma)$ is E-complete, for some witness variable a in $\mathsf{var}(\Gamma)$, $\neg A(a) \in \mathsf{form}(\Gamma)$ so by consistency, $A(a) \notin \mathsf{form}(\Gamma)$. By the induction hypothesis, $\mathcal{M}, \Gamma \nVdash A(a)$, and it follows that $\mathcal{M}, \Gamma \nVdash \forall x A(x)$.

□

Theorem 10.42 (FOLP Completeness) *Suppose closed formula A of $\mathcal{L}$ is not* FOLP *provable using constant specification* CS, *which is axiomatically appropriate and variant closed. Then A is not* FOLP(CS) *valid, in particular, A fails at some possible world of the canonical model meeting* CS.

Proof Assume A is not FOLP provable using CS. Then $\{\neg A\}$ is $\mathcal{L}$ consistent using CS. Let $V \subseteq \mathbf{V}$ contain countably many members of $\mathbf{V}$, while also omitting countably many. Applying Proposition 10.40 with $W_1 = \emptyset$ and $W_2 = V$, extend $\{\neg A\}$ to a set M that is maximally $\mathcal{L}^M(V)$ consistent using $\mathsf{CS}(V)$, and E-complete. Let $\Gamma = \langle M, V \rangle$. This is a possible world in the canonical model, and by the Truth Lemma, 10.41, A will be false in it. □

10.3.8 Fully Explanatory Models

For propositional justification logics, *fully explanatory* Fitting models were discussed earlier, Definition 4.4. This extends to the quantified setting very directly.

Definition 10.43 (Fully Explanatory) A Fitting model $\mathcal{M} = \langle \mathcal{G}, \mathcal{R}, \mathcal{D}, \mathcal{E}, \mathcal{I} \rangle$

is *fully explanatory* if it meets the following condition. Let A be an arbitrary formula with no free individual variables, but with constants from the domain of the model $\mathcal{M}$, Section 10.3.2, and let Γ be an arbitrary possible world in $\mathcal{G}$ in which A lives. If $\mathcal{M}, \Delta \Vdash A$ for every $\Delta \in \mathcal{G}$ such that $\Gamma \mathcal{R} \Delta$ then $\mathcal{M}, \Gamma \Vdash t{:}_X A$ for some justification term t, where X is the set of domain constants appearing in A.

This is the direct extension of the propositional version and informally says the same thing. If A is necessary at a possible world of a fully explanatory model, in the sense that it is true at all accessible worlds, then there is a reason for it—A has a justification at that world. The following establishes that FOLP is complete with respect to the class of fully explanatory Fitting models.

Theorem 10.44 *The canonical model, meeting an axiomatically appropriate, variant closed constant specification, is fully explanatory.*

Proof We show this in the contrapositive form. Let $\mathcal{M} = \langle \mathcal{G}, \mathcal{R}, \mathcal{D}, \mathcal{A}, \mathcal{I} \rangle$ be the canonical Fitting model for FOLP, with an axiomatically appropriate, variant closed constant specification CS. Suppose there is some $\Gamma \in \mathcal{G}$ and there is some formula A that lives in Γ and has no free individual variables, though it can have members of the domain (witness variables), and suppose that $\mathcal{M}, \Gamma \nVdash t{:}_X A$ for every justification term t, where X is the set of witness variables appearing in A. We show there is some $\Delta \in \mathcal{G}$ with $\Gamma \mathcal{R} \Delta$ such that $\mathcal{M}, \Delta \nVdash A$.

By our supposition and the Truth Lemma, 10.41, $t{:}_X A \notin \mathsf{form}(\Gamma)$ for every justification term t. Using this we first show the set $(\mathsf{form}(\Gamma))^\sharp \cup \{\neg A\}$ is $\mathcal{L}(\mathsf{var}(\Gamma))$ consistent using $\mathsf{CS}(\mathsf{var}(\Gamma))$, then we use the set to build an appropriate possible world Δ at which A is false.

Suppose $(\mathsf{form}(\Gamma))^\sharp \cup \{\neg A\}$ is inconsistent using $\mathsf{CS}(\mathsf{var}(\Gamma))$; we derive a contradiction. Because of inconsistency there are $\forall \vec{y_1} B_1, \ldots, \forall \vec{y_n} B_n \in (\mathsf{form}(\Gamma))^\sharp$ so that

$$\forall \vec{y_1} B_1, \forall \vec{y_2} B_2, \ldots, \forall \vec{y_n} B_n \vdash A$$

where $\vec{y_i}$ are the free $\mathcal{L}$-variables of B_i. For each i, since $\forall \vec{y_i} B_i \in (\mathsf{form}(\Gamma))^\sharp$ then $u_i{:}_{X_i} B_i \in \mathsf{form}(\Gamma)$ for some u_i, where X_i is the set of witness variables in B_i. The $\mathcal{L}$-variables in $\vec{y_i}$ cannot occur in X_i so by repeated use of Axiom (B5) there is a proof term t_i such that $u_i{:}_{X_i} B_i \rightarrow t_i{:}_{X_i} \forall \vec{y_i} B_i$ is provable. (In fact, t_i consists of iterated applications of gen operators, but we do not need the details.)

By the Internalization Theorem, 10.10, there is a proof term t so that

$$t_1{:}_{X_1} \forall \vec{y_1} B_1, t_2{:}_{X_2} \forall \vec{y_2} B_2, \ldots, t_n{:}_{X_n} \forall \vec{y_n} B_n \vdash t{:}_{X_1 \cup X_2 \cup \ldots \cup X_n} A.$$

This, combined with the provability of each $u_i{:}_{X_i} B_i \rightarrow t_i{:}_{X_i} \forall \vec{y}_i B_i$, gives us provability of the following

$$u_1{:}_{X_1} B_1 \rightarrow u_2{:}_{X_2} B_2 \rightarrow \ldots \rightarrow u_n{:}_{X_n} B_n \rightarrow t{:}_{X_1 \cup X_2 \cup \ldots \cup X_n} A.$$

For each i, $u_i{:}_{X_i} B_i \in \mathsf{form}(\Gamma)$, so by maximal consistency, $t{:}_{X_1 \cup X_2 \cup \ldots \cup X_n} A \in \mathsf{form}(\Gamma)$. It follows using Axioms (A2) and (A3) that $t{:}_X A \in \mathsf{form}(\Gamma)$, where X is exactly the set of witness variables appearing in A. But this contradicts the assumption that $t{:}_X A \notin \mathsf{form}(\Gamma)$ for every proof term t.

We have finished showing that $(\mathsf{form}(\Gamma))^\# \cup \{\neg A\}$ is consistent using $\mathsf{CS}(\mathsf{var}(\Gamma))$. We now apply Proposition 10.40. Let $\mathsf{var}(\Delta)$ extend $\mathsf{var}(\Gamma)$ with the addition of a countable set of members of $\mathbf{V}$, so that countably many members are still omitted. Extend the consistent set $(\mathsf{form}(\Gamma))^\# \cup \{\neg A\}$ to a set $\mathsf{form}(\Delta)$ that is maximally consistent and E-complete with respect to $\mathcal{L}(\mathsf{var}(\Delta))$, with members of $\mathsf{var}(\Delta)$ as witnesses. Then $\Delta = \langle \mathsf{form}(\Delta), \mathsf{var}(\Delta) \rangle \in \mathcal{G}$ and $\Gamma \mathcal{R} \Delta$. Because $\neg A \in \mathsf{form}(\Delta)$, by the Truth Lemma $\mathcal{M}, \Delta \not\Vdash A$, which completes the proof. □

Corollary 10.45 FOLP *is complete with respect to the class of* fully explanatory *models.*

10.3.9 Mkrtychev Models

Fitting models for FOLP have both a semantic and a syntactic component. They use the semantic machinery of first-order modal models and also have an evidence function that depends on syntactic details of formulas. Possible world semantics is flexible and provides a plausible intuition, but in fact the syntactic-based approach of Mkrtychev models, Section 3.5, extends to admit quantifiers. We sketch the details.

Definition 10.46 (Mkrtychev Model) An *Mkrtychev* FOLP *model* is a structure, $\mathcal{M} = \langle\langle \mathcal{D}, \mathcal{I}\rangle, \mathcal{E}\rangle$ where $\langle \mathcal{D}, \mathcal{I}\rangle$ is a classical first-order model, and $\mathcal{E}$ is an evidence function which must have the following closure properties.

$\mathcal{D}$ is the *domain* of the model, a non-empty set.

$\mathcal{I}$ is an *interpretation*, assigning to each n-place relation symbol of language $\mathcal{L}$ some n-ary relation on $\mathcal{D}$.

$\mathcal{E}$ is an *evidence function*, mapping proof term t and formula A with domain constants from $\mathcal{D}$ to a Boolean truth value, $\mathcal{E}(t, A)$. It must meet the following conditions.

- $\cdot$ **Condition** $\mathcal{E}(s, A \rightarrow B) \wedge \mathcal{E}(t, A) \rightarrow \mathcal{E}((s \cdot t), B)$.
- $+$ **Condition** $\mathcal{E}(s, A) \vee \mathcal{E}(t, A) \rightarrow \mathcal{E}((s + t), A)$.

! Condition $\mathcal{E}(t, A) \to \mathcal{E}(!t, t{:}_X A)$ where X is the set of all members of $\mathcal{D}$ that appear in A.

Instantiation Condition If $a \in \mathcal{D}$ then $\mathcal{E}(t, A(x)) \to \mathcal{E}(t, A(a))$.

gen_x Condition $\mathcal{E}(t, A) \to \mathcal{E}(\mathsf{gen}_x(t), \forall x A)$.

Let A be a formula of the language $\mathcal{L}(\mathcal{D})$ containing no free occurrences of $\mathcal{L}$ variables We write $\mathcal{M} \Vdash A$ to symbolize that A is true in Mkrtychev model $\mathcal{M}$. The truth conditions are as follows.

Atomic For an n place relation symbol Q and $k_1, \ldots, k_n \in \mathcal{D}$,
$\mathcal{M} \Vdash Q(k_1, \ldots, k_n) \iff \langle k_1, \ldots, k_n \rangle \in \mathcal{I}(Q)$.

Propositional $\mathcal{M} \Vdash (A \to B) \iff \mathcal{M} \nVdash A$ or $\mathcal{M} \Vdash B$, and similarly for other connectives.

Quantifier $\mathcal{M} \Vdash \forall x A(x) \iff \mathcal{M} \Vdash A(a)$ for every $a \in \mathcal{D}$.

Justification Term Assume $t{:}_X A(\vec{x})$ has no free $\mathcal{L}$ variables, where $\vec{x}$ are all the free variables of A. (Then members of X must be domain constants from $\mathcal{D}$.)
$\mathcal{M} \Vdash t{:}_X A(\vec{x}) \iff \mathcal{E}(t, A(\vec{x}))$ and $\mathcal{M} \Vdash A(\vec{a})$ for all $\vec{a}$ in $\mathcal{D}$.

Example 10.47 Example 10.26 showed $t{:}_{\{x,y\}} Q(x, y) \to t{:}_{\{x\}} Q(x, y)$ was not valid, where $Q(x, y)$ is atomic and the individual variables are the ones displayed. It did so with a two-world model. Here is an Mkrtychev model that also shows nonvalidity, but in the Mkrtychev sense.

Let $\mathcal{M} = \langle\langle \mathcal{D}, \mathcal{I} \rangle, \mathcal{E}\rangle$ be the following Mkrtychev model, where $\mathcal{D} = \{a, b\}$. $\mathcal{I}(Q) = \{\langle a, b\rangle\}$. For every proof term t, $\mathcal{E}(t, A)$ is true if A is a formula with domain constants from $\mathcal{D}$ but no free individual variables from $\mathcal{L}$, and false if there are individual variables from $\mathcal{L}$ in A.

First we note that $\mathcal{E}$ meets the conditions required of an evidence function in Mkrtychev models. Here are two representative cases. For the $\cdot$ Condition, if $\mathcal{E}((s \cdot t), B)$ is false B contains free $\mathcal{L}$ variables, but then so does $A \to B$ so $\mathcal{E}(s, A \to B)$ is false and the implication is true. For the ! Condition, $t{:}_X A$ contains no free $\mathcal{L}$ variables if members of X are all domain constants from $\mathcal{D}$, so the consequent of the implication is true.

Now $\mathcal{M} \Vdash Q(a, b)$ because $\langle a, b\rangle \in \mathcal{I}(Q)$. Then $\mathcal{M} \Vdash t{:}_{\{a,b\}} Q(a, b)$ because $\mathcal{E}(t, Q(a, b))$ is also true. But $\mathcal{M} \nVdash t{:}_{\{a\}} Q(a, y)$ because $\mathcal{E}(t, Q(a, y))$ is false.

Example 10.27 actually gives another example of a Mkrtychev model.

Mkrtychev models are essentially one-world FOLP models, and so we have soundness with respect to them. Each possible world in the canonical model is an Mkrtychev model, and completeness with respect to Mkrtychev models follows.

10.4 Arithmetical Semantics for FOLP

Disclaimer Despite our best efforts, this section is not easy reading and may require a certain proficiency in reasoning "within" formal theories, in particular, PA, Peano arithmetic (PA has been widely used already, e.g., in Chapter 9). Several variants of arithmetical provability semantics for FOLP were sketched in Artemov and Yavorskaya (Sidon) (2011), and a detailed exposition of all their aspects goes far beyond the scope of this book. The study of arithmetical provability semantics for FOLP, first-order S4, and first-order intuitionistic logic HPC is currently work in progress.

According to the classical Gödel approach as realized for the propositional case in Chapter 9, the main idea is to interpret FOLP formulas as arithmetical formulas, proof terms as proofs in PA, and proof assertions $t{:}_X F$ as arithmetical proof predicates *t is a proof of F in* PA *with parameters from X*.

The big difference from the propositional case, however, is the intrinsic incompleteness of the first-order logic of proofs, Section 10.4.3, which makes explorations in the first-order case an open-ended endeavor.

10.4.1 Free Variables in First-Order Derivations

The role of X in $t{:}_X F$ is to provide substitutional access to derivation t and formula F for all variables from X, in a sense, to keep variables in X "global," i.e., open for substitution in $t{:}_X F$. Other free variables in t and F are treated as "local": They are syntactic objects serving their purposes but not available for substitutions in $t{:}_X F$. This idea is implemented in the corresponding definitions below.

We assume the following convention about free occurrences of variables in derivations. By a PA-proof d of arithmetic formulas $F_1, F_2, \ldots, F_n$, we mean a finite collection of tree-like PA-derivations of formulas F_i's for every i. Free variables in these $F_1, F_2, \ldots, F_n$ are called *free variables of d*. We assume that if x is free in a proof d of F and b is a term then $d(b/x)$ is a proof of $F(b/x)$ (with the natural renaming of bound variables if necessary).

Let X be a finite set of individual variables, which we call "global variables," and let d be a PA-proof. We write $d(X)$ to reflect the fact that free variables in d that are from X are marked as *global* and other free variables in d as *local*.

10.4.2 Basic Semantics: Derivations from Hypotheses in PA

The intended interpretation of $t{:}_X F$ is:

$$t(X) \textit{ is a proof of } F(X).$$

There are different ways to interpret $t{:}_X F$ as a proof assertion in Peano arithmetic. The most straightforward approach would be to read t as a PA-derivation of F with variables from X accessible for substitution in both. The soundness of this interpretation is warranted, but there is a catch: Such a reading does not properly distinguish between the quantifiers $\forall$ and $\exists$ and hence remains conceptually propositional. Indeed, using this interpretation, $\exists x t{:}_X F$ yields $\forall x t{:}_X F$.

For a truly first-order arithmetical semantics of FOLP, we need a more sophisticated reading of $t{:}_X F$. The next natural idea is to consider a nonuniform interpretation $*$ of $t{:}_X F$, namely, t^* is an arithmetical proof, F^* is an arithmetical formula, X^* is a set of numbers N and

$$t^* \textit{ is a proof of } F^*$$

with X substituted by N. For example, $\{x = x\}$ with $X = \{x\}$ becomes a proof of infinitely many formulas $n = n$ for $n = 0, 1, 2, \ldots$ for appropriate evaluations $*$. This "infinity of proofs" effect complicates the technical side and here we shall look for a lighter approach.

In this section, we will model $t{:}_X F$ by *derivations from hypotheses*. So, t will be interpreted as a derivation of F from hypotheses in PA; without loss of generality we assume that each t has only one hypothesis H_t, or just H where H is an arithmetical formula, possibly with free variables. For technical convenience, e.g., to keep proof assertions decidable, we will limit such hypotheses to primitive recursive formulas. Furthermore, we assume that all true instances of primitive recursive formulas are axioms of PA. Therefore, for some values of X the hypothesis H is true, hence an axiom, hence t is a legitimate PA-derivation. Under this reading, it is possible to have $\exists x t{:}_X F$ without $\forall x t{:}_X F$.

Let $X = \{x_{i_1}, \ldots, x_{i_n}\}$ be a set of individual variables. For each arithmetical formula F, by $[F(\underline{X})]$ we understand a natural arithmetical term for the primitive recursive function that, for each value $K = (k_1, \ldots, k_n)$ of X, returns the Gödel number of the result of substituting K for X in all free occurrences of $x_1, \ldots, x_n$ in F. Formally,

$$[F(\underline{X})] \text{ is a natural arithmetical term for } \lambda K \ulcorner F(K/X) \urcorner.$$

We extend this formalism to arithmetical proofs and let

$$[d(\underline{X})] \text{ be a natural arithmetical term for } \lambda K \ulcorner d(K/X) \urcorner.$$

We assume that if a variable y is not free in F, then $[F(\underline{Xy})]$ and $[d(\underline{Xy})]$ coincide with $[F(\underline{X})]$ and $[d(\underline{X})]$, respectively. We will also systematically skip brackets "[ ]" in $[F(\underline{X})]$ and $[d(\underline{X})]$ whenever it is safe.

As a notational example, consider arithmetical formulas $F(x, y)$ and $G(x)$. A natural arithmetical term for

$$\lambda n \ulcorner F(n, \ulcorner G(n) \urcorner) \urcorner$$

in the full notation will be

$$[F(\underline{x}, [G(\underline{\underline{x}})])],$$

and in our simplified notation

$$F(\underline{x}, G(\underline{\underline{x}})).$$

We assume that Peano arithmetic PA contains all primitive recursive terms and their defining equalities, cf. Smoryński (1985). A *primitive recursive formula* is a formula $f(\vec{x}) = 0$ for some primitive recursive term f. A formula is *provably primitive recursive* if it is provably in PA equivalent to some primitive recursive formula. The class of provably primitive recursive formulas is closed under Boolean connectives, bounded quantifiers, and bounded μ-operator. Primitive recursive formulas constitute a decidable class: there is an algorithm, which for any given arithmetical formula decides whether this formula is primitive recursive.

Definition 10.48 A *filtered derivation d* is a derivation (tree) with a primitive recursive hypothesis H in PA. In particular, d may be a PA-derivation without hypotheses. A *filtered proof* p is a finite set of filtered derivations, p is a *filtered proof of a formula* F if F is the root formula of some filtered derivation occurring in p.

Comment. Due to the so-called Σ-completeness of PA, any instance of a primitive recursive formula $H(\vec{x})$ is provable if true. We make the convenient assumption that all true instances of primitive recursive formulas are axioms of PA.

Proofs are *sets* of filtered derivations and this reflects the multiconclusion character of FOLP proof objects. Technically, we need sets to interpret the operation "+," Lemma 10.51.

PA proves some simple combinatorial facts about filtered derivations, e.g., a filtered derivation d with hypothesis H is a PA-proof iff H is true.

Let us fix a natural multiconclusion Gödel proof predicate $\mathsf{Proof}(x, y)$ stating that x *is the Gödel number of a* PA*-derivation of a formula with Gödel number* y. By construction, $\mathsf{Proof}(x, y)$ is primitive recursive.

Lemma 10.49 *For each filtered proof d, arithmetical formula F, and set of* *variables X,* PA *proves*

$$\mathsf{Proof}(d(\underline{X}), F(\underline{X})) \rightarrow \mathsf{Proof}(d(\underline{Xy}), F(\underline{Xy}))$$

Proof A representative special case of this claim is: if

$$\mathsf{PA} \vdash \mathsf{Proof}(\ulcorner d(y) \urcorner, \ulcorner F(y) \urcorner)$$

holds with y being a local variable, then making y global preserves the derivation:

$$\mathsf{PA} \vdash \mathsf{Proof}(d(\underline{y}), F(\underline{y}))$$

with a free variable y. The proof is essentially a straightforward formalization of the property that a substitution of any y for a free variable in a derivation d of a formula F is a legitimate derivation $d(y)$ of $F(y)$. □

Lemma 10.50 *For each filtered proof d, arithmetical formula F, and set of variables X,* PA *proves*

$$\mathsf{Proof}(d(\underline{X}), F(\underline{X})) \rightarrow F.$$

Proof By external induction on the construction of d from assumptions to conclusions. Without loss of generality, assume that d is itself a filtered derivation and reason in PA. Suppose $\mathsf{Proof}(d(\underline{X}), F(\underline{X}))$ and F is the leaf node of d. If F is an instance of a PA-axiom, then F holds. Let F be a hypothesis H. Reason in PA. By the construction of a proof predicate, $\mathsf{Proof}(d(\underline{X}), F(\underline{X}))$ holds only if X contains all free variables of H and $H(X)$ is true, hence H given X. All steps down derivation tree in d are made according to the derivation rules in first-order logic, which are respected by proofs in PA, hence all F's from d hold. □

We define natural computable operations on filtered proofs that correspond to the functional symbols of FOLP.

Lemma 10.51 *There exist computable operations on filtered proofs* $\cdot$, $+$, $!$, *and* gen_x *such that for any filtered proofs d and e, formulas F and G, and a set of individual variables X, the following are provable in* PA*:*

(1) $\mathsf{Proof}(d(\underline{X}), (F \rightarrow G)(\underline{X})) \rightarrow (\mathsf{Proof}(e(\underline{X}), F(\underline{X})) \rightarrow \mathsf{Proof}((d \cdot e)(\underline{X}), G(\underline{X}))$;
(2) $\mathsf{Proof}(d(\underline{X}), F(\underline{X})) \vee \mathsf{Proof}(e(\underline{X}), F(\underline{X})) \rightarrow \mathsf{Proof}((d+e)(\underline{X}), F(\underline{X}))$;
(3) $\mathsf{Proof}(d(\underline{X}), F(\underline{X})) \rightarrow \mathsf{Proof}(\mathsf{gen}_x(d)(\underline{X}), \forall x F(\underline{X}))$, $x \notin X$;
(4) $\mathsf{Proof}(d(\underline{X}), F(\underline{X})) \rightarrow \mathsf{Proof}(!d(\underline{X}), \mathsf{Proof}(d(\underline{\underline{X}}), F(\underline{\underline{X}})))$.

Proof

(1) Given filtered proofs d and e as shown, build $d \cdot e$ as follows. Take any filtered derivation $\mathcal{T}_1 \in d$ of $Y \to Z$, any filtered derivations $\mathcal{T}_2 \in e$ of Y and construct a new filtered derivation consisting of $\mathcal{T}_1$ and $\mathcal{T}_2$ followed by Z (obtained by modus ponens); all these new derivations constitute $d \cdot e$. Let us check the desired property. Reason in PA. Without loss of generality, assume that d and e are themselves filtered derivations with the same hypothesis H. Because $\mathsf{Proof}(d(\underline{X}), (F \to G)(\underline{X}))$ and $\mathsf{Proof}(e(\underline{X}), F(\underline{X}))$ the hypothesis $H(X)$ holds. Because H is also the hypothesis of $d \cdot e$, we conclude $\mathsf{Proof}((d \cdot e)(\underline{X}), G(\underline{X}))$.

(2) Put $d+e$ to be the union of d and e, then use straightforward reasoning in PA.

(3) Given a variable x, by $\mathsf{gen}_x(d)$ we mean the union of d and the collection of tree-like arithmetical derivations that consists of a tree $\mathcal{T} \in d$ followed by $\forall x F$ where F is a root formula of d. By construction, $\mathsf{gen}_x(d)$ has the same hypothesis as d. Therefore, because $\mathsf{Proof}(d(\underline{X}), F(\underline{X}))$, $\mathsf{gen}_x(d)$ with given X is also a PA-proof. In particular, the last step from F to $\forall x F$ is a legitimate PA-proof step because $x \notin X$ hence x is a local variable and the subject of generalization. It remains to note that the root formula of $\mathsf{gen}_x(d)$ is $\forall x F$ and conclude $\mathsf{Proof}(\mathsf{gen}_x(d)(\underline{X}), \forall x F(\underline{X}))$.

(4) Reason in PA. Let $\mathsf{Proof}(d(\underline{X}), F(\underline{X}))$ and, without loss of generality assume that d is itself a filtered derivation and let its hypothesis be H. First, as in Lemma 10.49, we conclude that $H(X)$ holds. Formalizing in PA the comment after Definition 10.48, we find a PA-derivation d_1 such that

$$\mathsf{PA} \vdash \mathsf{Proof}(d_1, \forall X(H(X) \to \mathsf{Proof}(d(\underline{X}), F(\underline{X})))),$$

and then, by stripping the universal quantifier, find d_2 such that

$$\mathsf{PA} \vdash \mathsf{Proof}(d_2, \ulcorner H(X) \to \mathsf{Proof}(d(\underline{X}), F(\underline{X}))\urcorner).$$

By Lemma 10.49,

$$\mathsf{PA} \vdash \mathsf{Proof}(d_2(\underline{X}), H(\underline{X}) \to \mathsf{Proof}(d(\underline{\underline{X}}), F(\underline{\underline{X}}))). \tag{10.1}$$

As in Lemma 10.49, because $\mathsf{Proof}(d(\underline{X}), F(\underline{X}))$, the hypothesis $H(X)$ is a PA-axiom, hence is a PA-proof of itself

$$\mathsf{PA} \vdash \mathsf{Proof}(d(\underline{X}), F(\underline{X})) \to \mathsf{Proof}(H(\underline{X}), H(\underline{X})). \tag{10.2}$$

From 10.1 and 10.2 we get

$$\mathsf{PA} \vdash \mathsf{Proof}(d(\underline{X}), F(\underline{X})) \to \mathsf{Proof}((d_2 \cdot H)(\underline{X}), \mathsf{Proof}(d(\underline{\underline{X}}), F(\underline{\underline{X}}))).$$

Note that both d_2 and H are built from d and take $d_2 \cdot H$ as $!d$.

□

Example 10.52 In this example we illustrate how local and global parameters work in filtered proofs, and we see how the operation gen_x behaves in part (3) of Lemma 10.51. Let d be a filtered proof (derivation) $\{x = 0\}$, where this itself is a legitimate primitive recursive hypothesis. Then the filtered proof $\mathsf{gen}_x(d)$ will consist of two filtered derivations

$$x = 0, \quad \frac{x = 0}{\forall x(x = 0)}.$$

The formula $\mathsf{Proof}(\mathsf{gen}_x(d), F)$ without global variables is false for any F, since neither $x = 0$ nor

$$\frac{x = 0}{\forall x(x = 0)}$$

is a legitimate PA-derivation because $x = 0$ is not a PA axiom.

The formula $\mathsf{Proof}(\mathsf{gen}_x(d)(\underline{x}), F(\underline{x}))$ already depends on x, which is now regarded as a global parameter in $\mathsf{gen}_x(d)$. However, x is not a free variable in the derivation tree

$$\frac{x = 0}{\forall x(x = 0)}$$

hence $\mathsf{gen}_x(d)(\underline{0})$ is the Gödel number of the filtered derivation

$$0 = 0, \quad \frac{x = 0}{\forall x(x = 0)}$$

which is a PA-proof of $0 = 0$, but not of $\forall x(x = 0)$, for the same reasons as before. Therefore, $\mathsf{Proof}(\mathsf{gen}_x(d)(\underline{x}), (x = 0)(\underline{x}))$ is true only for $x = 0$, $\mathsf{Proof}(\mathsf{gen}_x(d)(\underline{x}), \forall x(x = 0)(\underline{x}))$ is false for any x.

Definition 10.53 A *filtered arithmetical interpretation* for the language FOLP is defined by operations $+$, $\cdot$, $!$ and gen_x which satisfy Lemma 10.51, and an evaluation $*$ that

(1) maps proof variables and constants to filtered arithmetical proofs,
(2) maps predicate symbols of arity n to arithmetical formulas with n free variables, and
(3) commutes with the renaming of individual variables.

Now we can define the interpretation t^* of an FOLP-term t as follows: for proof variables and constants t^* is given by the evaluation $*$, and we take $(s{\cdot}t)^*$ to be $s^*{\cdot}t^*$, $(s + t)^* = s^* + t^*$, $(\mathsf{gen}_x(t))^* = \mathsf{gen}_x(t^*)$, and $(!t)^* = !(t^*)$.

For formulas, $*$ commutes with the Boolean connectives and quantifiers and

$$(t{:}_X F)^* = \mathsf{Proof}(t^*(\underline{X}), F^*(\underline{X})).$$

Each derivation in FOLP generates a finite *constant specification*, which is a set of formulas $c{:}A$ introduced by the axiom necessitation rule (R3). We say that interpretation $*$ respects constant specification CS, if all formulas from CS are provable in PA under interpretation $*$.

Theorem 10.54 (Arithmetical soundness) *If* FOLP $\vdash A$ *with a constant specification* CS, *then for every filtered arithmetical interpretation* $*$ *respecting* CS, PA $\vdash A^*$.

Proof Induction on the proof of A in FOLP.

All instances of logical axioms (A1) and rules (R1) and (R2) are trivially provable in PA. Validity of axioms (B2)–(B5) is secured by the definition of operations $\cdot$, $+$, $!$, and gen_x; see Lemma 10.51. Rule (R3) holds once we assume that the constant specification CS is respected. Provability of (A3), and (B1) is guaranteed by Lemma 10.49 and Lemma 10.50.

It remains to verify (A2), i.e., that given $y \notin FVar(F)$

$$\mathsf{PA} \vdash \mathsf{Proof}(d(\underline{Xy}), F(\underline{Xy})) \to \mathsf{Proof}(d(\underline{X}), F(\underline{X})).$$

Reason in PA. Given $\mathsf{Proof}(d(\underline{Xy}), F(\underline{Xy}))$ pick a specific derivation $d_1(Xy)$ of $F(Xy)$ in $d(Xy)$. Note that because $y \notin FVar(F)$, y is not free in d_1. Therefore, by an earlier convention, $d_1(Xy)$ coincides with $d_1(X)$ and $F(Xy)$ coincides with $F(X)$. Hence $\mathsf{Proof}(d(\underline{X}), F(\underline{X}))$. □

Example 10.55 It is well known that the Barcan formula

$$\forall x \Box A(x) \to \Box \forall x A(x)$$

is not valid in the arithmetical semantics in which $\Box F$ stands for "*F is provable in* PA". Intuitively, even if $A(n)$ is provable for all n, this does not guarantee that $\forall x A(x)$ is provable.

The explicit version of Barcan formula can be formulated as

$$\forall x\, u{:}_{\{x\}} A(x) \to f(u){:}\forall x A(x) \tag{10.3}$$

for some function f on proofs.

We show that (10.3) fails in the provability (filtered) semantics for any f. To do this, we provide a specific arithmetical evaluation $*$ under which (10.3) is false in the standard model of arithmetic and hence not provable in PA.

Put $A^*(x)$ to be $\neg\mathsf{Proof}(x, \bot)$ and u^* to be $\{A^*(x)\}$. Note that for each n, $\neg\mathsf{Proof}(n, \bot)$ is provable in PA as a true primitive recursive formula, hence $A^*(n)$ is an axiom of PA for each n. On the other hand, $\forall x \neg\mathsf{Proof}(x, \bot)$ is the consistency formula, which is not provable in PA.

Furthermore, $u^*(x)$ is a filtered derivation from the hypothesis $A^*(x)$, which is a legitimate PA-derivation of $A^*(x)$ for each specific $x = n$. Therefore

$$(\forall x\, u{:}_{\{x\}}A(x))^*$$

is true in the standard model of arithmetic.

On the other hand,

$$(f(u){:}\forall x A(x))^*$$

claims that $(f(u))^*$ is a PA-proof of the consistency formula: this is false in the standard model of arithmetic for any f. Therefore, (10.3) under the interpretation $*$ is false and hence not provable in PA.

The same argument demonstrates arithmetical falsity of a "synthetic" explicit-implicit Barcan formula

$$\forall x\, u{:}_{\{x\}}A(x) \to \Box\forall x A(x).$$

10.4.3 Completeness Is Not Attainable

In this section, to simplify formulations but without a loss of generality, we consider the language of FOLP without proof constants, and the logic FOLP without the axiom necessitation rule. Let $\mathcal{F}$ be the set of FOLP-formulas valid under the filtered arithmetical semantics,

$$\mathcal{F} = \{X \mid \textit{for all } *,\ \mathsf{PA} \vdash X^*\}$$

We now show that, assuming consistency of PA, $\mathcal{F}$ is not recursively enumerable and hence cannot be effectively axiomatized. We note that the same argument works for any other natural arithmetical semantics.

Theorem 10.56 (Incompleteness Theorem) *$\mathcal{F}$ is not recursively enumerable.*

Proof Let Y be a pure first-order sentence (no proof terms, no free variables) and p be a proof variable.

Lemma 10.57 *$\neg p{:}Y \in \mathcal{F}$ if and only if for all $*$, PA $\nvdash Y^*$.*

Proof of Lemma Indeed, let $\neg p{:}Y \in \mathcal{F}$. Then for all $*$, PA $\vdash (\neg p{:}Y)^*$. Suppose PA $\vdash Y^*$ for some $*$ and let d be a corresponding derivation of Y^* in PA. Then for all $*$, PA $\vdash$ Proof(d, Y^*). Put $p^* = d$ and obtain PA $\vdash$ Proof(p^*, Y^*), i.e., for this $*$, PA $\vdash (p{:}Y)^*$ and PA $\vdash \neg(p{:}Y)^*$, which would make PA inconsistent.

Now assume that for all $*$, PA $\nvdash Y^*$. Then, by reflexivity, for all $*$, PA $\nvdash$ $(p{:}Y)^*$. Because $(p{:}Y)^*$ is a primitive recursive arithmetical formula, for all $*$, PA $\vdash \neg(p{:}Y)^*$, i.e., for all $*$, PA $\vdash (\neg p{:}Y)^*$, hence $\neg p{:}Y \in \mathcal{F}$. □

Continuing the proof of Theorem 10.56, using Lemma 10.57 it now suffices to show that the set

$$\mathcal{G} = \{Y \mid Y \text{ is a first-order formula and for all } * \ \mathsf{PA} \nvdash Y^*\}$$

is not recursively enumerable. For this, it suffices to demonstrate that

$$\mathcal{H} = \{Y \mid Y \text{ is a first-order formula and for all } * \ \mathsf{PA} \nvdash \neg Y^*\}$$

is not recursively enumerable.

Let FO be the set of first-order formulas that are valid in all models (this set coincides with first-order logic). Let FIN be the set of formulas that are valid in all finite models and $\overline{\mathsf{FIN}}$ be its complement. It is obvious that $\mathsf{FO} \subset \mathsf{FIN}$, hence $\mathsf{FO} \cap \overline{\mathsf{FIN}} = \emptyset$. The following lemma holds, cf. Ershov and Palyutin (1984):

Lemma 10.58 FO *and* $\overline{\mathsf{FIN}}$ *are recursively inseparable.*

Proof of Lemma Note that $\mathsf{FO} \subseteq \mathcal{H} \subseteq \mathsf{FIN}$. Indeed, if Y is a first-order theorem, then, for each interpretation $*$, $\mathsf{PA} \vdash Y^*$, hence $\mathsf{PA} \nvdash \neg Y^*$ and $Y \in \mathcal{H}$, yielding $\mathsf{FO} \subseteq \mathcal{H}$.

Suppose $Y \notin \mathsf{FIN}$. Then Y is false in some finite model $\mathcal{M}$. Because each finite model can be represented in PA (e.g., Artemov and Dzhaparidze, 1990), this yields an arithmetical interpretation $*$ for which $\mathsf{PA} \vdash \neg Y^*$, hence $Y \notin \mathcal{H}$. We conclude that $\mathcal{H} \subseteq \mathsf{FIN}$.

Therefore $\mathcal{H}$ separates FO and $\overline{\mathsf{FIN}}$ □

By Lemma 10.58, $\mathcal{H}$ is not decidable. It is easily seen from the definition of $\mathcal{H}$ that its complement $\overline{\mathcal{H}}$ is recursively enumerable. Therefore $\mathcal{H}$ is not recursively enumerable. □

Corollary 10.59 FOLP *is not complete with respect to filtered arithmetical semantics.*

10.4.4 Limitations of Provability Semantics

Note that in the propositional case, the arithmetical completeness of LP yields proof realizability of each finite constant specification, Theorem 9.22, which is essential for provability BHK semantics. This route, however, is impossible for FOLP because FOLP is not arithmetically complete in a rather strong sense, cf. Section 10.4.3.

Due to Incompleteness Theorem 10.56, the set of tautologies in any provability semantics of FOLP contains principles not derivable in FOLP.

One limitation of the arithmetical semantics stems from its commitment to the specific proof predicate $\mathsf{Proof}(x, y)$ and hence to a specific (Gödel) numbering of PA-axioms. In particular, the usual Gödel numbering is monotonic and the number of a formula is always greater than the number of any term that is part of it. Because the number of a proof of a formula is not less than a number of the formula, this makes all nested constructions of the type $u{:}A(u)$ vacuously false. This produces scores of "identities" $\neg u{:}A(u)$ not justified by their intended provability reading, cf. Chapter 9 in which this phenomenon was discussed for the propositional logic of proofs. We again refer to Artemov and Yavorskaya (Sidon) (2011) for approaches to mitigate this defect.

11
Going Past Modal Logic

In the history of classical justification logic there have been two conceptual breakthroughs. The first was the origin itself: *Artemov's logic of proofs*, the discovery of explicit counterparts for modal logics and proof semantics. The second was *the Mkrtychev–Fitting semantics*, at which point justification logic departed from arithmetical semantics and embraced general epistemic models, adding justification components to Kripke models for modalities.

A third conceptual breakthrough appears to be happening now: *going past (explicit counterparts of) modal logic*. We must take into consideration very general evidence tracking, and this naturally leads us beyond a purely modal framework because some epistemic scenarios include reasoning about reasons that is basically not of a modal nature. For such scenarios, a modeling that directly uses justification logic tools without intermediate modal logic steps could be appropriate.

There have already been significant developments in this direction. The papers (Krupski, 1997, 2001) described the logic of proofs for single-conclusion proof predicates (note Example 9.2). This departs from modal logic but is still within the arithmetical provability paradigm. Fitting's semantical realization methods, Chapter 7, actually produce quasi-realizations of modal logics in "+"-free versions of justification logics, which themselves are not counterparts of standard modal logics. Modular models (Artemov, 2012; Kuznets and Studer, 2012; cf. also Chapter 3) offer semantics of justifications using possible worlds constructions with justifications and closure conditions, in which modalities can be defined *a posteriori*. Artemov (2018) drops "+" and considers single-conclusion justification epistemic logics and models, that is, where for every justification term t there is at most one F such that $\models t{:}F$.

This chapter offers a refined version of Artemov (2018). It analyzes a well-known epistemic scenario, the Russell Prime Minister Example, using justification logic with additional predicates representing "t is accepted as a justification for F" and "t is a knowledge-producing justification for F." We argue that these predicates are not closed under sum on justifications and hence do not correspond to modalities.

11.1 Modeling Awareness

We will introduce *Justification Awareness Models*, or *JAMs*, in Section 11.3. These incorporate two principal ideas:

(1) *justifications are prime objects of the model*; thus knowledge and belief are defined evidence-based concepts;

(2) *awareness restrictions are applied to justifications* rather than to propositions, which allows for the maintaining of desirable closure properties.

JAMs naturally include major justification models, Kripke models and, in addition, can represent situations with multiple possibly fallible justifications. As an example, we will build a *JAM* for the Russell Prime Minister Example, which brings an attention to the details that was missing in previous epistemic modeling.

For formalizing epistemic scenarios one needs specific *domain-dependent models*, with additional features that are not necessary for the usual soundness and completeness analysis of proof systems.

Awareness is an important concept in epistemic modeling, but when applied to propositions directly, it may seriously diverge from intuition due to the lack of natural closure properties, see Fagin and Halpern (1988); Fagin et al. (1995); and Meyer and van der Hoek (1995). We propose applying awareness directly to justifications

agent is aware/unaware of a justification t for a proposition F

rather then to propositions, "agent is aware/unaware of a proposition F." As we will show, this approach allows for maintaining natural closure properties.

In *JAMs*, justifications are primary objects, and a distinction is made between *accepted* and *knowledge-producing* justifications. Belief and knowledge become derived notions, which depend on the status of supporting justifications. We argue that *JAMs* can work in a wide range of situations in which standard nonhyperintensional tools fail to fairly represent the corresponding epistemic structure.

Some samples of justification logic analysis of epistemic situations (specifically Gettier examples and the Red Barn example) are presented in Artemov

(2008) using Fitting justification models (Fitting, 2005), though due to the relative simplicity of those examples the analysis could be replicated in a bi-modal language (cf. Williamson, 2015). However, one cannot go much farther without adopting a justification framework because the situation changes when we have to represent several conflicting pieces of evidence for a stated fact. We now begin our examination of the promised example, due to Bertrand Russell in 1912 (Russell, 2001).

Russell Prime Minister Example *If a man believes that the late Prime Minister's last name began with a "B," he believes what is true, since the late Prime Minister was Sir Henry Campbell Bannerman.*[1] *But if he believes that Mr. Balfour was the late Prime Minister, he will still believe that the late Prime Minister's last name began with a "B," yet this belief, though true, would not be thought to constitute knowledge.*

To keep it simple, from here on P is the following proposition.

The late Prime Minister's last name began with a "B."

According to Russell's description, there are two justifications for P: the right one, which we call r, and the wrong one, which we call w. The agent chooses w as a reason to believe that P holds, and therefore cannot be said to know P.

One might suggest that the shortcomings of justifications in the Russell Prime Minister Example have to do with "false premises." To avoid such a reduction consider another Russell example, from 1912.

Russell True Premises Example *If I know that all Greeks are men and that Socrates was a man, and I infer that Socrates was a Greek, I cannot be said to-know-that Socrates was a Greek, because, although my premisses and my conclusion are true, the conclusion does not follow from the premisses.*

This example illustrates that "false premises" in the story told in the Russell Prime Minister Example is an instance of a more general phenomenon, namely an erroneous justification, which, in principle, can fail for many different reasons not limited to unreliable premises: hidden assumptions, deduction errors, an erroneous identification of the goal sentence, etc.[2]

There is a mathematical version of the story with a true proposition and two justifications, where one is correct and the other not.

[1] Which was true in 1912.

[2] Moreover, one can easily imagine knowledge-producing reasoning from a source with false beliefs (both an atheistic and a religious scientist can produce reliable knowledge products though one of them has false beliefs), so "false premises" are neither necessary nor sufficient for a justification to fail.

Arithmetic Example *Consider the display:*

$$\frac{1\not{6}}{\not{6}4} = \frac{1}{4}.$$

The true proposition is "16/64 = 1/4," the right justification is dividing both the numerator and the denominator by 16*, and the wrong (but shorter and thus more attractive) justification is simplifying as in the display.*

Given these considerations, we prefer speaking about *erroneous justifications* in a general setting, without reducing them to propositional entities such as "false premises." To be specific, we'll continue with the Russell Prime Minister Example.

To formalize Russell's scenario in modal logic, we introduce two modalities: **K** for knowledge and **J** for justified belief. In the real world,

(1) P holds,
(2) $\mathbf{J}P$ holds, because the agent has a justification w for P,
(3) $\mathbf{K}P$ does not hold.

This yields the following set of assumptions

$$\Gamma = \{P, \mathbf{J}P, \neg\mathbf{K}P\}.$$

However, Γ doesn't do full justice to Russell's scenario: The right justification r is not represented and Γ instead corresponds to an oversimplified Russell scenario in which the right justification r is absent. The epistemic structure of the example is not respected.

Within the *JAM* framework we will provide a model for Russell Prime Minister Example in Section 11.4, which, we think, fairly represents its intrinsic epistemic structure.

11.2 Precise Models

We introduce here a set of convenient tools for building *JAMs*, which will be formally introduced in Section 11.3. Our base logic for this section is J, with operations application "·" and sum "+" and its basic models, Definition 3.1, in which

$$s^* \triangleright t^* \subseteq (s\cdot t)^* \quad \text{and} \quad s^* \cup t^* \subseteq (s+t)^*.$$

The basic models for J(CS) are the basic CS-models for J_0.

Definition 11.1 (Precise Basic Models) A *precise* basic model of J(CS) is one in which the interpretation of constants is *fair to* CS, that is,

$$A \in c^* \ \textit{iff} \ c{:}A \in \mathsf{CS}$$

and closure conditions for operations are *exact*, that is,

$$[s{\cdot}t]^* = s^* \rhd t^* \ \text{ and } \ (s + t)^* = s^* \cup t^*.$$

Note that a precise model is completely defined by CS, evaluations of atomic justifications, and atomic propositions. There is no room for what might be called uncontrolled extraneous justifications. For example, in a basic model where $\mathsf{CS} = \{c_1{:}A_1, c_2{:}A_2\}$, it is possible to have $c_1^* = c_2^* = Fm$ and hence these constants become indistinguishable. In a precise model we must have $c_1^* = \{A_1\}$ and $c_2^* = \{A_2\}$.

11.3 Justification Awareness Models

We need more expressive power to capture epistemic differences between justifications and their consequent use by the knower. Some justifications are knowledge-producing, some are not. The agent makes choices about which justifications should serve as a base for beliefs and knowledge and which justifications should be ignored. These actions are commonly present in epistemic scenarios—we will continue to focus on Russell Prime Minister Example, where the central points are these:

- there are justifications w (Balfour was the late prime minister) and r (Bannerman was the late prime minister) for P;
- r is knowledge-producing for P whereas w is not;
- the agent opts to base belief that P on w and ignores r;
- the resulting belief is evidence-based, but is not knowledge.

We are about to discuss and adopt some natural properties of *accepted* and *knowledge-producing* justifications for a given sentence in a given model. First note that these sets are conceptually different: both Russell and Gettier examples, as well as their modifications, are built on this observation. Agents do not necessarily accept only knowledge-producing justifications as the basis for the agent's belief, and some knowledge-producing justifications may be left not accepted due to unawareness, ignorance, and/or other reasons. This combination of acceptance and knowledge-producing predicates for justifications allows us to represent awareness (which has suggested the name *Justification*

Awareness Models) in a way that is intuitive and maintains desirable closure conditions.

In our analysis of the Russell Prime Minister Example we will assume that *justification constants* are both knowledge-producing and accepted.

A fundamental natural assumption concerning the basic logical intelligence of agents is that if justifications s and t are accepted (or knowledge-producing) then their product $s \cdot t$ is also accepted (respectively, knowledge-producing) for the corresponding formulas. A more elaborate discussion of closure conditions of acceptance and knowledge-producing predicates will be found in Section 11.4.1.

Definition 11.2 A (basic) *Justification Awareness Model* (*JAM*) is a triple $\mathcal{M} = (*, \mathcal{A}, \mathcal{KP})$ where

(1) $*$ is a basic J(CS)-model for some axiomatically appropriate constant specification CS.

(2) $\mathcal{A}$ is an *acceptance* predicate, "t is accepted as a justification for F," such that

$\mathcal{A}$ is fair to CS: $\mathcal{A}(c, A)$ *iff* $c{:}A \in \mathsf{CS}$,

$\mathcal{A}$ closed under application:

if $\mathcal{A}(s, F \to G)$ *and* $\mathcal{A}(t, F)$ *then* $\mathcal{A}(s \cdot t, G)$,

$\mathcal{A}$ is consistent: *not* $\mathcal{A}(t, \bot)$ *for any* t;

(3) $\mathcal{KP}$ is a *knowledge-producing* predicate, "t is knowledge-producing for F," such that

$\mathcal{KP}$ is fair to CS: $\mathcal{KP}(c, A)$ *iff* $c{:}A \in \mathsf{CS}$,

$\mathcal{KP}$ is closed under application:

if $\mathcal{KP}(s, F \to G)$ *and* $\mathcal{KP}(t, F)$ *then* $\mathcal{KP}(s \cdot t, G)$,

$\mathcal{KP}$ is factive: $\mathcal{KP}(t, F)$ *yields* $\models_* F$ *for every* t *and* F.

We say a sentence F is *believed* if it has an accepted justification, i.e., $\mathcal{A}(t, F)$ holds for some t. Sentence F is *known* if it has an accepted knowledge-producing justification, i.e., $\mathcal{A}(t, F)$ and $\mathcal{KP}(t, F)$ both hold for the some t.

In models intended to represent beliefs but not knowledge, the component $\mathcal{KP}$ can be dropped and the corresponding *JAM* is just $(*, \mathcal{A})$.

This definition builds in the assumption that constants in a model are both knowledge-producing and accepted, though one could envision a more refined analysis.

It is important to understand that *JAMs* do not analyze *why* certain justifications are knowledge-producing or accepted, but rather *JAMs* assume knowledge-producing and accepted justifications to be given and provide a formal framework for reasoning about them.

11.4 The Russell Scenario as a *JAM*

We now use the *JAM* semantics just introduced to formally analyze the Russell Prime Minister Example. Consider a version of J(CS) in a language with two justification variables w and r, one propositional letter P, and an axiomatically appropriate single-conclusion constant specification CS, i.e., each axiom A has a constant c such that $c{:}A \in \mathsf{CS}$ and each c has at most one A for which $c{:}A \in \mathsf{CS}$.

Define a precise basic model $*$ by setting $P^* = 1$ and $w^* = r^* = \{P\}$. Define predicates $\mathcal{A}$ and $\mathcal{KP}$ by setting $\mathcal{A}(w, P)$ and $\mathcal{KP}(r, P)$, and (deterministically) extending them to the minimal predicates that are fair to CS and closed under application. We now have a candidate *JAM* model $\mathcal{R}$, defined to be $(*, \mathcal{A}, \mathcal{KP})$. We proceed to show it meets the conditions of Definition 11.2.

Lemma 11.3 *Each $t \in Tm$ is factive, that is, $\models_* t{:}F \to F$.*

Proof By structural induction on t. By the definition of a model, all axioms of J(CS) hold at $*$. Factivity holds for constants because, by the fairness assumption, $\models_* c{:}F$ implies that F is an axiom of J(CS), which are all true in $*$. Factivity holds for atomic justifications r and w because $r^* = w^* = \{P\}$ and P is true in $*$.

The induction step corresponding to application. Suppose $\models_* [s{\cdot}t]{:}F$, i.e., $F \in (s{\cdot}t)^*$. Because the model $*$ is precise, $(s{\cdot}t)^* = s^* \triangleright t^*$ and, by the definition of $\triangleright$, there should be $X \in t^*$ such that $X \to F \in s^*$. This means that $\models_* s{:}(X \to F)$ and $\models_* t{:}X$. By the induction hypothesis both s and t are factive, hence $\models_* X \to F$ and $\models_* X$, therefore $\models_* F$.

The induction step corresponding to sum. Suppose $\models_* [s + t]{:}F$, i.e., $F \in (s + t)^*$. Because the model $*$ is precise, $(s + t)^* = s^* \cup t^*$ and either $F \in s^*$ or $F \in t^*$, i.e., either $\models_* s{:}F$ or $\models_* t{:}F$. By the induction hypothesis, both s and t are factive, hence $\models_* F$. □

Informally, in the precise basic model $*$, P is true, both r, w are justifications of P, w is accepted for P and r is knowledge-producing for P.

It follows from Lemma 11.3 that $\mathcal{A}$ is consistent and $\mathcal{KP}$ is factive. Therefore, $\mathcal{R}$, defined to be $(*, \mathcal{A}, \mathcal{KP})$, is a *JAM*.

Lemma 11.4 *For any formula F, r does not occur in any t such that $\mathcal{A}(t, F)$, and w does not occur in any t such that $\mathcal{KP}(t, F)$.*

Proof First, we observe that $\mathcal{A}(t, F)$ holds iff $\mathcal{A}(t, F)$ can be obtained from assumptions

- $\mathcal{A}(c, A)$ for $c{:}A \in \mathsf{CS}$,
- $\mathcal{A}(w, P)$

by the rule

$$\mathcal{A}(s, F \to G) \text{ and } \mathcal{A}(t, F) \text{ yield } \mathcal{A}(s{\cdot}t, G). \tag{11.1}$$

Indeed, the "if" direction follows from the closure of $\mathcal{A}$ with respect to applications. Because the set D of "derived" truths $\mathcal{A}(t, F)$ is closed under application, by the minimality assumption on $\mathcal{A}$, there are no other truths $\mathcal{A}(t, F)$ outside D. The claim that r does not occur in any t such that $\mathcal{A}(t, F)$ trivially holds for the base cases in D and the use of application rule (11.1) obviously supports this property: once r occurs neither in s nor in t, their product $s{\cdot}t$ does not contain r either.

The same argument applies to $\mathcal{KP}$. □

Corollary 11.5 *If $\mathcal{A}(t, F)$ and $\mathcal{KP}(t, G)$ hold for the same t, then t is a ground term, i.e., built from constants only.*

Theorem 11.6 *In the model $\mathcal{R}$, sentence P is true, justified and believed, but not known.*

Proof In model $\mathcal{R}$ the sentence P is true because $\models_* P$. Also P is justified because $\models_* w{:}P$. And finally, P is believed because $\mathcal{A}(w, P)$ holds. We have yet to show that P is not known, i.e., that there is no justification g such that $\mathcal{A}(g, P)$ and $\mathcal{KP}(g, P)$ both hold.

To do this we build an auxiliary precise model $\bullet$ by flipping the truth value of P in $*$ from "true," $P^* = 1$ to "false" $P^\bullet = 0$, and leaving all evaluations of atomic justifications (i.e., r, w, and constants) the same as in $*$.

Lemma 11.7 *For each justification term t we have $t^* = t^\bullet$.*

Proof The inductive process (based on the operations in a precise model) of evaluating all justifications, given evaluations of atomic justifications, operates only with formulas of type $t{:}F$ and starts with the same initial set of such formulas in $*$ and $\bullet$, hence the results of these processes coincide. □

Lemma 11.8 *Each term g such that $\mathcal{A}(g, F)$ and $\mathcal{KP}(g, G)$ hold is factive in $\bullet$, i.e., $\models_\bullet g{:}X \to X$ for each X.*

Proof By Corollary 11.5, g is a ground term, i.e., built from constants. So, it suffices to verify factivity of ground terms only.

Base case: $\models_\bullet g{:}X$ and g is a justification constant c. Then, by Lemma 11.7, $X \in c^*$ and, by the fairness of $*$, X is an axiom of J(CS) and hence $\models_\bullet X$.

Application step: $g = s \cdot t$. Let $\models_\bullet [s \cdot t]{:}X$ for some formula X. By Lemma 11.7, $X \in (s \cdot t)^*$ and because $*$ is a precise model, there is $Y \in t^*$ such that $Y \to X \in s^*$. Therefore, by Lemma 11.7, $Y \in t^\bullet$ and $Y \to X \in s^\bullet$, i.e.,

$\models_\bullet s{:}(Y \to X)$ and $\models_\bullet t{:}Y$. By the induction hypothesis, both s and t are factive in $\bullet$ hence $\models_\bullet Y \to X$ and $\models_\bullet Y$, which yields $\models_\bullet X$.

The step corresponding to sum is treated similarly. Let $g = s + t$ and $\models_\bullet [s + t]{:}X$ for some X. By Lemma 11.7, $X \in (s + t)^*$ and because $*$ is a precise model, $X \in s^*$ or $X \in t^*$. By Lemma 11.7, $X \in s^\bullet$ or $X \in t^\bullet$, i.e., $\models_\bullet s{:}X$ or $\models_\bullet t{:}X$. Because, by the induction hypothesis, both s and t are factive in $\bullet$, $\models_\bullet X$. □

To complete the proof of Theorem 11.6, suppose $g{:}P$ is true in $\mathcal{R}$, i.e., $\models_* g{:}P$ for some g such that $\mathcal{A}(g, P)$ and $\mathcal{KP}(g, P)$. By Lemma 11.7, $\models_\bullet g{:}P$, and, by Lemma 11.8, $\models_\bullet P$, which is not the case. □

11.4.1 Closure Conditions for Acceptance and Knowledge-Producing Predicates

In this section we discuss why acceptance predicates $\mathcal{A}$ and knowledge-producing predicates $\mathcal{KP}$ are assumed closed under application "·" and not assumed closed under sum "+."

Acceptance is a subjective act by an agent who possesses a certain amount of rationality. In particular, the well-established epistemic closure principle for beliefs in its explicit form suggests that if s is accepted as belief-producing evidence for $F \to G$ and t is accepted as belief-producing evidence for F, then s applied to t should be accepted as belief-producing evidence for G.

Similar considerations justify the closure of $\mathcal{KP}$. If a justification s is knowledge-producing for $F \to G$ and t is knowledge-producing for F, then $s{\cdot}t$ is a knowledge-producing justification for G: a procedure of knowledge-producing for G could produce knowledge for $F \to G$ and F and then conclude that G.

As for the sum operation, within the Russell example it is easy to show that $\mathcal{A}$ and $\mathcal{KP}$ cannot both be closed under "+." Indeed, otherwise the justification $r + w$ would be both accepted, and knowledge-producing for P thus making P known, which is not the case.

We argue that actually *none* of $\mathcal{A}$ and $\mathcal{KP}$, generally speaking, is closed under "+," each for its own reasons.

Acceptance is subjective, and it may happen that the agent accepts s as a sufficient justification for believing that F, but considers a broader argument $s + t$ less trustworthy. In particular, the additional component t may yield something that is incompatible with F, as in the Russell example.

> On the basis of available observations (justification s) we could believe that a doctor recommends medicines objectively (sentence F), exclusively based on the patient's health condition. However, a broader justification $s + t$, in which s is supplemented

with additional evidence t stating that this doctor is a paid lobbyist of one of the relevant companies could well undermine our trust that F.

Knowledge-producing is more objective and thus more detached from the agent. A justification s may be knowledge-producing for F, for example, s could be a mathematically precise proof of F. However with another, possibly not proof-grade justification t added, the combined justification $s + t$ may no longer be a solid mathematical proof of anything let along F. In addition, some kind of consistency argument appears to work here as well: t might be an argument for $\neg F$; then s and t are incompatible and their sum $s + t$ cannot be viewed as knowledge-producing.

One can ask why in the logic of proofs LP the operation + is legitimate. Our answer is that mathematical proofs themselves are both knowledge-producing and assumed accepted by a rational agent, hence $\mathcal{A}(t, F)$ and $\mathcal{KP}(t, F)$ are both equivalent to $t{:}F$ and this guarantees their closure w.r.t. "+."

11.4.2 Can Russell's Scenario Be Made Modal?

One could try to express Russell Prime Minister Example in a modal language by introducing the justified belief modality

$$\mathbf{J}F \quad \Leftrightarrow \quad \models t{:}F \textit{ for some } t \textit{ accepted as a justification for } F$$

and the knowledge-producing modality

$$\mathbf{E}F \quad \Leftrightarrow \quad \models t{:}F \textit{ for some } t \textit{ which is knowledge-producing for } F,$$

and by stipulating that F is known iff F is both accepted and supported by a knowledge-producing justification:

$$\mathbf{K}F \quad \Leftrightarrow \quad \mathbf{J}F \wedge \mathbf{E}F.$$

This, however, fails because both $\mathbf{J}P$ and $\mathbf{E}P$ hold in $\mathcal{R}$, but $\mathbf{K}P$ does not. This is the essence of a Gettier-style phenomenon, when a proposition is supported by a knowledge-producing justification (hence true) and believed, but not known because knowledge-producing and accepted justifications for P are different. This illustrates the limitations of a purely modal language in tracking and sorting justifications.

11.5 Kripke Models and Master Justification

An epistemic reading of Kripke models relies on a hidden assumption of (common) knowledge of the model. This observation was made in Artemov (2016a) and leads to the following presentation of a Kripke model as a multiworld *JAM*.[3] The Kripkean accessibility relation between worlds, uRv, can be recovered by the usual rule: *What is believed at u, holds at v.*

The informal construction is as follows. Let $\mathcal{K}$ be a Kripke model. We have to find a justification $m{:}F$ for each knowledge/belief assertion $\Box F$ in $\mathcal{K}$. We claim that the model $\mathcal{K}$ itself is such a justification. Indeed, let $u \Vdash \Box F$ in $\mathcal{K}$. Then a complete description of $\mathcal{K}$ yields that, at state u, the agent knows/believes F *because the agent knows the model* $\mathcal{K}$ *and knows that* F *holds at all possible worlds.* So, the knowledge/belief-producing evidence for F is delivered by $\mathcal{K}$ itself, assuming the agent is aware of $\mathcal{K}$.

Syntactically, we consider a very basic justification language in which the set of justification terms consists of just one term m, called *master justification.* Think of m as representing a complete description of model $\mathcal{K} = (W, R, \Vdash)$. Specifically, we extend the truth evaluation in $\mathcal{K}$ to justification assertions by stipulating that at each $u \in W$

$$\mathcal{K}, u \Vdash m{:}X \quad \textit{iff} \quad \mathcal{K}, v \Vdash X \textit{ for any } v \in R(u) \quad \textit{iff} \quad \mathcal{K}, u \Vdash \Box X.$$

This reading provides a meaningful justification semantics of epistemic assertions in $\mathcal{K}$ via the master justification m representing the whole of $\mathcal{K}$. Because a Kripkean agent is logically omniscient, then along with $\mathcal{K}$ the agent knows all its logical consequences. Technically, we can assume that the description of $\mathcal{K}$ is closed under logical consequence and hence m is idempotent w.r.t. application, $m \cdot m = m$. This condition manifests itself in a special form of the application principle

$$m{:}(A \rightarrow B) \rightarrow (m{:}A \rightarrow m{:}B).$$

On the technical side, a switch from $\Box X$ to $m{:}X$ is a mere transliteration that does not change the epistemic structure of a model. Finally, for each $u \in W$, we define a basic model—the maximal consistent set Γ_u in the propositional language with $Tm = \{m\}$:

$$\Gamma_u = \{X \mid u \Vdash X\}.$$

So, from a justification perspective, a Kripke model is a collection of basic models with master justification that represents (common) knowledge of the model.

[3] In which we suppress the knowledge-producing component $\mathcal{KP}$ to capture beliefs.

11.6 Conclusion

JAMs do not offer a complete self-contained analysis of knowledge but rather reduce knowledge to knowledge-producing justifications accepted by the agent. This, however, constitutes meaningful progress because it decomposes knowledge in a way that moves justification objects to the forefront of epistemic modeling. Note that the Gettier and Russell examples clearly indicate which justifications are knowledge-producing or accepted. So *JAMs* fairly model situations in which the corresponding properties of justifications (knowledge-producing, accepted) are given.

There are many natural open questions that indicate possible research directions. Are justification assertions checkable, or decidable for an agent? Is the property of a justification to be knowledge-producing checkable by the agent? In multiagent cases, how much do agents know about each other and about the model? Do agents know each other's accepted and knowledge-producing justifications? What is the complexity of these new justification logics and what are their feasible fragments that make sense for epistemic modeling?

References

For convenience, each reference item ends with a list of pages on which that item is cited.

Achilleos, Antonis. 2014a. A complexity question in justification logic. *Journal of Computer and System Sciences*, **80**(6), 1038–1045. pages xix

Achilleos, Antonis. 2014b. On the complexity of two-agent Justification Logic. Pages 1–18 of: Bulling, N., van der Torre, L., Villata, S., Jamroga, W., and Vasconcelos, W. (eds.), *Computational Logic in Multi-Agent Systems. CLIMA 2014*. Lecture Notes in Computer Science, vol. 8624. Cham, Switzerland: Springer. pages xx

Achilleos, Antonis. 2015a. *Interactions and Complexity in Multi-Agent Justification Logic*. Ph.D. thesis, City University of New York. pages xx

Achilleos, Antonis. 2015b. Tableaux and complexity bounds for a multiagent Justification Logic with interacting justifications. In: Bulling, N. (ed.), *Multi-Agent Systems*. Lecture Notes in Computer Science, vol. 8953. Cham, Switzerland: Springer. pages xx

Alt, Jesse, and Artemov, Sergei. 2001. Reflective λ-calculus. Pages 22–37 of: Kahle, Reinhard, Schroeder-Heister, Peter, and Stärk, Robert (eds.), *Proof Theory in Computer Science, International Seminar, PTCS 2001, Dagstuhl Castle, Germany, October 7–12, 2001, Proceedings*. Lecture Notes in Computer Science, vol. 2183. Berlin: Springer. pages xix

Antonakos, Evangelia. 2007. Justified and common knowledge: Limited conservativity. Pages 1–11 of: Artemov, Sergei, and Nerode, Anil (eds.), *Logical Foundations of Computer Science, International Symposium, LFCS 2007, New York, NY, USA, June 4–7, 2007, Proceedings*. Lecture Notes in Computer Science, vol. 4514. Springer. pages xx

Antonakos, Evangelia. 2013. Explicit generic common knowledge. Pages 16–28 of: Artemov, Sergei, and Nerode, Anil (eds.), *Logical Foundations of Computer Science, International Symposium, LFCS 2013, San Diego, CA, USA, January 6–8, 2013, Proceedings*. Lecture Notes in Computer Science, vol. 7734. Berlin: Springer. pages xx

Artemov, Sergei. 1994. Logic of Proofs. *Annals of Pure and Applied Logic*, **67**(1–3), 29–59. pages xviii

Artemov, Sergei. 1995. *Operational Modal Logic*. Tech. rept. MSI 95–29. Ithaca, NY: Cornell University Press. pages xiv, xv, xviii, xix, 7, 46, 75, 103, 139, 160, 161

Artemov, Sergei. 2001. Explicit provability and constructive semantics. *Bulletin of Symbolic Logic*, **7**(1), 1–36. pages xiv, xv, xix, 7, 26, 46, 75, 139, 161, 181

Artemov, Sergei. 2002. Unified semantics for modality and λ-terms via proof polynomials. Pages 89–118 of: Vermeulen, Kees, and Copestake, Ann (eds.), *Algebras, Diagrams and Decisions in Language, Logic and Computation*. CSLI Lecture Notes, vol. 144. Stanford: CSLI Publications. pages xix, xx

Artemov, Sergei. 2004. Kolmogorov and Gödel's approach to intuitionistic logic: Current developments. *Russian Mathematical Surveys*, **59**(2), 203–229. Originally published in Russian. pages xx

Artemov, Sergei. 2006. Justified common knowledge. *Theoretical Computer Science*, **357**(1–3), 4–22. pages xix, xx, 46

Artemov, Sergei. 2008. The Logic of Justification. *The Review of Symbolic Logic*, **1**(4), 477–513. pages xv, xvi, 46, 224

Artemov, Sergei. 2012. The ontology of justifications in the logical setting. *Studia Logica*, **100**(1–2), 17–30. pages xv, 31, 44, 46, 51, 222

Artemov, Sergei. 2016a. Knowing the model. *arXiv preprint arXiv:1610.04955*. pages 231

Artemov, Sergei. 2016b. On aggregating probabilistic evidence. Pages 27–42 of: Artemov, Sergei, and Nerode, Anil (eds.), *Logical Foundations of Computer Science, International Symposium, LFCS 2016, Deerfield Beach, FL, USA, January 4-7, 2016, Proceedings*. Lecture Notes in Computer Science, vol. 9537. Cham, Switzerland: Springer. pages xx

Artemov, Sergei. 2018. Justification awareness models. Pages 22–36 of: Artemov, Sergei, and Nerode, Anil (eds.), *Logical Foundations of Computer Science, International Symposium, LFCS 2018, Deerfield Beach, FL, USA, January 8–11, 2018, Proceedings*. Lecture Notes in Computer Science, vol. 10703. Cham, Switzerland: Springer. pages 31, 46, 222

Artemov, Sergei, and Bonelli, Eduardo. 2007. The Intensional Lambda calculus. Pages 12–25 of: Artemov, Sergei, and Nerode, Anil (eds.), *Logical Foundations of Computer Science, International Symposium, LFCS 2007, New York, NY, USA, June 4–7, 2007, Proceedings*. Lecture Notes in Computer Science, vol. 4514. Berlin: Springer. pages xix, xx

Artemov, Sergei, and Dzhaparidze, Giorgie. 1990. Finite Kripke models and predicate logics of provability. *The Journal of Symbolic Logic*, **55**(3), 1090–1098. pages 220

Artemov, Sergei, and Fitting, Melvin. 2012. Justification Logic. In: Zalta, Edward N. (ed.), *The Stanford Encyclopedia of Philosophy*. The Metaphysics Research Lab, Center for the Study of Language and Information, Stanford University. pages xv

Artemov, Sergei, and Kuznets, Roman. 2006. Logical omniscience via proof complexity. Pages 135–149 of: Ésik, Zoltán (ed.), *Computer Science Logic, 20th International Workshop, CSL 2006, 15th Annual Conference of the EACSL, Szeged, Hungary, September 25–29, 2006, Proceedings*. Lecture Notes in Computer Science, vol. 4207. Berlin: Springer. pages xx

Artemov, Sergei, and Kuznets, Roman. 2009. Logical omniscience as a computational complexity problem. Pages 14–23 of: Heifetz, Aviad (ed.), *Theoretical Aspects of Rationality and Knowledge, Proceedings of the Twelfth Conference (TARK 2009)*. Stanford University, California: ACM. pages xx

Artemov, Sergei, and Kuznets, Roman. 2014. Logical omniscience as infeasibility. *Annals of Pure and Applied Logic*, **165**(1), 6–25. pages xx

Artemov, Sergei, and Nogina, Elena. 2005. Introducing justification into epistemic logic. *Journal of Logic and Computation*, **15**(6), 1059–1073. pages 46

Artemov, Sergei, and Straßen, Tyko. 1992. *The Basic Logic of Proofs*. Tech. rept. IAM 92–018. Institute of Computer Science and Applied Mathematics, University of Bern. pages xviii

Artemov, Sergei, and Straßen, Tyko. 1993a. The basic logic of proofs. Pages 14–28 of: Börger, E., Jäger, G., Kleine Büning, H., Martini, S., and Richter, M. M. (eds.), *Computer Science Logic, 6th Workshop, CSL'92, San Miniato, Italy, September 28–October 2, 1992, Selected Papers*. Lecture Notes in Computer Science, vol. 702. Berlin: Springer. pages xviii

Artemov, Sergei, and Straßen, Tyko. 1993b. *Functionality in the Basic Logic of Proofs*. Tech. rept. IAM 93–004. Institute of Computer Science and Applied Mathematics, University of Bern. pages xviii

Artemov, Sergei, and Yavorskaya, Tatiana. 2001. On first order Logic of Proofs. *Moscow Mathematical Journal*, **1**(4), 475–490. pages xix, 181

Artemov, Sergei, and Yavorskaya (Sidon), Tatiana. 2011. *First-Order Logic of Proofs*. Tech. rept. TR–2011005. City University of New York. pages xvii, xix, 181, 212, 221

Artemov, Sergei, Kazakov, E., and Shapiro, D. 1999. *Logic of Knowledge with Justifications*. Tech. rept. CFIS 99–12. Ithaca, NY: Cornell University Press. pages xx, 29, 139

Baltag, Alexandru, Renne, Bryan, and Smets, Sonja. 2014. The logic of justified belief, explicit knowledge, and conclusive evidence. *Annals of Pure and Applied Logic*, **165**(1), 49–81. pages xx

Blackburn, Patrick, de Rijke, Maarten, and Venema, Yde. 2001. *Modal Logic*. Cambridge, UK: Cambridge University Press. pages 29, 57

Borg, Annemarie, and Kuznets, Roman. 2015. Realization theorems for Justification Logics: Full modularity. Pages 221–236 of: De Nivelle, Hans (ed.), *Automated Reasoning with Analytic Tableaux and Related Methods*. Cham, Switzerland: Springer Lecture Notes in Computer Science, vol. 9323. TABLEAUX 2015. pages 140

Brezhnev, Vladimir N. 2000. *On Explicit Counterparts of Modal Logics*. Tech. rept. CFIS 2000–05. Ithaca, NY: Cornell University Press. pages 139

Brezhnev, Vladimir N., and Kuznets, Roman. 2006. Making knowledge explicit: How hard it is. *Theoretical Computer Science*, **357**(1-3), 23–34. pages xix, 108, 139, 180

Bucheli, Samuel. 2012. *Justification Logics with Common Knowledge*. Ph.D. thesis, Universität Bern. pages xx

Bucheli, Samuel, Kuznets, Roman, and Studer, Thomas. 2010. Two ways to common knowledge. Pages 83–98 of: Bolander, Thomas, and Braüner, Torben (eds.), *Proceedings of the 6th Workshop on Methods for Modalities (M4M–6 2009), Copenhagen, Denmark, 12–14 November 2009*. Electronic Notes in Theoretical Computer Science, no. 262. Amsterdam: Elsevier. pages xx

Bucheli, Samuel, Kuznets, Roman, and Studer, Thomas. 2011. Justifications for common knowledge. *Journal of Applied Non-Classical Logics*, **21**(1), 35–60. pages xx

Buss, Samuel R., and Kuznets, Roman. 2012. Lower complexity bounds in Justification Logic. *Annals of Pure and Applied Logic*, **163**(7), 888–905. pages xix

Chellas, Brian F. 1980. *Modal Logic, an Introduction*. Cambridge, UK: Cambridge University Press. pages 27, 55, 143, 145, 146

Cresswell, Max J. 1975. Hyperintensional logic. *Studia Logica*, **34**(1), 25–38. pages 8

Dean, Walter, and Kurokawa, Hidenori. 2010. From the Knowability Paradox to the existence of proofs. *Synthese*, **176**(2), 177–225. pages xx

Dean, Walter, and Kurokawa, Hidenori. 2016. Kreisel's theory of constructions, the Kreisel-Goodman paradox, and the second clause. In: Piecha, Thomas, and Schroeder-Heister, Peter (eds.), *Advances in Proof-Theoretic Semantics*. Trends in Logic (Studia Logica Library), vol. 43. Cham, Switzerland: Springer. pages 175, 179, 190

Dretske, Fred. 2005. Is knowledge closed under known entailment? The case against closure. Pages 13–26 of: Steup, M., and Sosa, E. (eds.), *Contemporary Debates in Epistemology*. Malden, MA: Blackwell. pages 2

Ershov, Yu. L., and Palyutin, E. A. 1984. *Mathematical Logic*. "Mir", Moscow. Translated from the Russian by Vladimir Shokurov. pages 220

Fagin, Ronald, and Halpern, Joseph Y. 1988. Belief, awareness, and limited reasoning. *Artificial Intelligence*, **34**, 39–76. pages 9, 223

Fagin, Ronald, Halpern, Joseph Y., Moses, Yoram, and Vardi, Moshe Y. 1995. *Reasoning about Knowledge*. Cambridge, MA: MIT Press. Paperback edition, 2004. pages 1, 223

Feferman, Solomon, Dawson Jr., John W., Kleene, Stephen C., Moore, Gregory H., Solovay, Robert M., van Heijenoort, Jean, Goldfarb, Warren D., Parsons, Charles, and Sieg, Wilfried (eds.). 1986–2003. *Kurt Gödel Collected Works*. Oxford, UK: Oxford. Five volumes. pages

Fitting, Melvin. 1996. *First-Order Logic and Automated Theorem Proving*. New York: Springer-Verlag. First edition, 1990. Errata at http://melvinfitting.org/errata/errata.html. pages 87

Fitting, Melvin. 1999. A simple propositional S5 tableau system. *Annals of Pure and Applied Logic*, **96**, 107–115. Originally in *The Parikh Project, Seven papers in honour of Rohit*, Uppsala Prints and Reprints in Philosophy, 1996 Number 18. pages 139

Fitting, Melvin. 2003. *A Semantics for the Logic of Proofs*. Tech. rept. TR-2003012. Ph.D. Program in Computer Science. City University of New York. pages xix, 31, 48, 51, 139

Fitting, Melvin. 2005. The Logic of Proofs, semantically. *Annals of Pure and Applied Logic*, **132**, 1–25. pages xv, xvi, xix, 31, 46, 48, 51, 62, 111, 112, 120, 136, 139, 224

Fitting, Melvin. 2009. Realizations and LP. *Annals of Pure and Applied Logic*, **161**(3), 368–387. pages 93, 140

Fitting, Melvin. 2010. Justification Logics and hybrid logics. *Journal of Applied Logic*, **8**(4), 356–370. pages 120, 139

Fitting, Melvin. 2011a. *Possible World Semantics for First Order* LP. Tech. rept. TR–2011010. City University of New York. pages xvii

Fitting, Melvin. 2011b. The Realization Theorem for S5, A simple, constructive proof. Chap. 4, pages 61–76 of: van Benthem, Johan, Gupta, Amitabha, and Pacuit, Eric (eds.), *Games, Norms and Reasons*. Synthese Library. New York: Springer. pages 139

Fitting, Melvin. 2013a. *Realization Implemented*. Tech. rept. TR–2013005. City University of New York. pages 93, 139

Fitting, Melvin. 2013b. Realization using the model existence theorem. *Journal of Logic and Computation*. pages 93, 111, 140

Fitting, Melvin. 2014a. *Justification Logics and Realization.* Tech. rept. TR-2014004. Ph.D. Program in Computer Science. City University of New York. pages xix, 112, 136, 139

Fitting, Melvin. 2014b. Possible world semantics for first order Logic of Proofs. *Annals of Pure and Applied Logic*, **165**, 225–240. pages xvii

Fitting, Melvin. 2015. Cut-free proof systems for Geach logics. *IfCoLog Journal of Logics and their Applications*, **2**(2), 17–64. pages xvi

Fitting, Melvin. 2016a. Modal logics, justification logics, and realization. *Annals of Pure and Applied Logic*, **167**, 615–648. pages xvi, 46, 112, 136

Fitting, Melvin. 2016b. Quasi-Realization. Pages 313–332 of: Hansen, Helle Hvid, Murray, Sarah E., Sadrzadeh, Mehrnoosh, and Zeevat, Henk (eds.), *Logic, Language, and Computation.* Lecture Notes in Computer Science, vol. 10148. New York: Springer. 11th International Tbilisi Symposium, TbiLLC, Tbilisi, Georgia, September 21-26, 2015. pages 112

Fitting, Melvin C. 2017. Paraconsistent logic, evidence, and justification. *Studia Logica*, **105**(6), 1149–1166. Special Issue: 40 Years of FDE. pages 30

Fitting, Melvin C., and Salvatore, Felipe. 2018. *First-order justification logic with constant domain semantics.* To appear in *Annals of Pure and Applied Logic.* pages xvii, xix, 181

Gabbay, Dov M. 1994. *Labeled Deductive Systems.* Oxford, UK: Oxford University Press. pages xv

Geach, Peter. 1973. On Certain modal systems connected with S4.3. *Bulletin of the Section of Logic*, **2**, 8–11. pages 29

Gettier, Edmund. 1963. Is knowledge justified true belief? *Analysis*, **23**(6), 121–123. pages 1, 24

Ghari, Meghdad. 2012. Cut elimination and realization for epistemic logics with justification. *Journal of Logic and Computation*, **22**(5), 1171–1198. pages xix

Ghari, Meghdad. 2014. Tableau proof systems for Justification Logics. *arXiv*, **preprint arXiv:1405.1828**(May). pages xix

Ghari, Meghdad. 2016a. Analytic tableaux for justification logics. *arXiv*, **preprint arXiv:1606.03562**. pages xix

Ghari, Meghdad. 2016b. Pavelka-style fuzzy justification logics. *Logic Journal of the IGPL*, **24**(5), 743–773. pages xx

Ghari, Meghdad. 2017. Labeled sequent calculus for justification logics. *Annals of Pure and Applied Logic*, **168**(1), 72–111. pages xix

Girard, J.-Y., Lafont, Y., and Taylor, P. 1989. *Proofs and Types.* Cambridge, UK: Cambridge University Press. pages 7

Gödel, Kurt. 1933. Eine Interpretation des intuistionistischen Aussagenkalkuls. *Ergebnisse eines mathematischen Kolloquiums*, **4**, 39–40. Translated as *An interpretation of the intuitionistic propositional calculus* in Feferman et al. (1986–2003) I, 296-301. *Kurt Gödel Collected Works.* Oxford, UK: Oxford University Press. pages xii, 5, 6, 175

Gödel, Kurt. 1938. *Vortrag bei Zilsel.* Translated as *Lecture at Zilsel's* in Feferman et al. (1986–2003) III, 62-113. *Kurt Gödel Collected Works.* Oxford, UK: Oxford University Press. pages xiv, xvii, 6, 175

Goodman, N. D. 1970. A theory of constructions is equivalent to arithmetic. Pages 101–120 of: Myhill, J., Kino, A., and Vesley, R. E. (eds.), *Intuitionism and Proof Theory.* Amsterdam: North-Holland. pages 7

Goris, Evan. 2008. Feasible operations on proofs: The Logic of Proofs for Bounded Arithmetic. *Theory of Computing Systems*, **43**(2), 185–203. pages xix

Hendricks, Vincent F. 2005. *Mainstream and Formal Epistemology*. Cambridge, UK: Cambridge University Press. pages 1

Heyting, Arend. 1934. *Mathematische Grundlagenforschung. Intuitionismus. Beweistheorie.* Berlin: Springer-Verlag. pages 4

Hintikka, Jaakko. 1962. *Knowledge and Belief: An Introduction to the Logic of the Two Notions*. Ithaca, NY: Cornell University Press. pages 1

Kazakov, E. L. 1999. *Logics of Proofs for S5*. M.Phil. thesis, Lomonosov Moscow State University, Faculty of Mechanics and Mathematics. In Russian. pages 139

Kleene, Stephen C. 1945. On the Interpretation of intuitionistic number theory. *Journal of Symbolic Logic*, **10**(4), 109–124. pages 175

Kokkinis, Ioannis, Ognjanović, Zoran, and Studer, Thomas. 2016. Probabilistic justification logic. Pages 174–186 of: Artemov, Sergei, and Nerode, Anil (eds.), *Logical Foundations of Computer Science, International Symposium, LFCS 2016, Deerfield Beach, FL, USA, January 4-7, 2016, Proceedings*. Lecture Notes in Computer Science, vol. 9537. New York: Springer. pages xx

Kolmogoroff, A. 1932. Zur Deutung der intuitionistischen Logik. *Mathematische Zeitschrift*, **35**, 58–65. In German. English translation in V. M. Tikhomirov (ed.), *Selected works of A.N. Kolmogorov. Volume I: Mathematics and Mechanics*, pages 151–158. Dordrecht: Kluwer, 1991. pages 4, 5

Kreisel, G. 1962. Foundations of Intuitionistic Logic. Pages 198–210 of: Nagel, E., Suppes, P., and Tarski, A. (eds.), *Logic, Methodology and Philosophy of Science. Proceedings of the 1960 International Congress*. Stanford, CA: Stanford University Press. pages 7, 175, 179, 190

Kreisel, G. 1965. Mathematical Logic. Pages 95–195 of: Saaty, T. L. (ed.), *Lectures in Modern Mathematics III*. New York: Wiley and Sons. pages 7

Krupski, Nikolai V. 2006a. On the complexity of the reflected Logic of Proofs. *Theoretical Computer Science*, **357**(1–3), 136–142. pages xix

Krupski, Nikolai V. 2006b. Typing in reflective combinatory logic. *Annals of Pure and Applied Logic*, **141**(1–2), 243–256. pages xx

Krupski, Vladimir N. 1997. Operational logic of proofs with functionality condition on proof predicate. Pages 167–177 of: Adian, Sergei, and Nerode, Anil (eds.), *Logical Foundations of Computer Science, LFCS'97, Yaroslav, Russia*. Lecture Notes in Computer Science, vol. 1234. New York: Springer. pages xviii, 160, 222

Krupski, Vladimir N. 2001. The single-conclusion proof logic and inference rules specification. *Annals of Pure and Applied Logic*, **113**(1–3), 181–206. pages xviii, 222

Krupski, Vladimir N. 2018. On the sharpness and the single-conclusion property of basic justification models. Pages 211–220 of: Artemov, Sergei, and Nerode, Anil (eds.), *Logical Foundations of Computer Science, International Symposium, LFCS 2018, Deerfield Beach, FL, USA, January 8–11, 2018, Proceedings*. Lecture Notes in Computer Science, vol. 10703. New York: Springer. pages xviii

Kurokawa, Hidenori. 2009. Tableaux and hypersequents for Justification Logic. Pages 295–308 of: Artemov, Sergei [N.], and Nerode, Anil (eds.), *Logical Foundations of Computer Science, International Symposium, LFCS 2009, Deerfield Beach, FL, USA, January 3–6, 2009, Proceedings*. Lecture Notes in Computer Science, vol. 5407. New York: Springer. Later version published as Kurokawa (2012). pages xix

Kurokawa, Hidenori. 2012. Tableaux and hypersequents for Justification Logics. *Annals of Pure and Applied Logic*, **163**(7), 831–853. pages xix, 239

Kurokawa, Hidenori. 2013. Prefixed tableau systems for Logic of Proofs and provability. Pages 203–218 of: Galmiche, D., and Larchey-Wendling, D. (eds.), *TABLEAUX 2013, Automated Reasoning with Analytic Tableaux and Related Methods*. Lecture Notes in Computer Science, vol. 8123. Berlin, Heidelberg: Springer. pages xix

Kuznets, Roman. 2000. On the complexity of explicit modal logics. Pages 371–383 of: Clote, Peter G., and Schwichtenberg, Helmut (eds.), *Computer Science Logic, 14th International Workshop, CSL 2000, Annual Conference of the EACSL, Fischbachau, Germany, August 21–26, 2000, Proceedings*. Lecture Notes in Computer Science, vol. 1862. New York: Springer. Errata concerning the explicit counterparts of $\mathcal{D}$ and $\mathcal{D}4$ are published as Kuznets (2009). pages xix, 46

Kuznets, Roman. 2009. Complexity through tableaux in justification logic. Page 121 of: *2008 European Summer Meeting of the Association for Symbolic Logic, Logic Colloquium '08, Bern, Switzerland, July 3–July 8, 2008*. Bulletin of Symbolic Logic, vol. 15(1). Poughkeepsie, NY: Association for Symbolic Logic. pages 180, 240

Kuznets, Roman, and Goetschi, Remo. 2012. Realization for justification logics via nested sequents: Modularity through embedding. *Annals of Pure and Applied Logic*, **163**(9), 1271–1298. pages xvi, 44, 46, 51, 140

Kuznets, Roman, and Studer, Thomas. 2012. Justifications, ontology, and conservativity. Pages 437–458 of: Bolander, Thomas, Braüner, Torben, Ghilardi, Silvio, and Moss, Lawrence (eds.), *Advances in Modal Logic, Volume 9*. College Publications. pages 31, 222

Kuznets, Roman, and Studer, Thomas. 2016. Weak arithmetical interpretations for the Logic of Proofs. *Logic Journal of the IGPL*, **24**(3), 424–440. pages xix

Lemmon, E. J., and Scott, Dana S. 1977. *The "Lemmon Notes": An Introduction to Modal Logic*. Amer. Phil. Quart., Monograph 11, Oxford. Blackwell. edited by Krister Segerberg. pages 142

Lewis, C. I. 1918. *A Survey of Symbolic Logic*. Berkeley: University of California Press. pages xiii

Lewis, C. I., and Langford, C. H. 1932. *Symbolic Logic*. Second edition 1959, Dover (New York) edn. London: Century. pages xiii

Lurie, Joseph. 2018. Probabilistic justification logic. *Philosophies*, **3**(1), 2. pages xx

McKinsey, J. C. C., and Tarski, A. 1948. Some theorems about the sentential calculi of Lewis and Heyting. *The Journal of Symbolic Logic*, **13**, 1–15. pages 6

Meyer, J.-J. Ch., and van der Hoek, W. 1995. *Epistemic Logic for AI and Computer Science*. Cambridge, UK: Cambridge University Press. pages 223

Milnikel, Robert [S.]. 2007. Derivability in certain subsystems of the Logic of Proofs is Π_2^p-complete. *Annals of Pure and Applied Logic*, **145**(3), 223–239. pages xix

Milnikel, Robert S. 2014. The logic of uncertain justifications. *Annals of Pure and Applied Logic*, **165**(1), 305–315. pages xx

Mkrtychev, Alexey. 1997. Models for the Logic of Proofs. Pages 266–275 of: Adian, Sergei, and Nerode, Anil (eds.), *Logical Foundations of Computer Science, 4th International Symposium, LFCS'97, Yaroslavl, Russia, July 6–12, 1997, Proceedings*. Lecture Notes in Computer Science, vol. 1234. New York: Springer. pages xv, xix, 31, 40, 46, 49

Negri, Sara. 2005. Proof analysis in modal logic. *Journal of Philosophical Logic*, **34**, 507–544. pages xvi

Negri, Sara, and von Plato, Jan. 2001. *Structural Proof Theory*. Paperback edition 2008 edn. Cambridge, UK: Cambridge University Press. pages xvi

Nogina, Elena. 2006. On Logic of Proofs and provability. *Bulletin of Symbolic Logic*, **12**(2), 356. pages xviii

Nogina, Elena. 2014a. On a hierarchy of reflection principles in Peano arithmetic. *arXiv preprint arXiv:1405.2558*. pages xviii

Nogina, Elena. 2014b. On Logic of formal provability and explicit proofs. *arXiv preprint arXiv:1405.2559*. pages xviii

Pacuit, Eric. 2005. A Note on some explicit modal logics. Pages 117–125 of: Dimitracopoulos, Costas (ed.), *Proceedings of the Fifth Panhellenic Logic Symposium*. pages xvi, 28, 56, 139, 241

Pacuit, Eric. 2006. *A Note on Some Explicit Modal Logics*. Tech. rept. PP–2006–29. Institute for Logic, Language and Computation, University of Amsterdam. Published as Pacuit (2005). pages 28, 46, 56, 139

Pouliasis, Konstantinos, and Primiero, Giuseppe. 2014. J-Calc: A typed lambda calculus for Intuitionistic Justification Logic. *Electronic Notes in Theoretical Computer Science*, **300**, 71–87. pages xx

Protopopescu, Tudor. 2016a. An arithmetical interpretation of verification and intuitionistic knowledge. Pages 317–330 of: Artemov, Sergei, and Nerode, Anil (eds.), *Logical Foundations of Computer Science, International Symposium, LFCS 2016, Deerfield Beach, FL, USA, January 4-7, 2016, Proceedings*. Lecture Notes in Computer Science, vol. 9537. New York: Springer. pages xx

Protopopescu, Tudor. 2016b. An arithmetical interpretation of verification and intuitionistic knowledge. *arXiv preprint arXiv:1601.03059*. pages xx

Renne, Bryan. 2004 (Mar.). *Tableaux for the Logic of Proofs*. Tech. rept. TR–2004001. Ph.D. Program in Computer Science. City University of New York. pages xix

Renne, Bryan. 2006. Semantic cut-elimination for two explicit modal logics. In: Huitink, J., and Katrenko, S. (eds.), *Proceedings of the Eleventh ESSLLI Student Session*. pages xix

Renne, Bryan. 2008. *Dynamic Epistemic Logic with Justification*. Ph.D. thesis. City University of New York. pages xx

Rose, Gene F. 1953. Propositional calculus and realizability. *Transactions of the American Mathematical Society*, **75**(1), 1–19. pages 7

Rubtsova, Natalia M. 2006a. Evidence reconstruction of epistemic modal logic S5. Pages 313–321 of: Grigoriev, Dima, Harrison, John, and Hirsch, Edward A. (eds.), *Computer Science – Theory and Application, First International Computer Science Symposium in Russia, CSR 2006, St. Petersburg, Russia, June 8–12, 2006, Proceedings*. Lecture Notes in Computer Science, vol. 3967. New York: Springer. pages xvi, 51, 52, 56

Rubtsova, Natalia M. 2006b. On realization of S5-modality by evidence terms. *Journal of Logic and Computation*, **16**(5), 671–684. pages xvi, 28, 46, 56, 139

Russell, Bertrand. 2001. *The Problems of Philosophy*. Oxford, UK: Oxford University Press. pages 224

Schwichtenberg, Helmut. *Fermat last theorem example*. Personal communication. pages 190

Sedlár, Igor. 2013. Justifications, awareness and epistemic dynamics. Pages 307–318 of: Artemov, Sergei [N.], and Nerode, Anil (eds.), *Logical Foundations of Computer Science, International Symposium, LFCS 2013, San Diego, CA, USA, January 6–8, 2013, Proceedings*. Lecture Notes in Computer Science, vol. 7734. New York: Springer. pages xx, 9

Shamkanov, Daniyar S. 2011. Strong normalization and confluence for reflexive combinatory logic. Pages 228–238 of: Beklemishev, Lev D., and de Queiroz, Ruy (eds.), *Logic, Language, Information and Computation, 18th International Workshop, WoLLIC 2011, Philadelphia, PA, USA, May 18–20, 2011, Proceedings*. Lecture Notes in Artificial Intelligence, vol. 6642. New York: Springer. pages xx

Shamkanov, Daniyar S. 2016. A Realization theorem for the Gödel-Löb provability logic. *Sbornik: Mathematics*, **207**(9), 1344. pages 62

Sidon, Tatiana L. 1997. Provability logic with operations on proofs. Pages 342–353 of: Adian, Sergei, and Nerode, Anil (eds.), *Logical Foundations of Computer Science, LFCS'97, Yaroslavl, Russia*. Lecture Notes in Computer Science, vol. 1234. New York: Springer. pages xviii, 160

Smoryński, Craig. 1985. *Self-Reference and Modal Logic*. New York: Springer-Verlag. pages 158, 214

Smullyan, Raymond M. 1963. A Unifying principle in quantification theory. *Proceedings of the National Academy of Sciences*, **49**(6), 828–832. pages 87

Smullyan, Raymond M. 1968. *First-Order Logic*. Berlin: Springer-Verlag. (Revised Edition, New York: Dover Press, 1994). pages 87, 96

Solovay, Robert M. 1976. Provability interpretation of modal logic. *Israel Journal of Mathematics*, **25**, 287–304. pages xiii

Takeuti, Gaisi. 1975. *Proof Theory*. Amsterdam: North-Holland. Second Edition, New York: Dover Publications, 2013. pages 158

Troelstra, Anne Sjerp. 1998. Realizability. Pages 407–474 of: Buss, S. (ed.), *Handbook of Proof Theory*. Amsterdam: Elsevier. pages 7

Troelstra, Anne Sjerp, and Schwichtenberg, Helmut. 1996. *Basic Proof Theory*. Amsterdam: Cambridge University Press. Second Edition 2000. pages 7, 187, 188

Troelstra, Anne Sjerp, and van Dalen, Dirk. 1988. Constructivism in Mathematics. Vol. I. *Studies in Logic and the Foundations of Mathematics*, 26. Amsterdam: North-Holland. pages 4

Twain, Mark. 1896. *Personal Recollections of Joan of Arc*. New York: Harper & Brothers. Real name: Samuel Langhorne Clemens. pages xxi

van Ditmarsch, Hans, van Der Hoek, Wiebe, and Kooi, Barteld. 2007. *Dynamic Epistemic Logic*. Vol. 337. Berlin: Springer Science & Business Media. pages 1

von Wright, Georg Henrik. 1951. *An Essay in Modal Logic*. Amsterdam: North-Holland. pages 1

Wang, Ren-June. 2011a. Knowledge, time, and the problem of logical omniscience. *Fundamenta Informaticae*, **106**(2–4), 321–338. pages xx

Wang, Ren-June. 2011b. *Non-Circular proofs and proof realization in modal logic*. Tech. rept. TR–2011012. City University of New York. pages 156

Wang, Ren-June. 2014. Non-circular proofs and proof realization in modal logic. *Annals of Pure and Applied Logic*, **165**(7-8), 1318–1338. pages 156

Ward, Artemus. 1861. Moses, the Sassy; or, the Disguised Duke. *Vanity Fair*, June 15. Real name: Charles Farrar Browne. pages x

Williamson, Timothy. 2015. A note on Gettier cases in epistemic logic. *Philosophical Studies*, **172**(1), 129–140. pages 224

Yavorskaya (Sidon), Tatiana. 2001. Logic of Proofs and provability. *Annals of Pure and Applied Logic*, **113**(1–3), 345–372. pages xviii

Yavorskaya (Sidon), Tatiana. 2008. Interacting Explicit Evidence Systems. *Theory of Computing Systems*, **43**(2), 272–293. pages xx

Yavorsky, Rostislav. 2001. Provability logics with quantifiers on proofs. *Annals of Pure and Applied Logic*, **113**(1–3), 373–387. pages xix

Yu, Junhua. 2014. Self-referentiality of Brouwer–Heyting–Kolmogorov semantics. *Annals of Pure and Applied Logic*, **165**(1), 371–388. pages 180

Index

www.ingramcontent.com/pod-product-compliance
Lightning Source LLC
LaVergne TN
LVHW101500200926
848483LV00001B/179
9781108424912